T0289623

Food Safety Handbook

Food Safety Handbook

Edited by Lisa Jordan

STATES
ACADEMIC PRESS
www.statesacademicpress.com

States Academic Press,
109 South 5th Street,
Brooklyn, NY 11249, USA

Visit us on the World Wide Web at:
www.statesacademicpress.com

© States Academic Press, 2023

This book contains information obtained from authentic and highly regarded sources. Copyright for all individual chapters remain with the respective authors as indicated. All chapters are published with permission under the Creative Commons Attribution License or equivalent. A wide variety of references are listed. Permission and sources are indicated; for detailed attributions, please refer to the permissions page and list of contributors. Reasonable efforts have been made to publish reliable data and information, but the authors, editors and publisher cannot assume any responsibility for the validity of all materials or the consequences of their use.

ISBN: 978-1-63989-722-3

Trademark Notice: Registered trademark of products or corporate names are used only for explanation and identification without intent to infringe.

Cataloging-in-Publication Data

Food safety handbook / edited by Lisa Jordan.
p. cm.
Includes bibliographical references and index.
ISBN 978-1-63989-722-3
1. Food--Safety measures. 2. Food. 3. Food industry and trade--Quality control.
4. Foodborne diseases--Prevention. I. Jordan, Lisa.
TX531 .F66 2023
363.192 6--dc23

Table of Contents

Preface

It is often said that books are a boon to mankind. They document every progress and pass on the knowledge from one generation to the other. They play a crucial role in our lives. Thus I was both excited and nervous while editing this book. I was pleased by the thought of being able to make a mark but I was also nervous to do it right because the future of students depends upon it. Hence, I took a few months to research further into the discipline, revise my knowledge and also explore some more aspects. Post this process, I begun with the editing of this book.

Food contains nutritional substances that are necessary for regulating vital processes as well as for the growth, maintenance and repair of body tissues. However, it can also transmit pathogens, which can cause illness or death in humans or other animals. The most common types of pathogens found in food are parasites, bacteria, fungi and viruses. Food safety refers to a scientific technique that is used for storing, handling and preparing food in order to reduce the risk of individuals becoming ill due to food-borne diseases. Food safety principles aim to avoid food from becoming contaminated, which may lead to food poisoning. This book explores all the important aspects of food safety in the present day scenario. It elucidates new techniques and their applications in a multidisciplinary manner. This book aims to equip students and experts with the advanced topics and upcoming concepts in this area of study.

I thank my publisher with all my heart for considering me worthy of this unparalleled opportunity and for showing unwavering faith in my skills. I would also like to thank the editorial team who worked closely with me at every step and contributed immensely towards the successful completion of this book. Last but not the least, I wish to thank my friends and colleagues for their support.

Editor

1

Integrated Ultrasonication and Microbubble-Assisted Enzymatic Synthesis of Fructooligosaccharides from Brown Sugar

Worraprat Chaisuwan [1,2], Apisit Manassa [1,2], Yuthana Phimolsiripol [2,3], Kittisak Jantanasakulwong [2,3], Thanongsak Chaiyaso [2,3], Wasu Pathom-aree [4], SangGuan You [5] and Phisit Seesuriyachan [2,3,*]

1 Interdisciplinary Program in Biotechnology, Graduate School, Chiang Mai University, Chiang Mai 50200, Thailand; worraprat_chai@cmu.ac.th (W.C.); apisit_man@cmu.ac.th (A.M.)
2 Faculty of Agro-Industry, Chiang Mai University, 155 Moo 2, Mae Hia, Mueang, Chiang Mai 50100, Thailand; yuthana.p@cmu.ac.th (Y.P.); kittisak.jan@cmu.ac.th (K.J.); thanongsak.c@cmu.ac.th (T.C.)
3 Cluster of Agro Bio-Circular-Green Industry (Agro BCG), Chiang Mai University, Chiang Mai 50200, Thailand
4 Department of Biology, Faculty of Science, Chiang Mai University, Chiang Mai 50200, Thailand; wasu.p@cmu.ac.th
5 Department of Marine Food Science and Technology, Gangneung-Wonju National University, Gangneung, Gangwon 210-702, Korea; umyousg@gwnu.ac.kr
* Correspondence: phisit.seesuriyachan@gmail.com or phisit.s@cmu.ac.th.

Abstract: Fructooligosaccharides (FOS) are considered prebiotics and have been widely used in various food industries as additives. Ultrasonication has been widely used to enhance food processes; however, there are few reports on ultrasound-assisted FOS synthesis. In the present study, FOS were produced from brown sugar using ultrasonication combined with microbubbles, and the production was optimised using a Box-Behnken experimental design. Here we showed that a combination of ultrasonication and microbubbles could boost the enzyme activity by 366%, and the reaction time was shortened by 60%. The reaction time was a significant variable affecting the FOS production. The optimum conditions were 5 min 45 s of ultrasonication and 7 min 19 s of microbubbles with a reaction time of 5 h 40 min. The maximum enzyme activity and total FOS yield were 102.51 ± 4.69 U·mL^{-1} and 494.89 ± 19.98 mg·g^{-1} substrate, respectively. In an enlarged production scale up to 5 L, FOS yields were slightly decreased, but the reaction time was decreased to 4 h. Hence, this technique offers a simple and useful tool for enhancing enzyme activity and reducing reaction time. We have developed a pilot technique as a convenient starting point for enhancing enzyme activity of oligosaccharide production from brown sugar.

Keywords: fructooligosaccharide; optimisation; Box–Behnken; ultrasound; microbubble; transfructosylation; enzyme activity enhancement

1. Introduction

Nowadays, people have attempted to improve their health and diet due to increased non-communicable diseases, including cancers, diabetes, metabolic syndrome, hypertension, stroke, and heart disease [1]. Therefore, many products, such as functional foods that have positive effects on health have been developed. Functional foods can be defined as foods that provide nutrients and energy, modulate an individual's health and physiological functions, and reduce diseases [2]. Oligosaccharides (short-chain carbohydrates) with prebiotic properties such as fructooligosaccharides

(FOS), galactooligosaccharides (GOS), xylooligosaccharides (XOS), inulooligosaccharides (IOS), soybean oligosaccharides (SOS), and cello-oligosaccharides (COS) have been interested from researchers because they have potential as ingredients in functional foods [3,4]. Among them, FOS are well-known non-digestible carbohydrates and have been extensively used in various food products [5].

It is safe to use FOS in food products, since FOS have been recognised by the Food and Drug Administration of the United States to be generally safe (GRAS) [6,7]. In the food industry, because FOS have favourable characteristics, including low calories, sweetening property, and unique physiochemical properties, they have been used as food additives in pastry, confectionery and dairy products [8,9]. Furthermore, FOS have many advantages for an individuals' health due to their biological activities, including prebiotic, anticancer, and immunomodulatory properties. FOS serve as carbon and energy sources for probiotics in the large intestine. FOS can promote the growth of prebiotic strains such as *Bifidobacterium* spp. and *Anaerostipes caccae*, butyrate-producing bacterium [10,11]. In contrast, FOS can inhibit the growth of some pathogenic bacteria such as *Clostridium* spp. [12]. As a prebiotic effect, gut microbiota can ferment FOS and then produce short-chain fatty acids (SCFAs), such as acetate, butyrate and propionate, which modulate intestinal epithelial functions and activate the host's immunity [13]. Buddington and colleagues [14] reported that FOS-fed mice had a low number of *Listeria monocytogenes* and *Salmonella enterica* subsp. *enterica* serovar Typhimurium (pathogenic bacteria). Taper and Roberfroid [15] also reported that FOS exhibited anti-proliferative effects on cancer cells in mice. However, consuming high levels of FOS (>20 g/day) might cause abdominal distension, abdominal rumbling, abnormal flatulence and abdominal pain because of excess gases from bacterial metabolism in the large intestine [16].

FOS are carbohydrates consisting of fructose monomers (2–10 residues) and can be naturally found in plant sources, such as onion, barley, asparagus, garlic, banana, and Jerusalem artichoke [17,18]. Moreover, FOS can be synthesised from low-cost materials such as sucrose and longan syrup, using an enzymatic reaction of transfructosylating enzymes, also called transfructosylation [19]. Enzymes with transfructosylation activity can transfer fructosyl groups, which are significant in FOS synthesis [20]. In the enzymatic synthesis, FOS can be produced using various enzymes, including β-fructofuranosidases (EC 3.2.1.26), β-fructosyltransferases (EC 2.4.1.9), and inulinases (EC 3.2.1.7) [8,20,21]. The commercial enzyme Pectinex Ultra SP-L was derived from *Aspergillus aculeatus* and is a complex enzyme, mainly consisting of polygalacturonase (pectinase) and others (cellulase, β-galactosidase and β-fructosyltransferases). This enzyme has been widely used for FOS production from sugar because it possesses high fructosyltransferase activity and was used in the present study [19,22]. After FOS synthesis, FOS molecules are a mixture of 1-kestose (1-kestotriose; GF2), nystose (1,1-ketstotetraose; GF3), and 1^F-β-fructofuranosylnystose (1,1,1-kestopentaose; GF4), which are categorised by the number of fructose monomers linking to glucose [8]. After production, some glucose and fructose contents remain as by-products, as well as excess sucrose, so it is necessary to purify the product. There are various techniques for purification of FOS, for example, chromatography, membrane filtration (such as nanofiltration), charcoal column and so on [23–25].

Various chemical, physical and biological approaches have been used to enhance enzyme activity for the improvement of FOS production. Among the physical methods, ultra-high pressure (UHP) is a useful technique that can activate or inactivate enzyme activity. In optimal levels of pressure, UHP enhanced fructosyltransferase activity of Pectinex Ultra SP-L and FOS contents could be increased 2.5-fold [5].

In the last few decades, an ultrasonication has been widely applied in various food and biotechnological processes. This technique has been performed as an enhancement for metabolite extraction from natural sources, such as fruits, microbial cells, and vegetables, among others, and can change the structure of food products [26,27]. This method has been used for enzyme inactivation; however, a large number of reports have been confirmed that the technique was able to activate some enzymes under suitable conditions [28]. Additionally, an ultrasonication has been approved that can increase enzyme activity at an appropriate frequency and intensity levels because ultrasound can act on conformation changes in enzymes or substrates [28,29]. Ultrasound can improve the activity of many enzymes including α-amylase, β-D-glucosidase, cellulose, dextranase and lipase [28,30–32]. Therefore, this method has been extensively used in various food industries, especially in fruit juice production, where it slightly affects the quality of fruit juice [33,34]. However, there are few reports on the use of ultrasonication for FOS production. This study, therefore, aimed to investigate the effects of ultrasonication and microbubble on enhancing enzyme activity for FOS production. Moreover, optimisation of the processes was done by a Box-Behnken design, and FOS yields (mg·g^{-1} substrate) were quantified using high-performance liquid chromatography (HPLC).

2. Materials and Methods

2.1. Enzymes and Chemicals

Pectinex Ultra SP-L, a commercial pectinolytic and cellulolytic enzyme derived from *Aspergillus aculeatus*, was purchased from Novozymes (Bagsværd, Denmark). Commercial brown sugar and white sugar derived from sugar canes were used as substrates in this study. 3,5-dinitrosalicylic acid (DNS) was purchased from Sigma-Aldrich (St. Louis, MO, USA). Standard carbohydrates including 1-kestose, nystose, 1^F-β-fructofuranosylnystose, sucrose, glucose and fructose were obtained from Wako Pure Chemical Industries (Osaka, Japan).

2.2. Enzymatic Assay

Enzyme activity was determined based on the generation of reducing sugar using sucrose as a substrate [19]. A concentration of Pectinex Ultra SP-L used in this study was 100 U·mL^{-1}, which was reported in a previous study [5]. Briefly, sucrose was dissolved in a sodium acetate buffer (pH 6.5) at a final concentration of 2% (*w*/*v*). Next, 0.2 mL of Pectinex Ultra SP-L and 1.8 mL of sucrose solution were mixed and then incubated at 55 °C for 15 min. Reducing sugars released from sucrose were analysed using the DNS assay [35]. After incubation, the sample (1 mL) and DNS reagent (4 mL) were mixed in a test tube and the test tube was placed in a boiling water bath for 5 min. The tube was placed in ice to instantly cool and then laid in a room temperature water bath (25 °C). Finally, the absorbance was measured at 540 nm using a Genesys 20 spectrophotometer (Thermo Fisher Scientific Inc., Waltham, MA, USA). One unit of Pectinex Ultra SP-L enzyme was defined as the amount of enzyme that produced 1 µmol·min^{-1} of fructose.

2.3. Production of FOS Using Ultrasonication and Microbubbles

Brown sugar was dissolved in 0.1 M sodium acetate buffer (pH 6.5) with a final concentration of 70% (*w*/*v*). The solution (98 mL) was incubated at 55 °C for 1–2 h, then Pectinex Ultra SP-L was added with a final volume of 100 mL in a 250-mL Erlenmeyer flask. The reaction was performed at 55 °C with ultrasonication (37 kHz) using an ultrasonic bath (Elmasonic S 30 H, Elma Schmidbauer GmbH, Singen, Germeny) and then microbubbles were applied direct into the solution using a microbubble generator with a flow rate of 100 mL·min^{-1} and pressure of 0.25–0.5 mPa. When the reaction was stopped by incubation in boiling water, the FOS content in each sample was analysed using HPLC. For large scale production, the reaction was done in 1.5- and 5-L batches, and samples were collected every 1 h for 6 h. The FOS content in samples were analysed using HPLC and compared to a 100-mL scale.

2.4. Optimisation of FOS Production

In this study, a Box-Behnken experimental design with three levels of three factors and response surface methodology was used to estimate optimal levels of variables. The three factors with three levels for FOS production included ultrasonication time (X_1, 2.00–15.00 min), microbubble time (X_2, 5.00–30.00 min), and reaction time (X_3, 2.00–6.00 h). Design expert 6.0.10 (Stat-Ease, Inc., Minneapolis, MN, USA) was used for experimental design and modelling analysis resulting in a total of 17 experiments, which were conducted in triplicate (Table 1). The model predicting the optimal values was expressed as Equation (1) given below:

$$Y = \beta_0 + \sum_{i=1}^{k} \beta_i X_i + \sum_{i=1}^{k} \beta_{ii} X_i^2 + \sum_{i \geq j}^{k} \sum_{i=1}^{k} \beta_{ij} X_{ij} X_j \quad (1)$$

where Y is the predicted response, β is the regression coefficient and X is the independent variable. The F-value determined the statistical significance of the equation. The accuracy of the polynomial model equation was expressed by the coefficient of determination (R^2).

Table 1. The Box-Behnken experimental design and experimental values of enzymatic activity and FOS yields produced from brown sugar.

Run	Variable			Enzyme Activity ($U \cdot mL^{-1}$)	Oligosaccharide Yields ($mg \cdot g^{-1}$)			
	X_1 (min) *	X_2 (min) **	X_3 (h) ***		1-Kestose	Nystose	1^F-β-Fructofuranosylnystose	Total FOS
1	2.00	5.00	4.00	102.03 ± 1.88	379.80 ± 23.15	112.84 ± 1.55	10.30 ± 0.06	502.94 ± 35.46
2	15.00	5.00	4.00	97.75 ± 4.10	371.04 ± 34.06	108.94 ± 2.33	12.24 ± 0.15	492.22 ± 26.12
3	2.00	30.00	4.00	91.44 ± 2.30	351.19 ± 28.85	95.27 ± 1.69	8.31 ± 0.25	454.77 ± 44.59
4	15.00	30.00	4.00	95.13 ± 4.50	362.96 ± 49.73	103.95 ± 2.03	9.14 ± 0.59	476.05 ± 36.21
5	2.00	17.50	2.00	64.82 ± 0.98	311.30 ± 32.11	41.36 ± 0.96	2.09 ± 0.06	354.74 ± 25.89
6	15.00	17.50	2.00	70.88 ± 2.10	330.06 ± 44.53	52.42 ± 1.12	3.02 ± 0.12	385.50 ± 38.01
7	2.00	17.50	6.00	102.67 ± 2.71	341.32 ± 25.85	146.26 ± 3.54	18.65 ± 1.06	506.26 ± 46.04
8	15.00	17.50	6.00	112.22 ± 2.50	352.25 ± 25.40	148.59 ± 3.59	24.61 ± 2.09	525.45 ± 56.09
9	8.50	5.00	2.00	73.95 ± 2.59	313.25 ± 35.33	55.56 ± 0.58	2.96 ± 0.08	371.77 ± 34.56
10	8.50	30.00	2.00	70.87 ± 3.50	313.25 ± 27.56	50.43 ± 0.49	2.65 ± 0.11	366.32 ± 23.15
11	8.50	5.00	6.00	104.07 ± 3.15	358.53 ± 16.81	143.10 ± 4.32	17.28 ± 0.56	518.92 ± 35.11
12	8.50	30.00	6.00	105.04 ± 4.97	356.39 ± 40.74	149.94 ± 2.51	18.67 ± 0.92	525.00 ± 43.65
13	8.50	17.50	4.00	94.47 ± 1.39	367.94 ± 32.00	100.38 ± 2.09	8.32 ± 0.49	476.64 ± 35.28
14	8.50	17.50	4.00	99.41 ± 6.83	361.44 ± 42.75	120.14 ± 3.05	12.75 ± 0.46	494.33 ± 43.19
15	8.50	17.50	4.00	95.86 ± 2.11	354.56 ± 41.56	111.06 ± 1.54	10.62 ± 0.54	476.24 ± 54.56
16	8.50	17.50	4.00	95.60 ± 3.25	365.72 ± 26.35	109.90 ± 4.56	12.26 ± 0.83	487.89 ± 35.21
17	8.50	17.50	4.00	98.66 ± 2.17	359.31 ± 25.50	122.59 ± 3.97	12.89 ± 0.49	494.79 ± 43.54

* X_1, ultrasonication time; ** X_2, Microbubble time; *** X_3, Reaction time.

2.5. Characterisation of FOS

The samples were filtered through a cellulose acetate membrane (0.22 μm; Sartorious, Göttingen, Germany) and then subjected to high-performance liquid chromatography (HPLC; HPLC 1260, Agilent Technology, Santa Clara, CA, USA) equipped with an Asahipak NH2P-50 4E column (5 μm, 250 × 4.6 mm; Showa Denko, Tokyo, Japan) and a refractive index (RI) detector. A mixture of acetonitrile and water (70:30) was used as a mobile phase at a flow rate of 1 $mL \cdot min^{-1}$ for 25 min. The column thermostat was 40 °C [36]. Peak identification and quantification of 1-kestose, nystose, 1^F-β-fructofuranosylnystose, sucrose, glucose, and fructose in samples were estimated from the calibration curve, which was built with standard reference sugars (Wako Pure Chemical Industries, Osaka, Japan) under the same HPLC conditions.

2.6. Statistical Analysis

All experiments were conducted in triplicate and the results were expressed as the mean ± standard deviations (SD). Statistical analysis of the results was performed using the SPSS statistical programme (version 17.0, IBM, Armonk, NY, USA). One-way analysis of variance (ANOVA), followed by Duncan's multiple range test, was carried out, and the differences between individual means were assessed at $p \leq 0.05$.

3. Results and Discussion

3.1. Production of FOS

FOS were synthesised from brown sugar using a combination of ultrasonication and microbubbles, in which the Box-Behnken design was used. Actual values of enzyme activity and FOS yield (1-kestose, nystose, 1^F-β-fructofuranosylnystose, and total FOS) obtained from an experimental design are shown in Table 1. Run 8 with 15 min of ultrasonication, 17.50 min of microbubbles and a 6.00 h reaction time led to the highest enzyme activity and a yield of 1^F-β-fructofuranosylnystose and total FOS, which were 112.22 ± 2.50 U·mL^{-1}, 24.61 ± 2.09 mg·g^{-1}, and 525.45 ± 56.09 mg·g^{-1}, respectively. Meanwhile, the highest yield of 1-kestose and nystose were obtained from Run 1 and Run 12, respectively, where the conditions were 2.00 min of ultrasonication, 5.00 min of microbubbles, and a 4.00 h reaction time for Run 1 and 8.50 min of ultrasonication, 30.00 min of microbubbles, and reaction time of 6.00 h for Run 12.

In addition, the relationship between the three variables (X_1, X_2 and X_3) and the responses were analysed by ANOVA, which are shown in Table 2. According to the analysis, the reaction time (X_3) was the most significant variable for Pectinex Ultra SP-L activity and production yield of 1-kestose, nystose, 1^F-β-fructofuranosylnystose and total FOS. Moreover, the quadratic term, reaction time (X_3^2) displayed the most significant effect on enzyme activity and yield of 1-kestose and total FOS. However, there were no interactive effects and quadratic terms that had a significant effect on the yield of nystose and 1^F-β-fructofuranosylnystose.

Table 2. Analysis of variance (ANOVA) for the response surface quadratic model for FOS production.

	Enzyme Activity		**1-Kestose**		**Nystose**		**1^F-β-Fructofuranosylnystose**		**Total FOS**	
	***F* Value**	***p* Value**	***F* Value**	***p* Value**	***F* Value**	***p* Value**	***F* Value**	***p* Value**	***F* Value**	***p* Value**
Model	24.57	0.0002	8.64	0.0048	75.78	<0.0001	54.54	<0.0001	42.45	<0.0001
X_1 *	2.10	0.1904	0.985	0.354	2.13	0.168	3.18	0.0979	1.58	0.248
X_2 **	2.18	0.1832	2.87	0.134	0.573	0.462	0.545	0.473	3.78	0.0931
X_3 ***	191.53	<0.0001	27.92	0.0011	224.63	<0.0001	159.89	<0.0001	310.96	<0.0001
X_1X_2	1.18	0.3135	2.03	0.197					1.91	0.210
X_1X_3	0.225	0.6494	0.173	0.690					1.76	0.226
X_2X_3	0.304	0.5984	0.0128	0.913					0.247	0.634
X_1^2	0.0859	0.7780	0.230	0.646					1.66	0.239
X_2^2	0.0307	0.8660	0.694	0.432					0.00004	0.995
X_3^2	23.34	0.0019	43.55	0.0003					58.68	0.0001
Lack of Fit	5.61	0.0645	6.06	0.0572	1.27	0.438	0.981	0.553	4.34	0.0951

* X_1, ultrasonication time (min); ** X_2, Microbubble time (min); *** X_3, Reaction time (h).

Considering the significant terms, enzyme activity and FOS yield could be described by the mathematical equations obtained in terms of coded variables. The predictive equations and statistical data are shown in Table 3. In the equations, a positive (+) coefficient indicates a synergistic effect, whereas a negative sign (–) refers to an antagonistic effect. According to the mathematical equations, the most significant variable was X_3 (reaction time) because it had a positive effect on all responses. Moreover, X_1 (ultrasonication time) also had a positive effect on the yield of nystose and 1^F-β-fructofuranosylnystose. However, the interaction of the three variables showed both positive and negative effects.

Table 3. Analysis of variance (ANOVA) of the model and coefficient estimates for response parameters.

Response	**Final Equation in Term of Actual Factors**	***P* ***	**LOF**	**R^2**	**Adj. R^2**	**AP**	**CV**
Enzyme activity	$35.10–0.20 \times X_1–0.59 \times X_2 + 24.95 \times X_3–0.012 \times X_1^2 + 0.002 \times X_2^2–2.16 \times X_3^2 + 0.024 \times X_1 \times X_2 + 0.07 \times X_1 \times X_3 + 0.04 \times X_2 \times X_3$	0.0002	0.0645	0.9693	0.9299	14.480	3.96
1-Kestose	$226.07–1.22 \times X_1–1.92 \times X_2 + 70.97 \times X_3 + 0.01 \times X_1^2 + 0.02 \times X_2^2–7.57 \times X_3^2 + 0.08 \times X_1 \times X_2–0.15 \times X_1 \times X_3–0.02 \times X_2 \times X_3$	0.0048	0.0572	0.9175	0.8113	9.140	2.69
Nystose	$12141 + 0.77 \times X_1–0.21 \times X_2 + 25.63 \times X_3$	<0.0001	0.4385	0.9459	0.9334	23.978	9.17
1^F-β-Fructofuranosyl-nystose	$-7.02 + 0.19 \times X_1–0.04 \times X_2 + 4.28 \times X_3$	<0.0001	0.5529	0.9264	0.9094	21.032	17.44
Total FOS	$165.75 + 4.35 \times X_1–1.94 \times X_2 + 125.68 \times X_3–0.17 \times X_1^2 + 0.0002 \times X_2^2–10.82 \times X_3^2 + 0.10 \times X_1 \times X_2–0.59 \times X_1 \times X_3 + 0.12 \times X_2 \times X_3$	<0.0001	0.0951	0.9820	0.9589	19.098	2.49

* *P*, probability of error; LOF, lack of fit; Adj. R^2, adjust R^2; AP, adequate precision; CV, coefficient of variance.

The relationships between three variables and responses (enzyme activity, 1-kestose, nystose, 1^F-β-fructofuranosylnystose and total FOS) are shown in Figure 1. The Pectinex Ultra SP-L activity (Figure 1a–c) and the yield of total FOS (Figure 1d–f) and 1-kestose (Figure 1g–i) increased when the reaction time (X_3) was increased to 4–5 h. Moreover, ultrasonication time (X_1) led to an increased yield of nystose (Figure 1j) and 1^F-β-fructofuranosylnystose (Figure 1k), while microbubbles (X_2) had no effect on them.

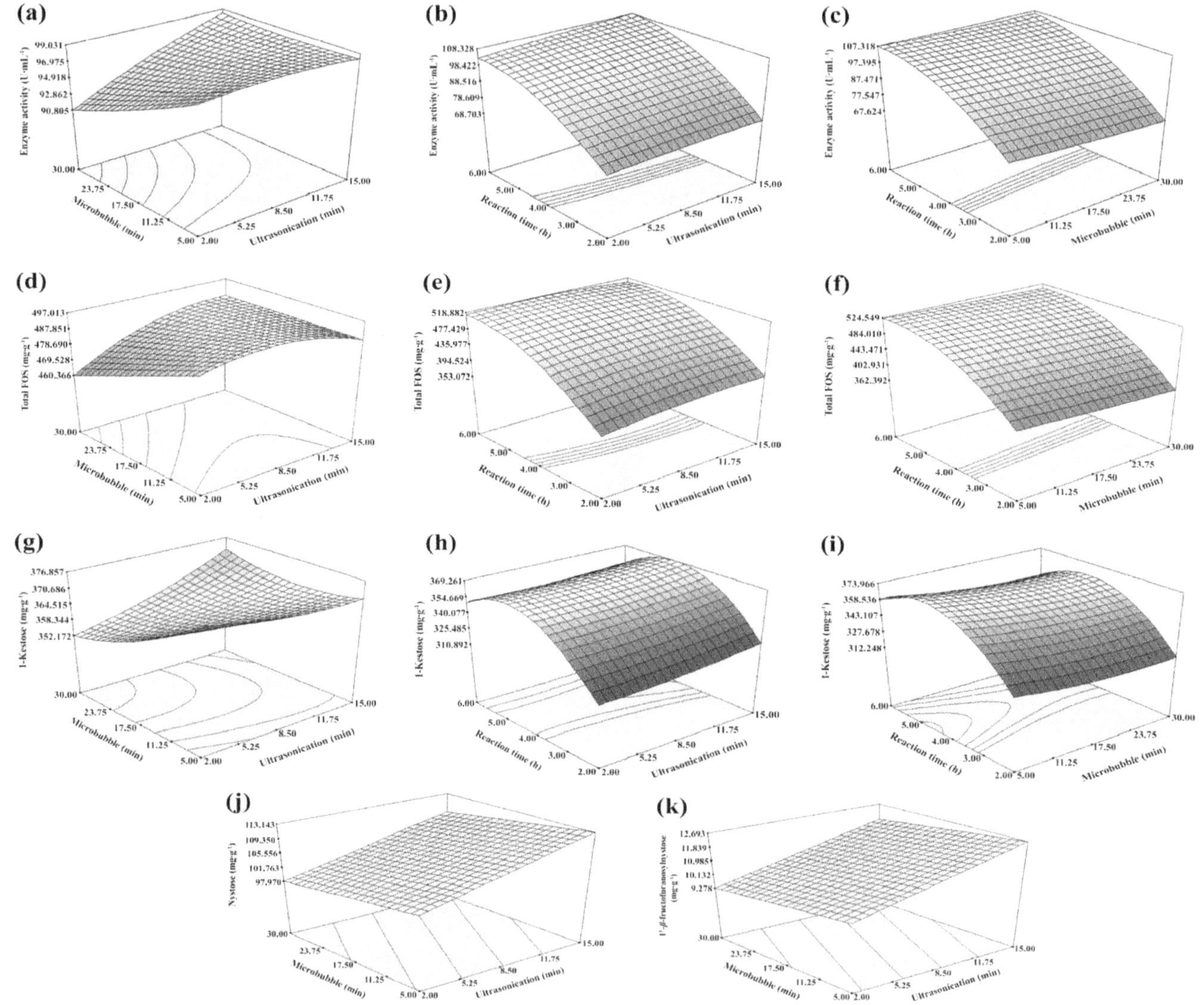

Figure 1. Response surface plots of the relationship between three variables (ultrasonication, microbubble and reaction time), (**a**–**c**) enzyme activity, (**d**–**f**) total FOS; (**g**–**i**) 1-kestose, (**j**) nystose, (**k**) 1^F-β-fructofuranosylnystose.

3.2. Optimisation of FOS Production

Based on the experimental designs, the optimum conditions for FOS production from brown sugar using ultrasonication and microbubbles were; (1) ultrasonication for 5 min 45 s, (2) microbubbles for 7 min 19 s, and (3) reaction time of 5 h 40 min. Under the optimum conditions, the predictive enzyme activity was 106.71 U·mL^{-1} and the predictive yield of kestose, nystose, 1^F-β-fructofuranosylnystose and total FOS were 365.25, 147.92, 17.92 and 525.44 mg·g^{-1}, respectively. To validate the optimum conditions, validation experiments were performed using white sugar and brown sugar as substrates. The actual values obtained from experiments under each set of conditions are shown in Table 4. The combination of ultrasonication and microbubbles enhanced the Pectinex Ultra SP-L activity up to 102.51 ± 4.69 U·mL^{-1}, which produced a total FOS yield of 494.89 ± 19.98 mg·g^{-1} substrate using brown sugar. For white sugar, the enzyme activity was 107.93 ± 5.26 U·mL^{-1} and total FOS yield was increased to 580.01 ± 21.35 mg·g^{-1} substrate. When the ultrasonication was performed without microbubbles, the enzyme activity and total FOS yield were decreased. This result confirmed that a microbubble step was important for the production. According to a previous report, Pectinex Ultra SP-L activity under normal conditions was 22 U·mL^{-1} [5]. Therefore, Pectinex Ultra SP-L activity was increased by 366% and 390% for brown sugar and white sugar, respectively, when the combined ultrasonication and microbubble were performed. Noticeably, white sugar led to a higher enzyme activity and total FOS yield than brown sugar (Table 4). Thus, white sugar was appropriate to use as a substrate for FOS production. Brown sugar contains sucrose, glucose, and fructose. Glucose in brown sugar could inhibit fructosyltransferase activity via feedback inhibition, so total FOS yield was lower [21]. According to the validation, the error was 5.18% between the predicted model and experimental values. Hu [37] reported that the error value should be less than 10%, so the model was acceptable to predict the production of FOS using Pectinex Ultra SP-L with a combination of ultrasonication and microbubbles.

Table 4. Fructooligosaccharide (FOS) yields produced under optimal conditions: (X_1) ultrasonication 5 min 45 s, (X_2) microbubble 7 min 19 s, and (X_3) reaction time of 5 h 40 min.

Factors	Pectinex Ultra SP-L Activity (U·mL^{-1})	Fructooligosaccharide Yields (mg·g^{-1})			
		1-Kestose	Nystose	1^F-β-Fructofurano-sylnystose	Total FOS
Brown sugar using a combination of ultrasonication and microbubble	102.51 ± 4.69	368.02 ± 20.22	110.53 ± 8.81	16.35 ± 1.49	494.89 ± 19.98
White sugar using a combination of ultrasonication and microbubbles	107.93 ± 5.26	415.02 ± 19.54	147.50 ± 9.84	17.49 ± 2.94	580.01 ± 21.35
White sugar using only ultrasonication	90.05 ± 3.54	399.48 ± 21.33	103.83 ± 12.33	10.23 ± 1.66	515.53 ± 20.43

3.3. *Production of FOS in a Large Scale*

The scale-up of FOS production was performed at 0.1, 1.5 and 5.0-L scales under the optimum conditions described above. The samples were collected every 1 h for a total of 6 h (Figure 2). At the 100-mL scale, the maximum yield of 1-kestose, nystose, 1^F-β-fructofuranosylnystose, and total FOS were obtained at 4, 6, 6, and 5 h reaction time, respectively. When a volume of 1.5-L production was used, the optimum reaction time was 4, 4, 3 and 4 h for 1-kestose, nystose, 1^F-β-fructofuranosyl-nystose, and total FOS, respectively. While the maximum yield of 1-kestose, nystose, 1^F-β-fructofuranosylnystose, and total FOS at the 5-L scale occurred at 4, 6, 6 and 4 h reaction time, respectively. In the 100-mL batch, the highest total FOS yield was 462.34 ± 24.22 mg·g^{-1} substrate at a 5 h reaction. While the optimum reaction time of the enlarge production scale (1.5 and 5 L) decreased to 4 h, the maximum yields of total FOS were 459.24 ± 25.64 and 452.55 ± 19.53 mg·g^{-1} substrate, respectively (Figure 2a). Thus, the enlarged production took a short reaction time than 100-mL production. Interestingly, the FOS yields produced in a 1.5-L batch were much higher than other batches at the optimum reaction time (Figure 2a,c,d). Therefore, 1.5-L scale might be appropriate for FOS production from brown sugar and the enlarged production scale decreased the reaction time, compared to the 100-mL scale.

In comparison, the total FOS yield from brown sugar was 0.45 g·g^{-1} substrate, which was lower than that of white sugar (0.58 g·g^{-1} substrate). Despite a low FOS yield, brown sugar had a shorter reaction time (4 h) than white sugar (5.67 h). Table 5 presents the total FOS yield obtained from this study in comparison to other studies. In previous studies, they used the conventional method and ultra-high pressure (UHP)-assisted method to synthesise FOS from sugar. The UHP.assisted method took the shortest reaction time (15 min) and had a total FOS yield of 0.57 g·g^{-1} substrate [5]. Among the conventional method, FOS yield ranged from 0.55 to 0.67 g·g^{-1} substrate within a reaction time of 5.6

to 16 h [38–43]. Although the FOS yield from this study was lower, the reaction time was reduced. The normal reaction time in the conventional method was ~10 h when using Pectinex Ultra SP-L in the synthesis. Here we found that the use of ultrasonication and microbubbles could decrease reaction time by 60%.

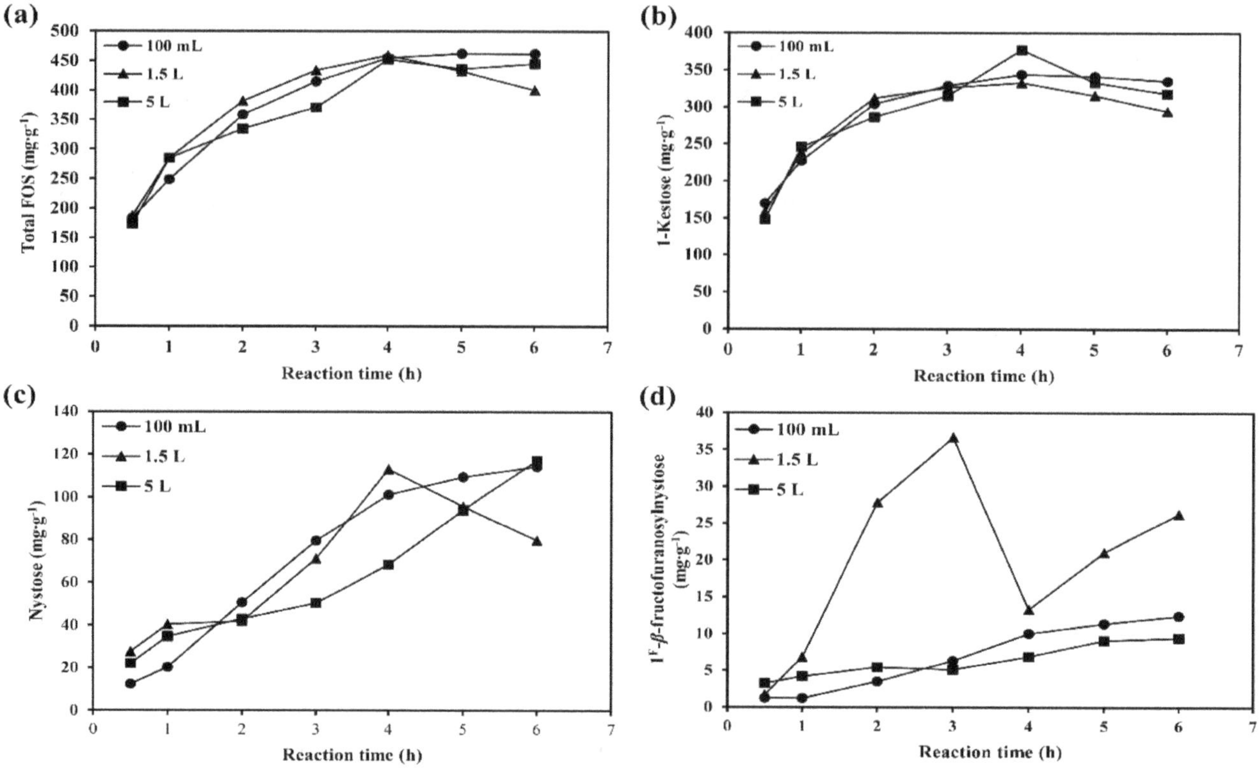

Figure 2. Fructooligosaccharide production from brown sugar in 100 mL, 1.5 L and 5 L batches. (**a**) Total FOS yield, (**b**) 1-Kestose yield, (**c**) Nystose yield, and (**d**) 1^F-β-fructofuranosylnystose yield.

Table 5. Comparison of FOS production from this study and other reports.

Substrate	Enzyme	Condition	Reaction Time (h)	FOS Yield ($g \cdot g^{-1}$ Substrate)	Reference
Brown sugar	Pectinex Ultra SP-L	Ultrasonication and microbubble, 55 °C, 700 $g \cdot L^{-1}$ substrate	4	0.45	This study
White sugar	Pectinex Ultra SP-L	Ultrasonication and microbubble, 55 °C, 700 $g \cdot L^{-1}$ substrate	5.67	0.58	This study
White sugar	GAPfopA_V1 (engineered β-fructofuronosidase)	Water bath, 62 °C, 120 rpm, 600 $g \cdot L^{-1}$ substrate	5. 6	0.55	[43]
White sugar	Pectinex Ultra SP-L	Ultra-high pressure at 300 MPa, 30 °C, 600 $g \cdot L^{-1}$ substrate	0.25	0.57	[5]
White sugar and Jerusalem artichoke	Inulinase	Water bath, 55 °C, 59.4 $g \cdot L^{-1}$ inulin and 598.7 $g \cdot L^{-1}$ refine sugar	9	0.67	[38]
White sugar	Pectinex Ultra SP-L	Water bath, 60 °C, 400 $g \cdot L^{-1}$ substrate	16	0.62	[39]
White sugar	Pectinex Ultra SP-L	Water bath, 55 °C, 500 $g \cdot L^{-1}$ substrate	10	0.60	[40]
White sugar	Pectinex Ultra SP-L	Water bath, 50 °C, 536.2 $g \cdot L^{-1}$ substrate	6	0.59	[41]
White sugar	Culture fluid of *Aspergillus oryzae* CFR202	Water bath, 55 °C, 64.55 $g \cdot L^{-1}$ substrate	12	0.58	[42]

This study demonstrated that the combination of ultrasonication and microbubbles could enhance FOS concentration and productivity compared to the conventional method (Table 6). In 4 h of reaction time, the combined methods gave the maximum concentration of 1-kestose (264.1 $g \cdot L^{-1}$) and total FOS (316.8 $g \cdot L^{-1}$), which also had the highest productivity (66.0 and 79.2 $g \cdot L^{-1} \cdot h^{-1}$, respectively).

Conversely, the concentration and productivity of nystose and 1^F-β-fructofuronosyl nystose were lower than in previous studies. This confirmed that using the combined methods had a higher productivity of FOS from brown sugar, despite a low FOS yield. Although the UHP-assisted approach produced the highest productivity of FOS, this method required complex high-pressure equipment systems and consumed a lot of energy due to extreme pressure [5]. In contrast, the ultrasonication and microbubble techniques were simple methods with mild conditions that consumed less energy. Additionally, these techniques have been accepted as cost-effective and eco-friendly processes [28]. Consequently, this method was suitable to apply on an industrial scale for FOS production. Many studies have reported that ultrasound at an appropriate frequency led to changes in enzyme conformation. The conformational changes had important roles in catalysis. A new conformation might enhance binding with substrates, therefore increasing enzyme activity [28,29,44].

Table 6. Fructooligosaccharide (FOS) yield and productivity obtained from this study compared to other reports.

Method	Reaction Time (h)	Fructooligosaccharide								Reference
		Concentration ($g{\cdot}L^{-1}$)				Productivity ($g{\cdot}L^{-1}{\cdot}h^{-1}$)				
		GF_2 *	GF_3	GF_4	Total FOS	GF_2	GF_3	GF_4	Total FOS	
Ultrasonic and microbubble methods	4	264.1	47.8	4.84	316.8	66.0	11.96	1.21	79.2	This study
Ultra-high pressure (UHP) method	0.25	189.72	110.88	28.2	328.8	758.88	443.52	112.8	1315.2	[5]
Conventional methods	5.6	220.8	321.6	60.0	330	39.4	57.4	10.7	58.9	[43]
	9	156.98	144.10	84.63	404.0	17.44	16.01	9.40	44.90	[38]
	10	155	130	15.0	300	15.5	13.0	1.50	30.0	[40]
	6	212.33	106.7	8.58	327.6	35.4	17.78	1.43	54.6	[41]
	12	174.2	96.6	26.8	297.7	14.5	8.05	2.24	24.8	[42]

* GF_2, 1-Kestose; GF_3, Nystose; GF_4, 1^F-β-fructofuronosylnystose.

With regard to microbubbles, the effect of microbubbles on enzymes and the production is still unclear and requires further investigation. Microbubbles might create turbulence in the system and increase the possibility of enzyme-substrate attachment. Based on the increased surface area of microbubbles, we speculated that microbubbles might adsorb sugar molecules on their surfaces and carry them to the catalytic sites of enzymes, thereby increasing the activity [45]. Furthermore, various reports have confirmed that high pressure has both positive and negative effects on enzyme activity. Adequate pressure can enhance enzyme activity [46], for the reason that microbubbles are unstable, are suddenly broken and gas is released. Spontaneously, pressure is generated and affects the enzyme. We suggested that the pressure from microbubbles affected enzyme activity, so FOS yields and enzyme activity were increased.

We found that the technique is useful and easy to use for FOS production and has the potential to be applied in industrial production. In the future, the combination of ultrasonication and microbubbles has the potential to increase production of other oligosaccharides, including galactooligosaccharides (GOS), xylooligosaccharides (XOS), inulooligosaccharides (IOS), and soybean oligosaccharides (SOS) and should be considered in future research.

4. Conclusions

The optimal conditions for FOS production from brown sugar were the ultrasonication for 5 min 45 s, microbubbles for 7 min 19 s, and a reaction time of 5 h 40 min. The combination of ultrasonication and microbubbles could enhance enzyme activity, and the Pectinex Ultra SP-L activity reached 102.51 ± 4.69 $U{\cdot}mL^{-1}$. The highest yield of total FOS, 1-kestose, nystose, and 1^F-β-fructofuranosylnystose were 494.89 ± 19.98, 368.48 ± 20.22, 110.53 ± 8.81, and 16.35 ± 1.49 mg·g^{-1} substrate, respectively. On a large scale (5 L), the maximum yield of total FOS, 1-kestose, nystose, and 1^{F}-β-fructofuranosylnystose were 452.55 ± 19.53, 377.28 ± 21.64, 68.35 ± 2.35 and 6.92 ± 0.09 mg·g^{-1} substrate, respectively within 4 h of reaction. Thus, the use of ultrasonication and microbubble in FOS production using brown sugar was able to increase Pectinex Ultra SP-L activity by 366% and reduce the reaction time by 60%, which productivity of FOS was enhanced.

This strategy has various advantages, such as easy and safe performance, and high efficacy. Additionally, this study uses a low-cost substrate (brown sugar) in which significantly reduces production costs and affects the economy of the process. Therefore, the use of brown sugar as a substrate with ultrasonication and microbubble assistants is an appropriate strategy for FOS production at an industrial scale.

Author Contributions: Conceptualisation, P.S., Y.P. and T.C.; funding Acquisition, P.S. and Y.P. methodology, W.C. and A.M.; software, Y.P. and K.J.; resources, P.S. and Y.P., validation, P.S., T.C., and S.Y.; formal analysis, Y.P., K.J. and T.C.; investigation, W.C. and A.M.; writing—original draft preparation, W.C.; writing—review and editing, P.S., W.P.-a. and S.Y. All authors have read and agreed to the published version of the manuscript.

Funding: The authors are grateful the financial support provided by the National Research Council of Thailand (NRCT) via the Royal Golden Jubilee PhD Programme, Thailand (Grant No. PHD/0185/2560) to Mr Worraprat Chaisuwan and Dr Phisit Seesuriyachan. P. Seesuriyachan also acknowledges partial financial supports and/or in-kind assistance from Program Management Unit – Brain Power (PMUB), The Office of National Higher Education Science Research and Innovation Policy Council (NXPO) in Global Partnership Project, Basic research fund, Thailand Science Research and Innovation (TSRI), Chiang Mai University (CMU), Faculty of Science (CoE64-P001) as well as Faculty of Agro-Industry (CMU-8392(10)/COE64), CMU.

Acknowledgments: The authors are grateful to Interdisciplinary Program in Biotechnology, Graduate School of Chiang Mai University; Cluster of Agro Bio-Circular-Green Industry (Agro BCG); Faculty of Agro-Industry and Chiang Mai University for the research infrastructure.

References

1. Noncommunicable Diseases. Available online: https://www.who.int/news-room/fact-sheets/detail/noncommunicable-diseases. (accessed on 17 April 2020).
2. Hasler, C.M. Functional foods: Benefits, concerns and challenges—A position paper from the American Council on science and health. *J. Nutr.* **2002**, *132*, 3772–3781. [CrossRef] [PubMed]
3. Al-Sheraji, S.H.; Ismail, A.; Manap, M.Y.; Mustafa, S.; Yusof, R.M.; Hassan, F.A. Prebiotics as functional foods: A review. *J. Funct. Foods* **2013**, *5*, 1542–1553. [CrossRef]
4. Karnaouri, A.; Matsakas, L.; Krikigianni, E.; Rov, U.; Christakopoulos, P. Valorization of waste forest biomass toward the production of cello-oligosaccharides with potential prebiotic activity by utilizing customized enzyme cocktails. *Biotechnol. Biofuels* **2019**, *12*, 285. [CrossRef]
5. Kawee-Ai, A.; Chaisuwan, W.; Manassa, A.; Seesuriyachan, P. Effects of ultra-high pressure on effective synthesis of fructooligosaccharides and fructotransferase activity using Pectinex Ultra SP-L and inulinase from Aspergillus niger. *Prep. Biochem. Biotechnol.* **2019**, *49*, 649–658. [CrossRef]
6. Fructo-Oligosaccharides, GRN No. 605. Available online: https://www.accessdata.fda.gov/scripts/fdcc/?set=GRASNotices&id=605&sort=GRN_No&order=DESC&startrow=1&type=basic&search=fructo. (accessed on 18 April 2020).
7. Fructooligosaccharide, GRN No. 44. Available online: https://www.accessdata.fda.gov/scripts/fdcc/?set=GRASNotices&id=44&sort=GRN_No&order=DESC&startrow=1&type=basic&search=fructo. (accessed on 18 April 2020).
8. Campbell, J.M.; Bauer, L.L.; Fahey, G.C., Jr.; Hogarth, A.J.C.L.; Wolf, B.W.; Hunter, D.E. Selected fructooligosaccharide (1-Kestose, Nystose, and 1F-β-Fructofuranosylnystose) composition of foods and feeds. *J. Agric. Food Chem.* **1997**, *45*, 3075–3082. [CrossRef]
9. Akalın, A.S.; Fenderya, S.; Akbulut, N. Viability and activity of bifidobacteria in yoghurt containing fructooligosaccharide during refrigerated storage. *Int. J. Food Sci. Tech.* **2004**, *39*, 613–621. [CrossRef]
10. Tanno, H.; Fujii, T.; Ose, R.; Hirano, K.; Tochio, T.; Endo, A. Characterization of fructooligosaccharide-degrading enzymes in human commensal Bifidobacterium longum and Anaerostipes caccae. *Biochem. Biophys. Res. Commun.* **2019**, *518*, 294–298. [CrossRef]

11. Liu, F.; Li, P.; Chen, M.; Luo, Y.; Prabhakar, M.; Zheng, H.; He, Y.; Qi, Q.; Long, H.; Zhang, Y.; et al. Fructooligosaccharide (FOS) and galactooligosaccharide (GOS) increase Bifidobacterium but reduce butyrate producing bacteria with adverse glycemic metabolism in healthy young population. *Sci. Rep.* **2017**, *7*, 11789. [CrossRef] [PubMed]
12. Rycroft, C.E.; Jones, M.R.; Gibson, G.R.; Rastall, R.A. A comparative in vitro evaluation of the fermentation properties of prebiotic oligosaccharides. *J. Appl. Microbiol.* **2001**, *91*, 878–887. [CrossRef] [PubMed]
13. Tang, C.; Ding, R.; Sun, J.; Liu, J.; Kan, J.; Jin, C. Impacts of natural polysaccharides on the intestinal microbiota and immune responses–A review. *Food Funct.* **2019**, *10*, 2290–2312. [CrossRef] [PubMed]
14. Buddington, R.K.; Kelly-Quangliana, K.; Buddington, K.K.; Kimura, Y. Non-digestible oligosaccharides and defense functions: Lessons learned from animal models. *Br. J. Nutr.* **2002**, *87*, S159–S162. [CrossRef]
15. Taper, H.; Roberfroid, M. Inulin/oligofructose and anticancer therapy. *Br. J. Nutr.* **2002**, *87*, S283–S286. [CrossRef] [PubMed]
16. Olesen, M.; Gudmand-Hoyer, E. Efficacy, safety, and tolerability of fructooligosaccharides in the treatment of irritable bowels syndrome. *Am. J. Clin. Nutr.* **2000**, *72*, 1570–1575. [CrossRef]
17. Yildiz, S. The metabolism of fructooligosaccharides and fructooligosaccharide-related compounds in plants. *Food Rev. Int.* **2010**, *27*, 16–50. [CrossRef]
18. Sangeetha, P.T.; Ramesh, M.N.; Prapulla, S.G. Recent trends in the microbial production, analysis and application of fructooligosaccharides. *Trends Food Sci. Technol.* **2005**, *16*, 442–457.
19. Surin, S.; Seesuriyachan, P.; Thakeow, P.; Phimolsiripol, Y. Optimization of enzymatic production of fructooligosaccharides from longan syrup. *J. Appl. Sci.* **2012**, *12*, 1118–1123. [CrossRef]
20. Karboune, S.; Appanah, N.; Khodaei, N.; Tian, F. Enzymatic synthesis of fructooligosaccharides from sucrose by endo-inulinase-catalyzed transfructosylation reaction in biphasic systems. *Process Biochem.* **2018**, *69*, 82–91. [CrossRef]
21. Silva, M.F.; Rigo, D.; Mossi, V.; Golunski, S.; Kuhn Gde, O.; Di Luccio, M.; Dallago, R.; de Oliveira, D.; Oliveira, J.V.; Treichel, H. Enzymatic synthesis of fructooligosaccharides by inulinases from Aspergillus niger and Kluyveromyces marxianus NRRL Y-7571 in aqueous-organic medium. *Food Chem.* **2013**, *138*, 148–153. [CrossRef]
22. Olwoch, I.P.; Greeff, O.B.W.; Jooné, G.; Steenkamp, V. The effects of the nastural enzyme, Pectinex Ultra SP-L, on human cell cultures and bacterial biofilms in vitro. *BMC Microbiol.* **2014**, *14*, 251. [CrossRef]
23. Yun, J.W. Fructooligosaccharides—Occurrence, preparation, and application. *Enzyme Microb. Technol.* **1996**, *19*, 107–117. [CrossRef]
24. Goulas, A.K.; Kapasakalidis, P.G.; Sinclair, H.R.; Rastall, R.A.; Grandison, A.S. Purification of oligosaccharides by nanofiltration. *J. Membrane Sci.* **2002**, *209*, 321–335. [CrossRef]
25. Nobre, C.; Teixeira, J.A.; Rodrigues, L.R. Fructo-oligosaccharides purification from a fermentative broth using an activated charcoal column. *New Biotechnol.* **2012**, *29*, 395–401. [CrossRef] [PubMed]
26. Jadwong, K.; Therdthai, N. Effects of ultrasonicand enzymatic treatment on cooking and eating quality of Sao Hai rice. *Food Appl. Biosci. J.* **2018**, *6*, 153–165.
27. Tanongkankit, Y.; Kalantakasuwan, S.; Varit, J.; Narkprasom, K. Ultrasonic-assiated extraction of allicin and its stability during storage. *Food Appl. Biosci J.* **2019**, *7*, 17–31.
28. Nadar, S.S.; Rathod, V.K. Ultrasound assisted intensification of enzyme activity and its properties: A mini-review. *World J. Microbiol. Biotechnol.* **2017**, *33*, 170. [CrossRef]
29. Wang, D.; Yan, L.; Ma, X.; Wang, W.; Zou, M.; Zhong, J.; Ding, T.; Ye, X.; Liu, D. Ultrasound promotes enzymatic reactions by acting on different targets: Enzymes, substrates and enzymatic reaction systems. *Int. J. Biol. Macromol.* **2018**, *119*, 453–461. [CrossRef]
30. Tran, T.T.T.; Nguyen, K.T.; Le, V.V.M. Effects of ultrasonication variables on the activity and properties of alpha amylase preparation. *Biotechnol. Prog.* **2018**, *34*, 702–710. [CrossRef]
31. Sun, Y.; Zeng, L.; Xue, Y.; Yang, T.; Cheng, Z.; Sun, P. Effects of power ultrasound on the activity and structure of β-D-glucosidase with potentially aroma-enhancing capability. *Food Sci. Nutr.* **2019**, *7*, 2043–2049. [CrossRef]
32. Subhedar, P.B.; Gogate, P.R. Enhancing the activity of cellulase enzyme using ultrasonic irradiations. *J. Mol. Catal. B Enzym.* **2014**, *101*, 108–114. [CrossRef]
33. O'Donnell, C.P.; Tiwari, B.K.; Bourke, P.; Cullen, P.J. Effect of ultrasonic processing on food enzymes of industrial importance. *Trends Food Sci. Technol.* **2010**, *21*, 358–367. [CrossRef]

34. Valero, M.; Recrosio, N.; Saura, D.; Muñoz, N.; Martí, N.; Lizama, V. Effects of ultrasonic treatments in orange juice processing. *J. Food Eng.* **2007**, *80*, 509–516. [CrossRef]
35. Miller, G.L. Use of dinitrosalicylic acid reagent for determination of reducing sugar. *Anal. Chem.* **1959**, *31*, 426–428. [CrossRef]
36. Endo, H.; Tamura, K.; Fukasawa, T.; Kanegae, M.; Koga, J. Comparison of fructooligosaccharides utilization by Lactobacillus and Bacteroides species. *Biosci. Biotechnol. Biochem.* **2012**, *76*, 176–179. [CrossRef]
37. Hu, R. *Food Product Design: A Computer-Aided Statistical Approach*; Technomic Publishing: Lancaster, PA, USA, 1999; p. 240.
38. Kawee-Ai, A.; Ritthibut, N.; Manassa, A.; Moukamnerd, C.; Laokuldilok, T.; Surawang, S.; Wangtueai, S.; Phimolsiripol, Y.; Regenstein, J.M.; Seesuriyachan, P. Optimization of simultaneously enzymatic fructo- and inulo-oligosaccharide production using co-substrates of sucrose and inulin from Jerusalem artichoke. *Prep. Biochem. Biotechnol.* **2018**, *48*, 194–201. [CrossRef] [PubMed]
39. Mutanda, T.; Wilhelmi, B.S.; Whiteley, C.G. Biocatalytic conversion of inulin and sucrose into short chain oligosaccharides for potential pharmaceutical applications. *Afr. J. Sci. Technol. Innov. Dev.* **2015**, *7*, 371–380. [CrossRef]
40. Kashyap, R.; Palai, T.; Bhattacharya, P.K. Kinetics and model development for enzymatic synthesis of fructooligosaccharides using fructosyltransferase. *Bioproc. Biosyst. Eng.* **2015**, *38*, 2417–2426. [CrossRef]
41. Vega-Paulino, R.J.; Zúniga-Hansen, M.E. Potential application of commercial enzyme preparations for industrial production of short-chain fructooligosaccharides. *J. Mol. Catal. B Enzym.* **2012**, *76*, 44–51. [CrossRef]
42. Sangeetha, P.T.; Ramesh, M.N.; Prapulla, S.G. Production of fructo-oligosaccharides by fructosyl transferase from Aspergillus oryzae CFR 202 and Aureobasidium pullulans CFR 77. *Process Biochem.* **2004**, *39*, 755–760. [CrossRef]
43. Coetzee, G.; van Rensburg, E.; Görgens, J.F. Evaluation of the performance of an engineered β-fructofuranosidase from Aspergillus fijiensis to produce short-chain fructooligosaccharides from industrial sugar streams. *Biocatal. Agric. Biotechnol.* **2020**, *23*, 101484. [CrossRef]
44. Gutteridge, A.; Thornton, J. Conformational change in substrate binding, catalysis and product release: An open and shut case? *FEBS Lett.* **2004**, *567*, 67–73. [CrossRef]
45. Park, J.S.; Kenji, K. Application of microbubbles to hydroponics solution promotes lettuce growth. *Horttechnology* **2009**, *19*, 212–215. [CrossRef]
46. Northrop, D.B. Effects of high pressure on enzymatic activity. *Biochim. Biophys. Acta Protein Struct. Mol. Enzymol.* **2002**, *1595*, 71–79. [CrossRef]

2

Physicochemical Effects of *Lactobacillus plantarum* and *Lactobacillus casei* Cultures on Soy–Wheat Flour Dough Fermentation

Bernadette-Emőke Teleky [1], Gheorghe Adrian Martău [1,2] and Dan Cristian Vodnar [1,2,*]

1 Institute of Life Sciences, University of Agricultural Sciences and Veterinary Medicine, Calea Mănăștur 3-5, 400372 Cluj-Napoca, Romania; bernadette.teleky@usamvcluj.ro (B.-E.T.); adrian.martau@usamvcluj.ro (G.A.M.)

2 Faculty of Food Science and Technology, University of Agricultural Sciences and Veterinary Medicine, Calea Mănăștur 3-5, 400372 Cluj-Napoca, Romania

* Correspondence: dan.vodnar@usamvcluj.ro.

Abstract: In contemporary food production, an important role is given to the increase in the nutritional quality of foodstuff. In the bakery industry, one of the main cereals used is wheat flour (WF), which creates bread with proper sensory evaluation but is nutritionally poor. Soy-flour (SF) has increased nutrient content, and its consumption is recommended due to several health benefits. Dough fermentation with lactic acid bacteria (LAB) increases bread shelf life, improves flavor, and its nutritional quality, mostly due to its high organic acid production capability. In the present study, the addition of SF to WF, through fermentation with the cocultures of *Lactobacillus plantarum* and *Lactobacillus casei* was analyzed. Three different batches were performed by using WF supplemented with SF, as follows: batch A consisting of 90% WF and 10% SF; batch B—95% WF and 5% SF; batch C—100% WF. The fermentation with these two LABs presented several positive effects, which, together with increased SF content, improved the dough's rheological and physicochemical characteristics. The dynamic rheological analysis exhibited a more stable elastic-like behavior in doughs supplemented with SF (G' 4936.2 ± 12.7, and G'' 2338.4 ± 9.1). Organic acid production changes were the most significant, especially for the lactic, citric, and tartaric content.

Keywords: dough fermentation; lactic acid bacteria; rheology; soy flour; wheat flour; organic acid

---

1. Introduction

The global food industry confronts serious challenges to provide consistent nutrition [1,2]. Cereals are classified among the primary food resources worldwide [3], wheat (*Triticum aestivum* L.) represents an essential cereal produced in the world, and it is also highly consumed, especially in Europe [4]. In the bakery industry, doughs are made mainly of refined wheat flour, owing to its particular viscoelastic features resulting mostly from the gluten network (gliadins and glutenins). Although bread produced from wheat flour (WF) has a high energy source, nutritionally, it is unsatisfactory with low mineral and fiber content [5]. Several studies search for alternative ingredients (i.e., amaranth, aleurone, potato, quinoa, etc.) as a substitute, in bakery products to increase its nutritional quality [5–8]. Soybean (*Glycine max* L.) is part of the Fabaceae family, with particularly high nutrient content like carbohydrates, lipids, proteins, minerals, fibers, vitamins, and has low saturated fat content [9,10]. Soybeans are rich in isoflavones (phytoestrogens) like daidzein and genistein, which specifically contribute to the biological activity of soybean that aroused considerable interest in functional food preparation [11]. As demonstrated, fermentation improves the sensory and physicochemical characteristics of soy-based products and has several health benefits [12].

An ancient biotechnological process is sourdough fermentation that is highly used in the food industry and was a unique bread leavening technique preceding the breakthrough of yeast fermentation [13]. Generally, sourdough fermentation has a positive effect on WF-based products, as a consequence of low pH on dough constituent's structure formation like arabinoxylans, starch, and gluten. Lactic acid bacteria (LAB) have a positive influence on bread quality, especially regarding shelf life, texture, nutrition, and flavor resulting in a product free of additives [14]. Through fermentation, carbohydrates are consumed being essential for the growth of LAB, and starch metabolization also leads to a drop in pH which enhances organic acid production [15]. LAB also have the ability to metabolize through hydrolysis the nondigestible oligosaccharides found in soy, like glycinin G1, β-conglycinin, and 2S albumin giving the final product flavors and aroma [16,17]. *Lactobacillus plantarum* (*LP*) is a facultative heterofermentative bacteria that use the Embden–Meyerhof–Parnas (EMP) pathway (Figure 1A) to break down glucose and generate lactic acid [18–20].

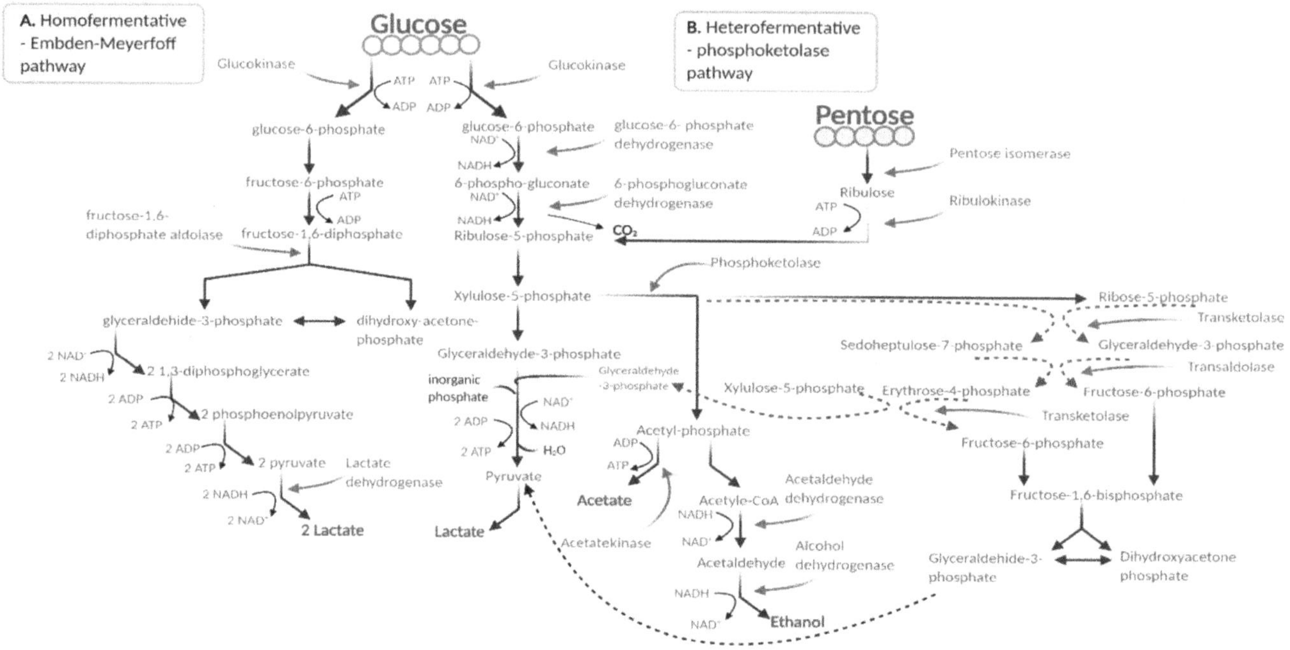

Figure 1. Glucose fermentation pathway. (**A**) Homofermentative metabolism (through Embden–Meyerhoff pathway); (**B**) heterofermentative metabolism (phosphoketolase pathway) (image created using BioRender application https://app.biorender.com).

This bacteria adapts easily, has an enhanced sourdough acidification rate, and is pliable with other bacteria or yeasts in cocultures [21]. *Lactobacillus casei* (*LC*) has the same adaptability, like *LP*, and can generate lactic acid through EMP (Figure 1A) and phosphoketolase pathways (Figure 1B), which leads to homo- and heterolactic fermentation [21]. *LC* is widely incorporated in foodstuffs owed to its appealing technological characteristics and health benefits [22,23]. Furthermore, dough fermentation has several beneficial effects, like the pretreatment of the substrate, through increasing and stabilizing the functional quality of the fermented dough. Moreover, it produces a diminished glycemic reaction, positively impacts allergic and intolerance effects in individuals, and increases the absorption of vitamins, minerals, and phytochemicals [24,25].

Through fermentation the rheological property of dough is influenced by many factors, like the content of sodium [26], the type of microorganism found in the fermented dough [16,27], the type of substrate used [14,28], and by the addition of different oleogels [29], proteins [30–32], or alternative sweeteners [33,34]. The impact of soy proteins on the wheat dough is highly studied [35,36]. The viscoelastic characteristics of wheat dough are given by its gluten network, and the consequence of incorporating different proteins might have a negative impact on bread properties [37]. The incorporation of soy-flour (SF) presents foam-stabilizing, moisture-holding potential, gelation, emulsifying activities, and through disulphide-linkages offers elasticity to bread products [32]. Additionally, a recent study proved that the enzyme β-glucosidase found in LAB can effectively increase the aglycone content of soybean products. An important aspect that should be analyzed through bread making is the property of dough after frozen storage, which can be effectively characterized through rheology [38].

The cocultivation of two LAB (*LP* and *LC*) on WF-SF substrates has not been studied only together with *Saccharomyces cerevisiae* (*Sc*) and especially on model media (MM) [21,39]. Fermentation with only LAB cocultures is important to see their competitiveness and organic acid production capabilities in bakery products. Additionally, the use of LAB as starter cultures in dough fermentation can increase the mineral availability, due to phytate degrading enzymes [40]. Only a few studies analyzed dough fermented exclusively by LAB, which could provide help to people suffering from various health problems [41]. An important aspect that needs to be better analyzed is the change during fermentation with LAB on different substrates and the effect of SF incorporation in WF doughs. Therefore, the present article has the objective to investigate the impact of SF incorporation in WF, through LAB coculture fermentation with *LP* and *LC*. Besides the dynamic rheological properties of three different concentrations of WF and SF doughs, the chemical properties, cell growth, and pH were also evaluated.

2. Materials and Methods

2.1. Strains and Culture Conditions

The fermentations involved two types of LAB, namely *Lb. plantarum* ATCC 8014 (*LP*) and *Lb. casei* ATCC 393 (*LC*) acquired from the University of Agricultural Science and Veterinary Medicine Cluj-Napoca. Culture media constituent elements and every reagent were of analytical grade. For both microorganisms MRS broth (casein peptone—tryptic digest 10 g/L; meat extract 10 g/L; yeast extract 5 g/L; glucose 20 g/L; Tween 80 1 g/L; K_2HPO_4 2 g/L; Na-acetate 5 g/L; $(NH_4)_3$ citrate 2 g/L; $MgSO_4 \times 7\ H_2O$ 0.2 g/L; $MnSO_4 \times H_2O$ 2 g/L; distilled water 1000 mL) was used.

2.2. Dough Preparation

The soy-flour utilized in this study originated from the Agricultural Research and Development Center Turda. The soybean type was Onix (*Glycine max* (L.) Merril), with a traditional cultivation method (tillage) using 60% of green fertilizer (vegetable debris). A commercially available wheat flour (type 000, in conformity with Romanian ash content categorization) was used, having 11.2% protein and 15.3% moisture content. Three types of flour concentrations were prepared for fermentation, composed of 90% WF enriched with 10% SF (batch A), 95% WF enriched with 5% SF (batch B), and 100% WF (batch C). The quantity of water added to the different flour concentrations (100 g of flour) was 60% of the whole formulations, while 40% was the inoculum composed of the two LABs used as coculture, with a total dough amount of 200 g. Before inoculation, the different WF-SF concentrations were sterilized, together with the necessary amount of distilled water, after which they were mixed thoroughly and inoculated with the activated and necessary amount of LAB as presented in our previous study Teleky et al. (2020) [39].

2.3. Model Media and Dough Fermentation

Model media (MM) cultured the same way as presented in a previous study [39], and the dough was inoculated with the microorganisms *LP* and *LC* in a concentration of 10^8 CFU/mL. The final volume of MM was 500 mL, and each dough combination had a final volume of 200 mL and was inoculated with the same amount of inoculum from both LAB. Sample extraction occurred at every 4 h (0, 4, 8, 12, 24 h) with 5 mL of sample for high-performance liquid chromatography (HPLC) analysis, 5 mL for pH, 1 mL for viability, and 5 mL for rheological measurements.

2.4. Testing Methods

2.4.1. pH

pH evaluation was established with a digital pH meter (InoLab 7110, Wellheim, Germany). Samples were dissolved with 45 mL of distilled water at room temperature and measured while the samples were homogenized continuously with a magnetic stirrer [42].

2.4.2. Cell Viability

LAB viability was established by diluting 1 mL of prelevated sample in 9 mL of sterile saline solution and processed with the spread plate method and incubated for 24 h at 37 °C [43]. The viability of LAB was evaluated through plate counting and colony-forming units/mL sample and was displayed in logarithmic values of colony-forming units/milliliter of the sample (log10 CFU/mL).

2.4.3. Rheological Properties

Samples for rheological analysis were stored at −20 °C, defrosted at room temperature after which they were analyzed with an Anton Paar MCR 72 rheometer (Anton Paar, Graz, Austria) [44,45]. The dynamic rheological measurements were performed with a Peltier plate-plate system (P-PTD 200/Air) supplied with temperature control, and with a smooth parallel plate geometry (PP-50-67300) of 50 mm. Shear strain (oscillating) was set at a constant value of 0.1% and angular frequency (ω) at a logarithmic ramp and set between the intervals of 0.628–628 rad/s. After the sample supplying, the gap between plates was set at 1 mm, the dough surplus was trimmed, and to prevent the sample from drying, silicone oil was distributed on the exterior.

2.4.4. HPLC

Organic acid consumption and production were measured by HPLC (Agilent 1200 series, Santa Clara, CA, USA) after the samples were filtered (for MM) or homogenized (for dough, 1 g of sample with 2 mL of distilled water), vortexed, sonicated, centrifuged, and filtered (0.45 µm pore size). The HPLC was provided with a solvent degasser, DAD detector coupled with a mass detector, quaternary pump, thermostat column, and an automatic injector (Agilent Technologies, Santa Clara, CA, USA). In the column, a volume of 20 µL of the sample was injected, with a 0.5 mL/min sample flow rate, and detection was carried out at 280 and 340 nm. The organic acid separation was accomplished on a reversed-phase chromatographic column Acclaim OA (5 µm, 4 × 150 mm Dionex), washed with NaH2PO4 50 mM concentration (pH 2.8) solution for 10 min, at 20 °C, and with a 0.5 mL/min flow rate. Chromatogram measurement was performed at a λ = 210 nm wavelength [46,47]. Organic acid standard stock solution (Merck) was prepared by mixing the components (lactic, tartaric, malic, citric, succinic, fumaric, butyric acids) (Supplementary Figure S1).

2.4.5. Statistical Analysis

Every measurement was performed in triplicate and expressed as mean value (±SD, n = 3). The statistical interpretation was performed with the help of Graph Prism Version 8.0.1. (GraphPad Software Inc., San Diego, CA, USA) with a one-way ANOVA test (Tukey multiple comparisons tests) [48]. Statistically significant differences of means were considered at a 5% level.

3. Results and Discussion

3.1. LAB Viability and pH

LABs are one of the most frequently used microorganisms in sourdough fermentation, and the use of appropriate starter cultures have great importance. In SF the carbon:nitrogen ratio is not satisfactory for the growth of LAB, and various carbon sources can influence the metabolism and growth of *LP* and *LC* [49]. Through fermentation on every WF and SF concentration, the growth of LAB cocultures increased from an average value of 6.1–6.4 log10 CFU/mL, which was very similar in every batch, to 10.8–10.9 log10 CFU/mL (Figure 2).

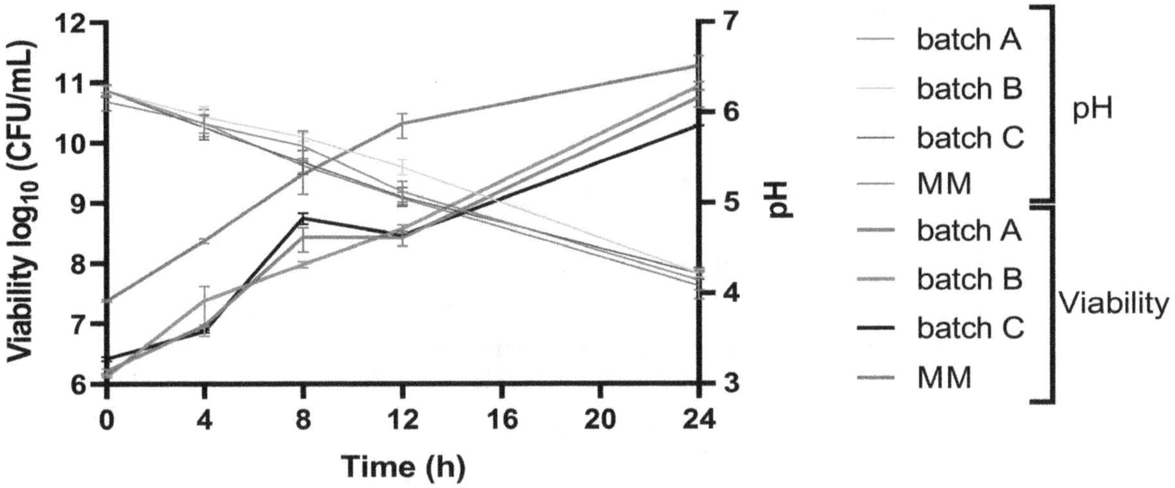

Figure 2. Cell viability (spread plate method) and pH profile of the fermentations at the three different substrate concentrations (wheat and soy-flour) and MM for 24 h with the cocultures. Values for LAB viable cell growth and pH are displayed as mean values ± SD, log10 CFU/mL, n = 3, GraphPad Prism Version 8.0.1 (Graph Pad Software, Inc., San Diego, CA, USA); batch A (90% WF + 10% SF addition); batch B (95% WF + 5% SF addition); batch C (100% WF), MM (model media); CFU/mL (colony-forming units/milliliter of sample).

Through the fermentation, the values gradually increased and reached a final concentration of 10.8–10.9 log10 CFU/mL on substrates where SF was added and 10.3 log10 CFU/mL where only WF was the sole substrate. In MM, the viability was considerably higher, with final values reaching 11.5 log10 CFU/mL. The same trend could be observed at viability with the cocultures of *LP*, *LC*, and *Sc* [39], where a final concentration of 10.4 log10 CFU/mL was reached in the case of flour as substrate and 12.5 log10 CFU/mL in the case of MM. Dough enriched with SF had a beneficial effect on viability which can be justified by the ability of LAB to degrade soy as substrate, particularly *LP* [50].

Throughout fermentation, the pH gradually decreased, but in the case of MM and batch C, the pH decreased more rapidly, owing to better accessibility of carbohydrates. A similar decline in pH could be observed in a study on different carbohydrates with starting values of about 6.1 and final values between 4.4 and 4.8 [21]. The decrease in pH is caused by the production of organic acids through fermentation by LAB especially lactic acid. With the continuous metabolic activities of the LAB, the pH reached lower values than <4.3 at the end of fermentation, which can also be seen in several studies led on different substrates [51–53].

3.2. Viscoelastic Behavior of Dough

Dough rheology became a field with a great deal of interest in the cereal industry, considering its implication in baked product quality [54]. An essential qualitative parameter of bread products is associated with dough's viscoelasticity, which is given by the gluten complex found in wheat. Through bread-production dough experiences a variety of strains and stresses of different degrees, which are essential in evaluating the viscoelastic behavior of dough. Particularly through fermentation, a significant element is extensional deformation, which has a major impact on the rheological characteristics and consistency of the end-product [32,54].

In Figure 3, and Supplementary Tables S1–S3, the viscoelastic properties of doughs at 30 °C in all three wheat–soy flour concentrations (batches A, B, and C) through coculture fermentation with *LP* and *LC* are presented. Every dough combination showed with the increase of angular frequency (ω) a rise in storage (G′) and loss modulus (G′′).

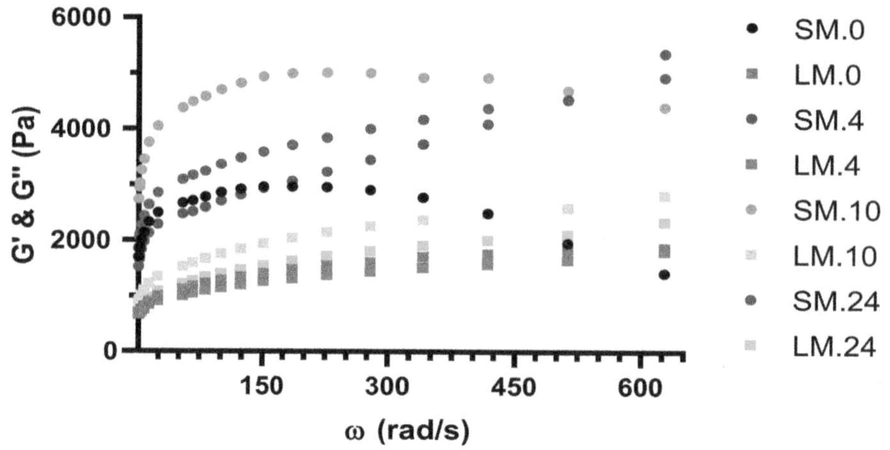

A.

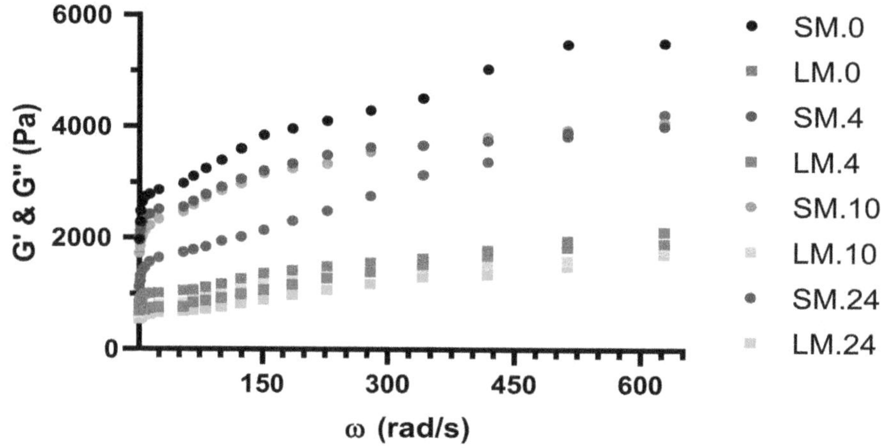

B.

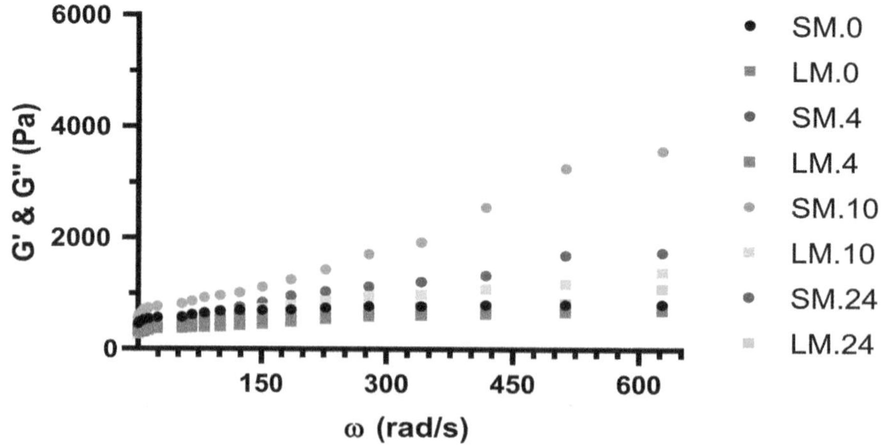

C.

Figure 3. Storage (SM—G′, filled dots •) and loss modulus (LM—G′′, filled squares ■) performed as a function of angular frequency (ω) in doughs at different strengths for (**A**) batch A (90% WF + 10% SF addition); (**B**) batch B (95% WF + 5% SF addition); (**C**) batch C (100% WF) fermented with the coculture of *LP* and *LC* through a 24 h period.

The moisture content of mixtures of WF and SF doughs have different water absorption capacities, which affects the dynamic moduli (with lower water content it amplifies) [55]. As a consequence, the water content of every batch was the same in every experiment. Considering that in every dough sample, the G′ was higher than the G″, and also increased with the increase of ω. Consequently, every dough sample presented a weak gel-like behavior (stable elastic behavior). The highest gel-like behavior was observed in the batches where SF was added, with final values at 24 h fermentation of G′ 4936.2 ± 12.7, and G″ 2338.4 ± 9.1 for batch A and G′ 4225.2 ± 11.9, and G″ 1726.7 ± 10.1 for batch B. The same behavior was reported in several studies where different types of flours were analyzed in dough production, like aleurone, durum, Psyllium, amaranth flour, corn starch, and pea isolate [5,7,56].

Although the addition of SF to WF has a negative impact on bread volume, the sour flavor given by the organic acids hides the odd flavor of soybean, and as a result, the consumer acceptance is favorable [57]. Similar reports stated that soy proteins, besides giving valuable nutritional quality to doughs, also display multiple functional characteristics like emulsifying, gelation action, moisture retention capability, and stabilize the foam activity, which has a good influence on dough products [32,36]. Furthermore, dough fermentation influenced the rheological property of wheat/wheat–soy doughs becoming flexible, easily extensible, with low elasticity [13]. In the batch where no SF was added, G′, and G″ had both very low values, although in these samples a small elastic behavior was also observed, with SF addition, this behavior was accentuated.

3.3. Organic Acid Production

Sourdough fermentation is an ancient bread-making technique, required for dough leavening. Through fermentation, LAB, which usually exists in flour, produces lactic acid. The production of an appropriate quantity of volatile compounds needs fermentation of approximately 12–24 h, whereas, with baker's yeast, it is sufficient for a couple of hours. For the acidification of dough, usually, the LAB is accountable. Lactic acid and acetic acid play an essential role in the general flavor perception of the produced bakery products. In the production of organic acids, an important role is given to the fermentation temperature, the utilized flour type, and especially the used starter culture (homo- or heterofermentative LAB) [58]. Through EMP, LAB ferment glucose, and carbohydrates are conveyed by the phosphotransferase system transporter systems. Besides glucose, sugars are metabolized under carbon-catabolite regression, fructose is entirely utilized as a carbon source, and pyruvate acts as the main connection point of metabolism [59].

Considering lactic acid production (Figure 4A), at the beginning of fermentation it was either elevated or with no production at all, after which the highest level was observed in batch A with 4661.1 mg/L, followed by batch B with 3440.9 mg/L and the lowest quantity was 2872.9 mg/L in batch C. This organic acid was found in a higher amount of 12,306.3 mg/L in a similar study, on amaranth flour dough fermentation with three LAB strains, *LP* RTa12, *Lb. sakei* RTa14, and *Pediococcus pentosaceus* RTa11 [60]. On different semolina and pistachio powder enriched WF sourdoughs with mixed LAB cocultures, lactic acid was produced between 5210 and 5890 mg/L without pistachio powder, and 7470–9800 mg/L in enriched flours, which also resulted in bread with better sensory analysis [53].

Heterofermentative LAB in the fermentation process achieves acidification by organic acid production, especially by lactic and acetic acids, which are also accountable for bread products' shelf-life extension [61]. Even though acetic acid production in the fermented dough was not observed in any of the three batches, in MM from an initial quantity of 3437.7 mg/L at 0 h, it decreased to 1892.3 mg/L at 24 h which is in accordance with similar studies [21,62].

Alongside lactic acid and acetic acid, several organic acids have antimicrobial activities, like citric acid, fumaric acid, malic acid (Figure 4D), tartaric acid (Figure 4F), and butyric acid. Through the decarboxylation of malic acid, lactic acid is produced [63], which in our studies decreased to values of 1366.107 (batch A), 786.589 (batch B), and 824.863 mg/L (batch C).

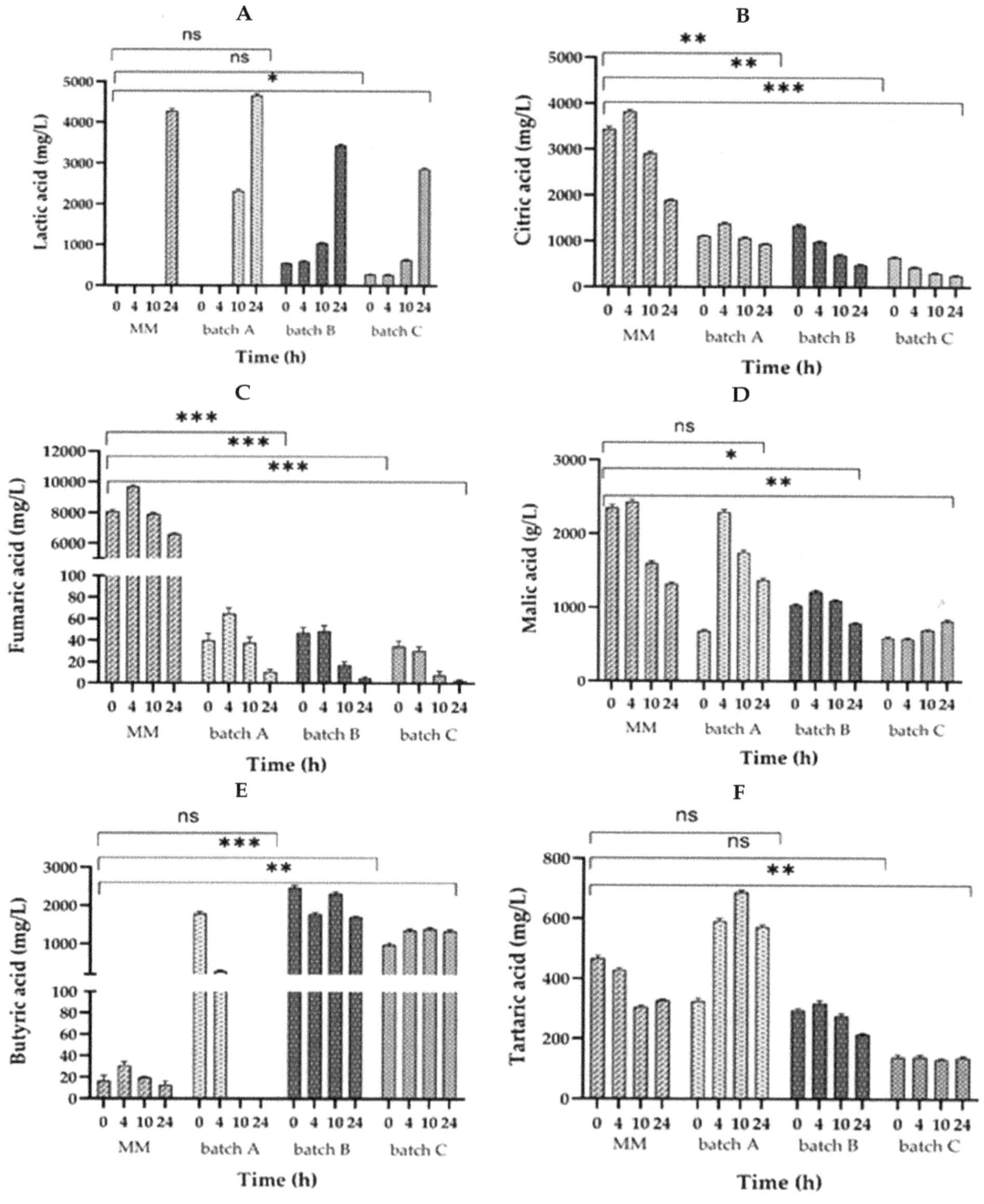

Figure 4. Organic acid production through 24 h fermentation with the cocultures *LP* and *LC* in MM (model media—control); batch A (90% WF and 10% SF), B (95% WF and 5% SF), and C (100% WF); (**A**) lactic acid, (**B**) citric acid, (**C**) fumaric acid, (**D**) malic acid, (**E**) butyric acid, and (**F**) tartaric acid. Values displayed as mean ± SD of the triple measurements and symbols (ns, *, **, ***) express the not-significant or significant differences ($p < 0.05$) between the MM and the three batches (one-way analysis of variance (ANOVA), multiple comparison tests).

In the three wheat–soy flour doughs, citric acid (Figure 4B) presented inconstant results through fermentation, because generally it is consummated when LAB are low in carbohydrates. Still, at 24 h, the quantity decreased to 931.7, 479.6, and 256.9 mg/L in the three batches. Simurina et al. (2014)

stated that the incorporation of citric acid substantially enhances bread quality, and has crumb softening aftermath [64]. Fumaric acid (Figure 4C), although recognized as an advantageous organic acid on human health as a result of its antioxidant activity [65], in these batches it was found in a low quantity. In the three batches from the beginning of the fermentation, it decreased from 35.051–47.510 to 2.673–10.835 mg/L at 24 h. Su et al. (2019) found that organic acids (lactic, citric, acetic, malic, and fumaric) have a good effect on specific volume, decrease the moisture content, pH, and bread hardness. On the other hand, organic acids also decreased the gas retention capacity and diminished the gluten network [65].

Butyric acid (Figure 4E) is especially beneficial for colon health and has the added capacity to prevent or to treat Crohn's disease, cancer, and distal ulcerative colitis [66]. The production of butyric acid decreased till the end of fermentation, reaching values of 1711.167 mg/L in batch B and 1349.320 mg/L in batch C. In batch A, the butyric acid quantity was not detected after 24 h of fermentation.

The highest concentration of organic acids (except butyric acid) at the end of fermentation was found in batch A, which indicates that SF addition improves considerably organic acid production. With a final concentration of tartaric acid of 572.508 mg/L, malic acid of 1366.107 mg/L, lactic acid of 4661.098 mg/L, citric acid of 931.690 mg/L, and fumaric acid of 10.835 mg/L, the addition of 10% of SF has an improved effect on dough quality. In comparison with our earlier results with single fermentation of *LP*, *LC*, or coculture fermentation of *LP* + *LC* + *Sc* [39], the coculture of *LP* + *LC* without yeast had the highest organic acid production. *Sc* generates high quantity of CO_2 which is accountable for dough leavening. LAB are also accelerated by CO_2, and inhibited by the aerobically produced reactive oxygen species [67]. The lower lactic acid concentration can be explained by the fact that *Sc* consumes this organic acid from dough, but this way [68] increases the growth of LAB by reducing the acidification of the substrate [68].

Cocultivation of *LP* and *LC* led to a high concentration of lactic acid, especially where SF was present. As in our earlier study carbohydrate metabolization was better with *LP* than with *LC*, especially in lactic acid production. In general, in single cultures, organic acid production was higher with *LP* which indicates that it has a dominant trait in carbohydrate consumption. However, on every substrate combination both LAB showed efficient synergy.

4. Conclusions

Dough fermentation with cocultures of *LP* + *LC* presented favorable results in organic acid development, especially with an increased lactic (4661.098 mg/L), citric (931.690 mg/L), and tartaric (572.508 mg/L) acid production. The viability of these two LABs, grown at 37 °C, increased to 10.28–11.51 log10 CFU/mL, and the pH decreased at values between the range of 4.14–4.23. These two LAB had good metabolite production efficiency and both *Lb.* strains were in symbiosis with favorable growth dynamics.

In the course of fermentation dough experiences important rheological alterations, which were investigated using dynamic rheological measurements. Besides, the effect of SF incorporation (up to 10%) in WF produced several positive changes in the rheological characteristics of dough, like increased elastic-like behavior. Additionally, with the increase of SF, an increase in organic acid content could be observed, which further supports the addition of SF in bakery products.

Further research should be run to evaluate the positive effects of other microorganisms like (i.e., *Lb. florum*, *Lb. bulgaricus*, *Oenococcus oeni*, etc.) through dough fermentation in single or cocultures, together or without the addition of baker's yeast (*S. cerevisiae*), and also the effect of higher substitution of SF to WF. As a future scope, the consequence of SF addition to WF through frozen storage in comparison with the unfrozen dough will be performed. The interferences of LAB in cocultures and their metabolic activity with the effect on dough characteristics is important.

Supplementary Materials: Figure S1. HPLC chromatogram of organic acid standards and retention times; Table S1. Cell viability and pH profile of the fermentations at the three different substrate concentrations (wheat and soy-flour) and MM for 24 h with the cocultures; Table S2. The standard deviation of storage (G′) and loss (G″) shear moduli for 100% wheat flour with *LP* + *LC* cocultures; Table S3. The standard deviation of storage (G′) and loss (G″) shear moduli for 95%wheat flour and 5% soy flour with *LP* + *LC* cocultures; Table S4. The standard deviation of storage (G′) and loss (G″) shear moduli for 90% wheat flour and 10% soy flour with *LP* + *LC* cocultures.

Author Contributions: D.C.V. and B.-E.T. conceived and designed the study. B.-E.T., G.A.M. performed the experiments. D.C.V., B.-E.T., and G.A.M. undertook the data analysis and cowrote the paper. All authors have read and agreed to the published version of the manuscript.

Acknowledgments: The authors would like to thank their colleagues from the Department of Food Science, for continued support.

References

1. Saget, S.; Costa, M.; Barilli, E.; Wilton, M.; Vasconcelos, D.; Sancho, C.; Styles, D.; Williams, M. Substituting wheat with chickpea flour in pasta production delivers more nutrition at a lower environmental cost. *Sustain. Prod. Consum.* **2020**, *24*, 26–38. [CrossRef]
2. Chis, M.S.; Paucean, A.; Man, S.M.; Bonta, V.; Pop, A.; Stan, L.; Beldean, B.V.; Pop, C.R.; Mureșan, V.; Muste, S.; et al. Effect of rice flour fermentation with *Lactobacillus spicheri* DSM 15429 on the nutritional features of gluten-free muffins. *Foods* **2020**, *9*, 822. [CrossRef]
3. Călinoiu, L.F.; Vodnar, D.C. Whole grains and phenolic acids: A review on bioactivity, functionality, health benefits and bioavailability. *Nutrients* **2018**, *10*, 1615. [CrossRef] [PubMed]
4. Wang, J.; Chatzidimitriou, E.; Wood, L.; Hasanalieva, G.; Markelou, E.; Iversen, P.O.; Seal, C.; Baranski, M.; Vigar, V.; Ernst, L.; et al. Effect of wheat species (*Triticum aestivum* vs. *T. spelta*), farming system (organic vs. conventional) and flour type (wholegrain vs. white) on composition of wheat flour—Results of a retail survey in the UK and Germany—2. Antioxidant activity, and phenoli. *Food Chem. X* **2020**, *6*, 100091. [CrossRef] [PubMed]
5. Guardianelli, L.M.; Salinas, M.V.; Puppo, M.C. Hydration and rheological properties of amaranth-wheat flour dough: Influence of germination of amaranth seeds. *Food Hydrocoll.* **2019**, *97*, 105242. [CrossRef]
6. Yamul, D.K.; Navarro, A.S. Effect of hydrocolloids on structural and functional properties of wheat/potato (50/50) flour dough. *Food Struct.* **2020**, *24*, 100138. [CrossRef]
7. Xu, M.; Hou, G.G.; Ma, F.; Ding, J.; Deng, L.; Kahraman, O.; Niu, M.; Trivettea, K.; Lee, B.; Wu, L.; et al. Evaluation of aleurone flour on dough, textural, and nutritional properties of instant fried noodles. *LWT* **2020**, *126*, 109294. [CrossRef]
8. Paucean, A.; Man, S.M.; Chis, M.S.; Mureşan, V.; Pop, C.R.; Socaci, S.A.; Muresan, C.C.; Muste, S. Use of pseudocereals preferment made with aromatic yeast strains for enhancing wheat bread quality. *Foods* **2019**, *8*, 443. [CrossRef]
9. Jayachandran, M.; Xu, B. An insight into the health benefits of fermented soy products. *Food Chem.* **2019**, *271*, 362–371. [CrossRef]
10. Hu, C.; Wong, W.-T.; Wu, R.; Lai, W.-F. Biochemistry and use of soybean isoflavones in functional food development. *Crit. Rev. Food Sci. Nutr.* **2019**, 1–15. [CrossRef]
11. Simmons, A.L.; Smith, K.B.; Vodovotz, Y. Soy ingredients stabilize bread dough during frozen storage. *J. Cereal Sci.* **2015**, *56*, 232–238. [CrossRef]
12. Bojňanská, T.; Šmitalová, J.; Vollmannová, A.; Tokár, M.; Vietoris, V. Bakery products with the addition of soybean flour and their quality after freezer storage of dough. *J. Microbiol. Biotechnol. Food Sci.* **2015**, *4*, 18–22. [CrossRef]
13. Angioloni, A.; Romani, S.; Pinnavaia, G.G.; Rosa, M.D. Characteristics of bread making doughs: Influence of sourdough fermentation on the fundamental rheological properties. *Eur. Food Res. Technol.* **2006**, *222*, 54–57. [CrossRef]
14. Zhang, Y.; Hong, T.; Yu, W.; Yang, N.; Jin, Z.; Xu, X. Structural, thermal and rheological properties of gluten dough: Comparative changes by dextran, weak acidification and their combination. *Food Chem.* **2020**, *330*, 127154. [CrossRef] [PubMed]
15. Chiş, M.S.; Păucean, A.; Man, S.M.; Vodnar, D.C.; Teleky, B.E.; Pop, C.R.; Stan, L.; Borsai, O.; Kadar, C.B.;

Urcan, A.C.; et al. Quinoa sourdough fermented with *Lactobacillus plantarum* ATCC 8014 designed for gluten-free muffins—A powerful tool to enhance bioactive compounds. *Appl. Sci.* **2020**, *10*, 7140. [CrossRef]
16. Masiá, C.; Jensen, P.E.; Buldo, P. Effect of *Lactobacillus rhamnosus* on physicochemical properties of fermented plant-based raw materials. *Foods* **2020**, *9*, 1182. [CrossRef] [PubMed]
17. Shirotani, N.; Bygvraa, A.; Lametsch, R.; Agerlin, M.; Rattray, F.P.; Ipsen, R. Proteolytic activity of selected commercial *Lactobacillus helveticus* strains on soy protein isolates. *Food Chem.* **2020**, *340*, 128152. [CrossRef]
18. Brizuela, N.; Tymczyszyn, E.E.; Semorile, L.C.; Valdes La Hens, D.; Delfederico, L.; Hollmann, A.; Bravo-Ferrada, B. *Lactobacillus plantarum* as a malolactic starter culture in winemaking: A new (old) player? *Electron. J. Biotechnol.* **2019**, *38*, 10–18. [CrossRef]
19. Collado-Fernández, M. Bread and dough fermentation. *Encycl. Food Sci. Nutr.* **2003**, *1997*, 647–655. [CrossRef]
20. Endo, A.; Dicks, L.M.T. Physiology of the LAB. In *Lactic Acid Bacteria: Biodiversity and Taxonomy*; Wiley Blackwell: Hoboken, NJ, USA, 2014; pp. 13–30.
21. Paucean, A.; Vodnar, D.C.; Socaci, S.A.; Socaciu, C. Carbohydrate metabolic conversions to lactic acid and volatile derivatives, as influenced by *Lactobacillus plantarum* ATCC 8014 and *Lactobacillus casei* ATCC 393 efficiency during in vitro and sourdough fermentation. *Eur. Food Res. Technol.* **2013**, *237*, 679–689. [CrossRef]
22. Saxami, G.; Ypsilantis, P.; Sidira, M.; Simopoulos, C.; Kourkoutas, Y.; Galanis, A. Distinct adhesion of probiotic strain *Lactobacillus casei* ATCC 393 to rat intestinal mucosa. *Anaerobe* **2012**, *18*, 417–420. [CrossRef] [PubMed]
23. Pop, O.L.; Dulf, F.V.; Cuibus, L.; Castro-Giráldez, M.; Fito, P.J.; Vodnar, D.C.; Coman, C.; Socaciu, C.; Suharoschi, R. Characterization of a sea buckthorn extract and its effect on free and encapsulated *Lactobacillus casei*. *Int. J. Mol. Sci.* **2017**, *18*, 2513. [CrossRef] [PubMed]
24. Gobbetti, M.; Rizzello, C.G.; Di Cagno, R.; De Angelis, M. How the sourdough may affect the functional features of leavened baked goods. *Food Microbiol.* **2014**, *37*, 30–40. [CrossRef] [PubMed]
25. Martău, G.A.; Coman, V.; Vodnar, D.C. Recent advances in the biotechnological production of erythritol and mannitol. *Crit. Rev. Biotechnol.* **2020**, *40*, 608–622. [CrossRef]
26. Voinea, A.; Stroe, S.-G.; Codină, G.G. The effect of sodium reduction by sea salt and dry sourdough addition on the wheat flour dough rheological properties. *Foods* **2020**, *9*, 610. [CrossRef] [PubMed]
27. Yildirim-Mavis, C.; Yilmaz, M.T.; Dertli, E.; Arici, M.; Ozmen, D. Non-linear rheological (LAOS) behavior of sourdough-based dough. *Food Hydrocoll.* **2019**, *96*, 481–492. [CrossRef]
28. Autio, K.; Flander, L.; Kinnunen, A.; Heinonen, R. Bread quality relationship with rheological measurements of wheat flour dough. *Cereal Chem.* **2001**, *78*, 654–657. [CrossRef]
29. Oh, I.; Lee, S. Rheological, microstructural, and tomographical studies on the rehydration improvement of hot air-dried noodles with oleogel. *J. Food Eng.* **2020**, *268*. [CrossRef]
30. Marco, C.; Rosell, C.M. Functional and rheological properties of protein enriched gluten free composite flours. *J. Food Eng.* **2008**, *88*, 94–103. [CrossRef]
31. Sozer, N. Rheological properties of rice pasta dough supplemented with proteins and gums. *Food Hydrocoll.* **2009**, *23*, 849–855. [CrossRef]
32. Zhou, J.; Liu, J.; Tang, X. Effects of whey and soy protein addition on bread rheological property of wheat flour. *J. Texture Stud.* **2018**, *49*, 38–46. [CrossRef] [PubMed]
33. Psimouli, V.; Oreopoulou, V. The effect of alternative sweeteners on batter rheology and cake properties. *J. Sci. Food Agric.* **2012**, *92*, 99–105. [CrossRef] [PubMed]
34. Asghar, A.; Anjum, F.M.; Butt, M.S.; Randhawa, M.A.; Akhtar, S. Effect of polyols on the rheological and sensory parameters of frozen dough pizza. *Food Sci. Technol. Res.* **2012**, *18*, 781–787. [CrossRef]
35. Omedi, J.O.; Huang, W.; Su, X.; Liu, R.; Tang, X.; Xu, Y.; Rayas-Duarte, P. Effect of five lactic acid bacteria starter type on angiotensin-I converting enzyme inhibitory activity and emulsifying properties of soy flour sourdoughs with and without wheat bran supplementation. *J. Cereal Sci.* **2016**, *69*, 57–63. [CrossRef]
36. Ammar, A.S.; Salem, S.A.; Badr, F.H. Rheological properties of wheat flour dough as affected by addition of whey and soy proteins. *Pakistan J. Nutr.* **2011**, *10*, 302–306. [CrossRef]
37. Crockett, R.; Ie, P.; Vodovotz, Y. Effects of soy protein isolate and egg white solids on the physicochemical properties of gluten-free bread. *Food Chem.* **2015**, *129*, 84–91. [CrossRef]
38. Yang, S.; Jeong, S.; Lee, S. Elucidation of rheological properties and baking performance of frozen doughs under different thawing conditions. *J. Food Eng.* **2020**, *284*, 110084. [CrossRef]
39. Teleky, B.E.; Martău, A.G.; Ranga, F.; Chețan, F.; Vodnar, D.C.; Gheorghe, A.; Chet, F. Exploitation of lactic

acid bacteria and baker's yeast as single or multiple starter cultures of wheat flour dough enriched with soy flour. *Biomolecules* **2020**, *10*, 778. [CrossRef] [PubMed]

40. De Angelis, M.; Rizzello, C.G.; Alfonsi, G.; Arnault, P.; Cappelle, S.; Cagno, R.D.; Gobbetti, M. Use of sourdough lactobacilli and oat fibre to decrease the glycaemic index of white wheat bread. *Br. J. Nutr.* **2007**, *98*, 1196–1205. [CrossRef]
41. Bottani, M.; Brasca, M.; Ferraretto, A.; Cardone, G.; Casiraghi, M.C.; Lombardi, G.; De Noni, I.; Cattaneo, S.; Silvetti, T. Chemical and nutritional properties of white bread leavened by lactic acid bacteria. *J. Funct. Foods* **2018**, *45*, 330–338. [CrossRef]
42. Mitrea, L.; Trif, M.; Vodnar, D.-C. The effect of crude glycerol impurities on 1,3-propanediol biosynthesis by *Klebsiella pneumoniae* DSMZ 2026. *Renew. Energy* **2020**, *153*, 1418–1427. [CrossRef]
43. Călinoiu, L.-F.; Catoi, A.-F.; Vodnar, D.C. Solid-state yeast fermented wheat and oat bran as a route for delivery of antioxidants. *Antioxidants* **2019**, *8*, 372. [CrossRef] [PubMed]
44. Szabo, K.; Teleky, B.E.; Mitrea, L.; Călinoiu, L.F.; Martău, G.A.; Simon, E.; Varvara, R.A.; Vodnar, D.C. Active packaging-poly (vinyl alcohol) films enriched with tomato by-products extract. *Coatings* **2020**, *10*, 141. [CrossRef]
45. Mitrea, L.; Călinoiu, L.-F.F.; Martău, G.-A.; Szabo, K.; Teleky, B.-E.E.; Mureșan, V.; Rusu, A.-V.V.; Socol, C.-T.T.; Vodnar, D.-C.C.; Mărtau, G.A.; et al. Poly (vinyl alcohol)-based biofilms plasticized with polyols and colored with pigments extracted from tomato by-products. *Polymers* **2020**, *12*, 532. [CrossRef]
46. Mitrea, L.; Vodnar, D.C. Klebsiella pneumoniae—A useful pathogenic strain for biotechnological purposes: Diols biosynthesis under controlled and uncontrolled pH levels. *Pathogens* **2019**, *8*, 293. [CrossRef]
47. Mitrea, L.; Leopold, L.F.; Bouari, C.; Vodnar, D.C. Separation and Purification of Biogenic 1,3-Propanediol from Fermented Glycerol through Flocculation and Strong Acidic Ion-Exchange Resin. *Biomolecules* **2020**, *10*, 1601. [CrossRef]
48. Szabo, K.; Diaconeasa, Z.; Catoi, A.; Vodnar, D.C. Screening of Ten Tomato Varieties Processing Waste for Bioactive Components and Their Related Antioxidant and Antimicrobial Activities. *Antioxidants* **2019**, *8*, 292. [CrossRef]
49. Zhang, B.; Yang, Z.; Huang, W.; Omedi, J.O.; Wang, F.; Zou, Q.; Zheng, J. Isoflavone aglycones enrichment in soybean sourdough bread fermented by lactic acid bacteria strains isolated from traditional Qu starters: Effects on in-vitro gastrointestinal digestion, nutritional, and baking properties. *Cereal Chem.* **2018**, *96*, 129–141. [CrossRef]
50. Aguirre, L.; Hebert, E.M.; Garro, M.S.; Savoy de Giori, G. Proteolytic activity of Lactobacillus strains on soybean proteins. *LWT Food Sci. Technol.* **2014**, *59*, 780–785. [CrossRef]
51. Hashemi, S.M.B.; Gholamhosseinpour, A.; Mousavi Khaneghah, A. Fermentation of acorn dough by lactobacilli strains: Phytic acid degradation and antioxidant activity. *LWT* **2019**, *100*, 144–149. [CrossRef]
52. Gerez, C.L.; Dallagnol, A.; Rollán, G.; Font de Valdez, G. A combination of two lactic acid bacteria improves the hydrolysis of gliadin during wheat dough fermentation. *Food Microbiol.* **2012**, *32*, 427–430. [CrossRef] [PubMed]
53. Gaglio, R.; Alfonzo, A.; Barbera, M.; Franciosi, E.; Francesca, N.; Moschetti, G.; Settanni, L. Persistence of a mixed lactic acid bacterial starter culture during lysine fortification of sourdough breads by addition of pistachio powder. *Food Microbiol.* **2020**, *86*, 103349. [CrossRef] [PubMed]
54. Lee, S.; Campanella, O. Impulse viscoelastic characterization of wheat flour dough during fermentation. *J. Food Eng.* **2013**, *118*, 266–270. [CrossRef]
55. Edwards, N.M.; Peressini, D.; Dexter, J.E.; Mulvaney, S.J. Viscoelastic properties of durum wheat and common wheat dough of different strengths. *Rheol. Acta* **2001**, *40*, 142–153. [CrossRef]
56. Mariotti, M.; Lucisano, M.; Ambrogina Pagani, M.; Perry, K.W.N. The role of corn starch, amaranth flour, pea isolate, and psyllium flour on the rheological properties and the ultrastructure of gluten-free doughs. *Food Res. Int.* **2009**, *42*, 963–975. [CrossRef]
57. Gobbetti, M.; De Angelis, M.; Di Cagno, R.; Calasso, M.; Archetti, G.; Rizzello, C.G. Novel insights on the functional/nutritional features of the sourdough fermentation. *Int. J. Food Microbiol.* **2019**, *302*, 103–113. [CrossRef]
58. Hansen, A.; Schieberle, P. Generation of aroma compounds during sourdough fermentation: Applied and fundamental aspects. *Trends Food Sci. Technol.* **2005**, *16*, 85–94. [CrossRef]
59. Gänzle, M.G. Lactic metabolism revisited: Metabolism of lactic acid bacteria in food fermentations and food

spoilage. *Curr. Opin. Food Sci.* **2015**, *2*, 106–117. [CrossRef]

60. Sterr, Y.; Weiss, A.; Schmidt, H. Evaluation of lactic acid bacteria for sourdough fermentation of amaranth. *Int. J. Food Microbiol.* **2009**, *136*, 75–82. [CrossRef]
61. Debonne, E.; Van Schoors, F.; Maene, P.; Van Bockstaele, F.; Vermeir, P.; Verwaeren, J.; Eeckhout, M.; Devlieghere, F. Comparison of the antifungal effect of undissociated lactic and acetic acid in sourdough bread and in chemically acidified wheat bread. *Int. J. Food Microbiol.* **2020**, *321*, 1–19. [CrossRef]
62. Robert, H.; Gabriel, V.; Lefebvre, D.; Rabier, P.; Vayssier, Y.; Fontagné-Faucher, C. Study of the behaviour of *Lactobacillus plantarum* and Leuconostoc starters during a complete wheat sourdough breadmaking process. *LWT Food Sci. Technol.* **2006**, *39*, 256–265. [CrossRef]
63. Kunkee, R.E. Some roles of malic acid in the malolactic fermentation in wine making. *FEMS Microbiol. Rev.* **1991**, *88*, 55–72.
64. Filipcev, B.; Simurina, O.; Bodroza-Solarov, M. Combined effect of xylanase, ascorbic and citric acid in regulating the quality of bread made from organically grown spelt cultivars. *J. Food Qual.* **2014**, 1–11. [CrossRef]
65. Su, X.; Wu, F.; Zhang, Y.; Yang, N.; Chen, F.; Jin, Z.; Xu, X. Effect of organic acids on bread quality improvement. *Food Chem.* **2019**, *278*, 267–275. [CrossRef]
66. Huda-Faujan, N.; Abdulamir, A.S.; Fatimah, A.B.; Muhammad Anas, O.; Shuhaimi, M.; Yazid, A.M.; Loong, Y.Y. The impact of the level of the intestinal short chain fatty acids in inflammatory bowel disease patients versus healthy subjects. *Open Biochem. J.* **2010**, *4*, 53–58. [CrossRef]
67. Stevens, M.J.A.; Wiersma, A.; de Vos, W.M.; Kuipers, O.P.; Smid, E.J. Improvement of *Lactobacillus plantarum* Aerobic Growth as Directed by Comprehensive Transcriptome Analysis. *Appl. Environ. Microbiol.* **2008**, *74*, 4776–4778. [CrossRef]
68. Sieuwerts, S.; Bron, P.A.; Smid, E.J. Mutually stimulating interactions between lactic acid bacteria and *Saccharomyces cerevisiae* in sourdough fermentation. *LWT Food Sci. Technol.* **2018**, *90*, 201–206. [CrossRef]

3

Comparison between Pressurized Liquid Extraction and Conventional Soxhlet Extraction for Rosemary Antioxidants, Yield, Composition, and Environmental Footprint

Mathilde Hirondart [1,2], Natacha Rombaut [1,2], Anne Sylvie Fabiano-Tixier [1,2], Antoine Bily [2,3] and Farid Chemat [1,2,*]

1 Avignon University, INRAE, UMR408, GREEN Team Extraction, F-84000 Avignon, France; mathilde.hirondart@sigma-clermont.fr (M.H.); rombaut.natacha@gmail.com (N.R.); anne-sylvie.fabiano@univ-avignon.fr (A.S.F.-T.)

2 ORTESA, LabCom Naturex-Avignon University, F-84000 Avignon, France; antoine.bily@givaudan.com

3 Naturex-Givaudan, 250 rue Pierre Bayle, BP 81218, CEDEX 9, F-84911 Avignon, France

* Correspondence: farid.chemat@univ-avignon.fr

Abstract: Nowadays, "green analytical chemistry" challenges are to develop techniques which reduce the environmental impact not only in term of analysis but also in the sample preparation step. Within this objective, pressurized liquid extraction (PLE) was investigated to determine the initial composition of key antioxidants contained in rosemary leaves: Rosmarinic acid (RA), carnosic acid (CA), and carnosol (CO). An experimental design was applied to identify an optimized PLE set of extraction parameters: A temperature of 183 °C, a pressure of 130 bar, and an extraction duration of 3 min enabled recovering rosemary antioxidants. PLE was further compared to conventional Soxhlet extraction (CSE) in term of global processing time, energy used, solvent recovery, raw material used, accuracy, reproducibility, and robustness to extract quantitatively RA, CA, and CO from rosemary leaves. A statistical comparison of the two extraction procedure (PLE and CSE) was achieved and showed no significant difference between the two procedures in terms of RA, CA, and CO extraction. To complete the study showing that the use of PLE is an advantageous alternative to CSE, the eco-footprint of the PLE process was evaluated. Results demonstrate that it is a rapid, clean, and environmentally friendly extraction technique.

Keywords: Pressurized liquid extraction; soxhlet; solvent extraction; green analytical chemistry; Rosemary

---

1. Introduction

In the field of raw material extraction, the first challenge consists of determining the potential of the plant matrix that means what can be extracted and valorized. The chemical composition of the plant material may highly vary depending on the local environmental conditions, development stages, plant part, harvesting season, the technique used for drying, and the storage condition. Therefore, for each batch of plant material used for industrial extraction, an analysis has to be performed to determine the amount of available extractives.

In general, an analytical procedure for antioxidants from plants or spices comprises two steps: Extraction (Soxhlet, maceration, percolation) followed by analysis (spectrophotometry, high performance liquid chromatography coupled or not to mass spectrometry (HPLC-MS), gas chromatography coupled or not to mass spectrometry (GC–MS)). Whereas the last step is finished after only 15 to 30 min, extraction takes at least several hours. Conventional Soxhlet extraction (CSE) is the most used method for solid-liquid extraction in natural product chemistry and is a reference procedure for the extraction of fat and oil according to International Organization for Standardization (ISO standards) [1–3]. It has several disadvantages such as long operation time requiring a minimum of hours or days, large solvent volumes involved, time and energy consuming for the concentration step by evaporation to recover the final extract, and inadequacy for thermolabile analytes.

Pressurized liquid extraction (PLE) has been intensively studied as an efficient extraction technique to substitute CSE [2,4]. It is based on the ability to perform rapid (less than 30 min) and clean extraction at high pressure and temperature. Various parameters of extraction can be modified to improve extraction performance (solvent, pressure, temperature, time of extraction, etc.) [5–9]. High temperature and pressure increase analytes' solubility and solvent diffusion rate, while solvent viscosity and surface tension decrease, resulting in a drained matrix after extraction [10]. With PLE, extractions can be programmed and automatically run, which is convenient for quality control.

In this study we focused on rosemary (*Rosmarinus Officinalis* L.), which is mostly studied and used in the food industry due to its richness in antioxidants' compounds [11,12], particularly rosmarinic acid (RA), carnosic acid (CA), and carnosol (CO) (Figure 1).

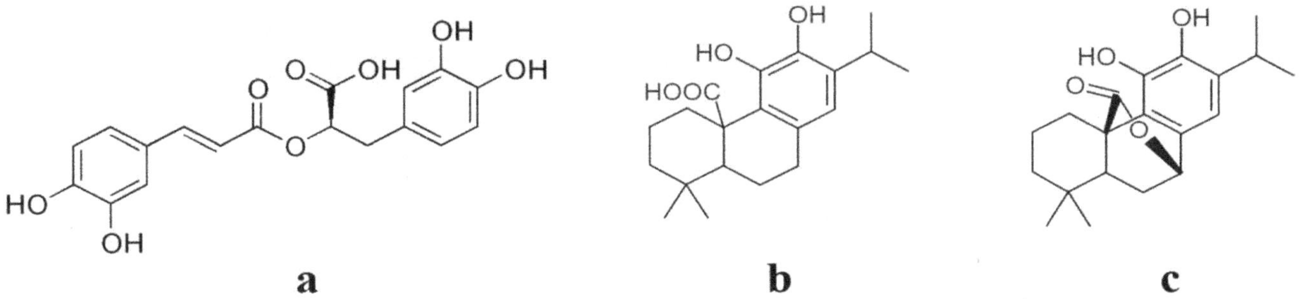

Figure 1. Structures of rosmarinic acid (**a**), carnosic acid, (**b**) and carnosol (**c**).

These compounds are extracted at industrial scale and are dedicated to food applications since the antioxidant extract of rosemary has been authorized in 2010 by the European Union as food additive E392 (directive No. 2010/69/EU). Throughout literature, extraction of chemical compounds from rosemary leaves has been investigated using PLE [13–16]. These studies were mainly focused on maximization of antioxidant activity of rosemary extracts and no complete parametric study of extraction of monitored compounds by PLE has been performed. Additionally, evaluation of the green aspects of PLE is not found in literature.

A major problem in the field of extraction remains the characterization of the raw material studied. The objective of our study was to propose a new method of raw material characterization by optimizing the extraction process in order to be sure to have exhausted the studied raw material. Numerous studies have already been carried out on the extraction of rosemary with innovative technologies such as supercritical fluid extraction (SFE) and pressurized liquid extraction (PLE) coupled with a new quantitative Ultra Performance Liquid Chromatography coupled to Tandem Mass Spectrometry (UPLC-MS/MS) method [13,14]. The difference with the work cited above is that we wanted to propose a green method that could replace Soxhlet in order to optimize the characterization of the raw material studied in the analytical laboratory. The procedure used minimizes the use of organic solvents, which makes it attractive in the analytical field.

In the present work, PLE was studied as a green alternative to Soxhlet extraction of antioxidants from rosemary leaves to extract qualitatively and quantitatively RA, CA, and CO. PLE was optimized via a response surface methodology and a desirability function, which simultaneously maximized extraction, was used. We ran statistical tests in order to check the reliability and the reproducibility of this new procedure. Finally, the eco-footprint of the PLE process was evaluated to demonstrate that it is a rapid, environmentally friendly, and clean extraction technique.

2. Materials and Methods

2.1. *Plant Material and Chemicals*

Rosemary leaves (*Rosmarinus officinalis* L.) were provided by the company Naturex (Avignon France), and rosemary leaves were collected in Morocco in 2015. Initial moisture was 8.2 ± 0.2%. Leaves were ground before extraction using a grinder (MF 10 basic, IKA, Staufen, Germany) with a 0.5-mm sieve. Granulometry of the rosemary powder was 610 ± 22 μm.

For the extraction solvent, food grade ethanol 96° *v*/*v* (Cristalco, FranceAlcools, Paris, France) and demineralized water were used. For HPLC analysis, solvent used were all HPLC grade: Methanol, water, acetonitrile, and tetrahydrofuran. Phosphoric acid 85% ACS grade (according to American Chemical Society specifications) and trifluoroacetic acid 99% were purchased from Sigma-Aldrich, USA. Standards used were rosmarinic acid (Extrasynthese, Genay, France) and carnosic acid (Sigma-Aldrich, St. Louis, MO, USA). Nitrogen used had a purity of 99.999% (Alphagaz 1 Smartop, Air Liquid, Paris, France).

2.2. *Extraction Procedures*

In this study, a procedure of PLE was developed and optimized for analytical determination of RA and CA contents in rosemary leaves. PLE performance was compared to the reference method of Soxhlet extraction. Those processes are illustrated in Figure 2.

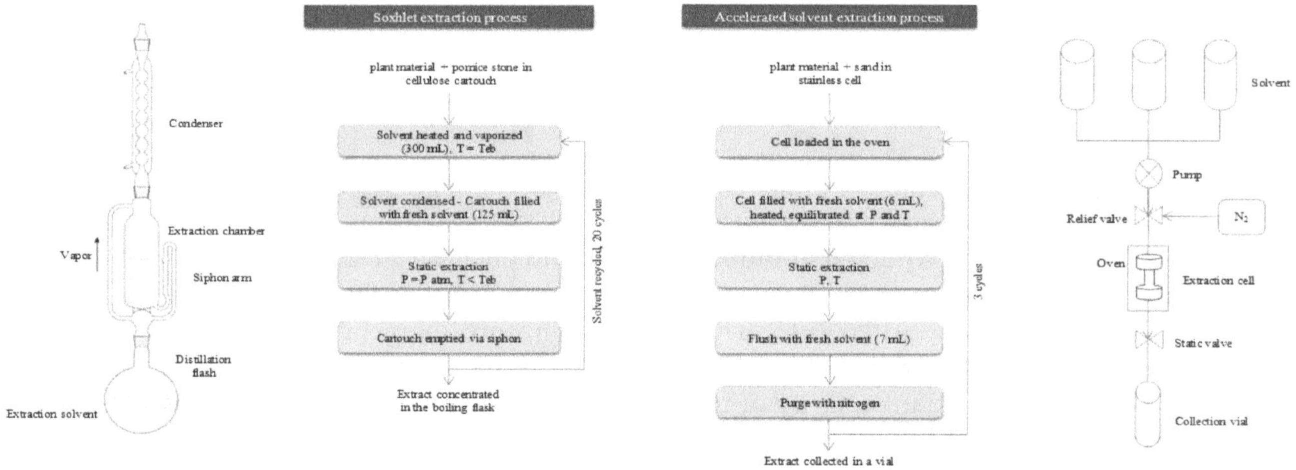

Figure 2. Comparison of Soxhlet and Accelerated Solvent Extraction (ASE) processes.

2.2.1. Reference Procedure: Conventional Soxhlet Extraction (CSE)

For CSE, 10 g of ground rosemary leaves and 5 g of pumice stone were mixed in a 34 × 130 mm cellulose thimble (plugged with cotton in order to avoid transfer of sample particles in the distillation flask) and placed in Soxhlet apparatus with flask containing 300 mL of solvent. Extractions were performed using a solid to liquid ratio of 1 to 12 (g/mL). Extraction was performed during 8 h. After extraction, the extract was concentrated under vacuum (Laborota 4001, Heidolph, Germany) and conserved at 4 °C before analysis. All extractions were done at least in duplicate and the mean values were reported.

2.2.2. Pressurized Liquid Extraction (PLE)

An accelerated solvent extractor ASE200 model was used (Dionex, Thermo Fisher Scientific, Waltham, MA, USA). This apparatus allows extraction of plant material at high pressure (up to 130 bar) and high temperature (up to 200 °C). Preliminary trials were made in order to determine the optimal parameters (loading of the cell, flushing volume, and percentage of dispersant), and will be discussed in the result section. Optimal loading was determined to be 3.1 g of ground rosemary leaves, and 7.3 g of

Fontainebleau sand (VWR Chemicals, Radnor, PA, USA) were homogenized in an 11-mL stainless-steel cell. The cells were equipped with stainless steel frits on both sides, and a cellulose filter at the bottom to obtain a filtered extract. The extraction procedure cycle was done as follows: First, the cell was filled with extraction solvent via an HPLC pump, pressurized, and placed into the preheated oven. Depending on the set extraction temperature, the cell preheating duration was between 5 and 9 min, followed by a static period of extraction. Then, the cell was flushed with fresh solvent (60% of the extraction cell volume) and purged with a flow of nitrogen during 1 min. Several cycles of extraction can be performed to drain active compounds from the plant matrix. Extracts were collected into a glass vial and analyzed without a concentration step. The dry matter content of each extract was determined by drying 5 mL of extract at 130 °C during 3 h, to calculate the mass extraction yield.

Preliminary trials were performed to evaluate the impact of some PLE parameters on extraction performance: Solvent, percentage of dispersant, and flushing volume. For these trials, the other extraction parameters were fixed according to literature [17]: Temperature (T) = 100 °C, Pressure (P) = 80 bar, static time of extraction = 5 min, and 3 cycles of extraction.

2.3. Statistical Analysis

2.3.1. Experimental Design

To investigate the influence of PLE extraction parameters on the extraction of rosemary antioxidants, a response surface methodology was used. Three independent factors, namely the temperature (A), the pressure (B), and the extraction time (C), were studied to evaluate their impact on several responses: The mass yield (%) and the contents in RA, CA, and CO (mg/g). The independent variables, given in Table 1, were coded according to Equation (1):

$$X_i = \frac{x_i - x_{i0}}{\Delta x_i} \quad (1)$$

where X_i and x_i are, respectively, the dimensionless and the actual values of the independent variable i, x_{i0} is the actual value of the independent variable i at the central point, and Δx_i is the step change of x_i corresponding to a unit variation of the dimensionless value. For the three variables, the design yielded randomized experiments with eight (2^3) factorial points, six axial points ($-\alpha$ and $+\alpha$ (in our case 1.68)) to form a central composite design, and six center points for replications and estimation of the experimental error and to prove the suitability of the model. Coded values of the independent variables are listed in Table 1.

$$Y = \beta_0 + \sum_{i=1}^{2} \beta_i X_i + \sum_{i=1}^{2} \beta_{ii} X_i^2 + \sum_{i} \sum_{j=i+1} \beta_{ij} X_i X_j \quad (2)$$

The responses are related to the coded independent variables X_i and X_j according to the second order polynomial expressed in Equation (2) with β_0 the interception coefficient, β_i the linear terms, β_{ii} the quadratic terms, and β_{ij} the interaction terms. Fisher's test for analysis of variance (ANOVA) performed on experimental data was used to assess the statistical significance of the proposed model. The experimental design was analyzed using the software Statgraphics (StatPoint Technologies, Inc., Warrenton, VA, USA) for Windows.

Table 1. Central composite design (CCD) matrix with experimental responses obtained (mass extraction yield and leaf content in rosmarinic acid, carnosic acid, and carnosol).

Variables						Responses			
Temperature		Pressure		Extraction Time		Mass Yield	RA Content	CA Content	CO Content
Actual Value (°C)	Coded Value	Actual Value (bar)	Coded Value	Actual Value (min)	Coded Value	%	mg/g	mg/g	mg/g
160	+1	111.8	+1	27	+1	44.0	11.56	21.76	2.01
160	+1	111.8	+1	9	−1	41.9	11.57	22.54	2.10
160	+1	58.2	-1	27	+1	44.9	11.18	22.32	2.07
160	+1	58.2	−1	9	−1	44.0	11.49	22.62	2.17
70	−1	111.8	+1	27	+1	31.3	11.13	20.55	2.22
70	−1	111.8	+1	9	−1	30.5	10.83	20.49	2.40
70	−1	58.2	−1	27	+1	31.3	11.84	21.20	2.16
70	−1	58.2	−1	9	−1	31.5	11.67	20.68	2.12
115	0	85	0	18	0	34.3	11.29	20.98	2.17
115	0	85	0	18	0	34.5	11.70	20.12	2.17
115	0	85	0	18	0	34.3	11.72	20.63	2.05
115	0	85	0	18	0	34.8	11.71	20.41	2.11
115	0	85	0	18	0	34.3	11.68	20.39	2.18
115	0	85	0	18	0	34.1	12.01	20.34	2.09
39.3	$-\alpha$	85	0	18	0	26.3	10.56	19.90	2.29
190.7	$+\alpha$	85	0	18	0	49.6	9.93	22.12	2.22
115	0	40	$-\alpha$	18	0	34.0	11.32	21.28	2.30
115	0	130	$+\alpha$	18	0	35.1	12.24	21.35	1.98
115	0	85	0	3	$-\alpha$	34.5	12.19	21.88	2.07
115	0	85	0	33	$+\alpha$	34.7	11.99	21.63	2.10

2.3.2. Reproducibility and Statistical Comparison

The optimized PLE method compared to the CSE method was performed for the extraction of antioxidants from rosemary. It consisted of a series of eight successive experiments performed for each extraction procedure. Then the statistical study was performed in two steps: First, the Fisher–Snedecor's test to compare the variability of the results and then the student test in order to compare the mean values obtained by the two different extraction procedures. Those two tests were performed with $\alpha = 0.05$.

2.4. HPLC Analysis

Analyses of RA, CA, and CO were performed by HPLC (Agilent 1100, Agilent Technologies, Santa Clara, CA, USA) equipped with a Diode Array Detector (DAD) detector. HPLC analyses were made according to previously reported procedures without further optimization and specific procedures for each compound are described below [18].

2.4.1. Rosmarinic Acid Analysis

The column used was a C_{18} column (5 µm, 4.6 mm × 250 mm, Zorbax SB, Agilent Technologies, Santa Clara, CA, USA). The mobile phase was composed of 32% acetonitrile and 68% water with 0.1% trifluoroacetic acid (mL/mL) and the flow rate was set at 1 mL/min. The column oven temperature was 20 °C and the run time was 10 min. Five µL were injected. Rosmarinic acid was detected at a wavelength of 328 nm. For quantification of rosmarinic acid in the extract, a calibration curve was calculated by linear regression analysis for rosmarinic acid standard.

2.4.2. Carnosic Acid and Carnosol Analysis

The column used was a C_{18} column (1.8 µm, 4.6 mm × 50 mm, Zorbax Eclipse XBD-C18, Agilent Technologies, France). The mobile phase was isocratic and composed of 0.5% H_3PO_4

(in water)/acetonitrile (35/65, mL/mL), and the flow rate was set at 1.5 mL/min. The column oven temperature was 25 °C and the run time was 15 min. Five µL were injected. Carnosic acid and carnosol were detected at a wavelength of 230 nm. For quantification of carnosic acid in the extract, a calibration curve was calculated by linear regression analysis for carnosic acid standard. Carnosol was expressed as carnosic acid.

2.5. Calculations

In order to assess the extraction performances of the evaluated processes, mass extraction yield, purity, and content in each compound of interest were calculated. Each mass included in equations below was expressed in dry weight.

$$mass\ extraction\ yield\left(\%, \frac{g}{100g}\right) = \frac{weight\ of\ extract}{weight\ of\ rosemary\ leaves} \times 100 \quad (3)$$

$$purity\left(\%, \frac{g}{100g}\right) = \frac{weight\ of\ RA,\ CA\ or\ CO}{weight\ of\ extract} \times 100 \quad (4)$$

$$content\ in\ RA,\ CA\ and\ CO\ (mg/g\ rosemary) = \frac{purity \times weight\ of\ extract}{weight\ of\ rosemary\ leaves} \quad (5)$$

3. Results and Discussion

3.1. Pressurized Liquid Extraction (PLE): Preliminary Study

3.1.1. Solvent Evaluation of Ethanol/Water Ratio on Extraction Efficiency

To determine the solvent that maximized the extraction of both RA and CA, PLE was performed with various percentages of ethanol in water: 0, 20, 40, 60, 80, and 100% (g/g). Hydro-alcoholic solutions as solvent offer many advantages. Indeed, they can solubilize both hydrophilic and lipophilic active compounds. To test different ethanolic solvent proportions, the flushing volume was fixed at 60% mL/mL of the extraction cell. The extraction cell was filled with 30% g/g of ground rosemary leaves and 70% g/g of Fontainebleau sand.

The influence of ethanol proportion in the extraction solvent on extract composition is reported in Figure 3.

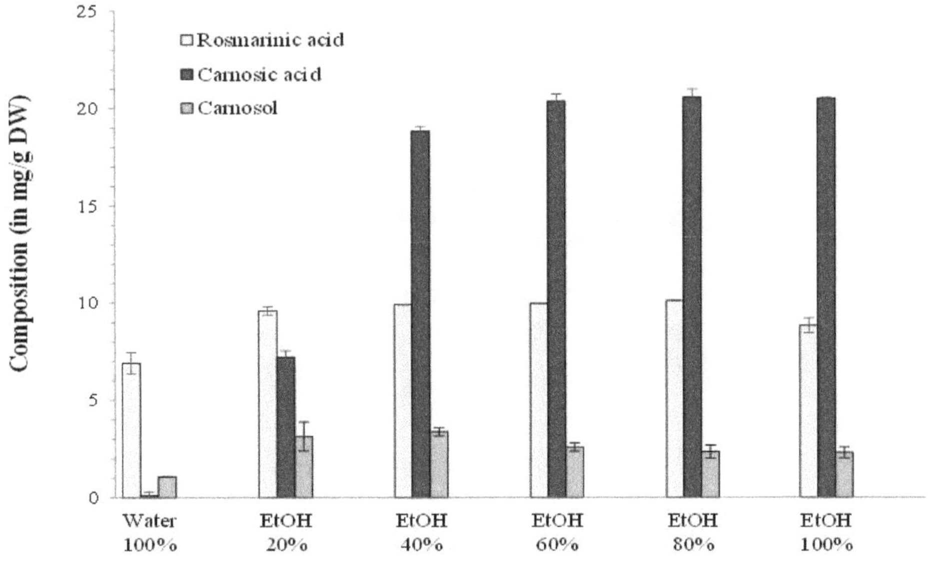

Figure 3. Influence of different ratios of ethanol/water as extraction solvent on the antioxidants composition of Pressurized Liquid Extraction (PLE) extracts of rosemary.

At low ethanol percentage (0 and 20%), RA was extracted but no CA, while from 40% ethanol, CA was extracted as well. These results showed that the solvent maximizing both RA and CA extraction was 80% ethanol, with 10.13 ± 0.02 mg RA/g rosemary and 20.6 ± 0.4 mg CA/g rosemary extracted. Maximal extraction and solubilization of both compounds was possible with 80% ethanol thanks to its intermediate polarity, despite the different chemical structures of RA and CA. Indeed, RA is a caffeic acid ester [19], rather hydrophilic, so preferentially extracted and solubilized in solvents that are relatively polar, as ethanol [20]. CA is a phenolic diterpene [21] and is relatively lipophilic, but still soluble in intermediate polarity solvents such as acetone or ethanol [22,23].

3.1.2. Dispersant

Fontainebleau sand was used as a dispersant in order to favor a uniform distribution of sample and maximize the extraction yield. It was mixed with ground rosemary leaves in different proportions to quantify the impact of dispersant on extraction yield. Trials were carried out with 30, 50, 70, and 90% g/g of dispersant. This parameter is usually fixed or not specified in literature, suggesting that it is not impacting the extraction performances. In this study, we measured its impact only on mass yield to verify this hypothesis. The flushing volume was fixed at 60% mL/mL of the extraction cell, and 80% ethanol was used as extraction solvent. It can be seen in Figure 4 that the proportion of dispersant in the cell had a small impact on extraction mass yield, which varied between 31 ± 2% and 37 ± 2%. The 70% dispersant must be selected to maximize the mass yield (37 ± 2%).

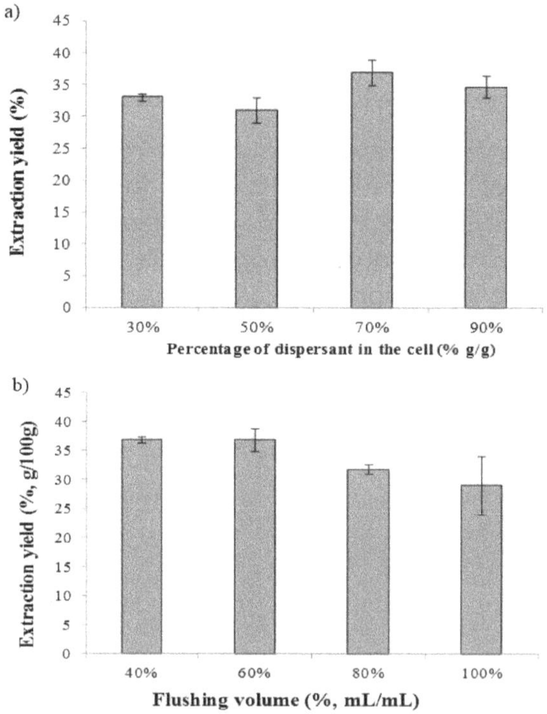

Figure 4. Influence of dispersant proportion in the extraction cell (**a**) and flushing volume (**b**) on the mass extraction yield of PLE of rosemary leaves. The bars with range mean standard deviation between three experiments.

3.1.3. Flushing Volume

The flushing volume is the amount of fresh solvent injected during PLE after static extraction (Figure 2). It is measured as a percentage mL/mL of the extraction cell volume (11 mL). Extractions were performed with 40, 60, 80, and 100% mL/mL (Figure 4).

As for the proportion of dispersant, the flushing volume is usually fixed in literature [14,15]. In this study we measured its impact on extraction mass yield. A flushing volume of 60% mL/mL maximized the extraction mass yield (36.9 ± 2.0%), while higher flushing volume decreased it. A 60% flushing volume was commonly applied throughout literature, which confirms the results obtained [14].

3.2. PLE Extraction: Experimental Design and Statistical Analysis

Three variables that could impact extraction efficiency of antioxidants from rosemary by PLE were studied in a central composite design, namely, temperature (A), pressure (B), and extraction time (C). The choice of the A and B variation domain was selected considering the limits of the ASE equipment. A range from 40 to 190 °C was chosen for the temperature (A), and a range from 40 to 130 bar for the pressure (B). This wide temperature range was chosen to thoroughly evaluate temperature impact on extraction. Within this range, thermal degradation of compounds could also be assessed. The total extraction duration depends on the duration of equilibration of the cell, which varies from 5 to 9 min according to the temperature of extraction. However, as the temperature is a factor impacting the extraction, we chose to consider the duration of static extraction as an independent variable (C). As we performed three cycles of PLE for each extraction cell (Figure 2), the extraction time (C) was the sum of the three static extraction periods. A range from $3 = 3 \times 1$ min to $33 = 3 \times 11$ min of static extraction was selected. The controlled variables were studied in a multivariate study with 20 experiments, as shown in Table 1.

Responses varied greatly as a function of the combination of parameter settings. Mass extraction yield ranged from 26.3 to 49.6%, RA content ranged from 9.93 mg/g rosemary to 12.24 mg/g rosemary, and CA content ranged from 19.90 mg/g rosemary to 22.12 mg/g rosemary (Table 1). Experimentally, the formation of Maillard reaction product at temperature of 150 °C or above occurred as evidenced by the brown color of the extract and the burnt smell. The presence of these toxic compounds in extracts must be avoided, so extractions carried out at temperature in the range of 150–200 °C are usually not recommended [15]. However, in this study extraction was investigated for analytical purpose and sensory characteristics of the extracts was not considered. Moreover, the degradation of targeted compounds did not occur because the RA and CA content in extracts were not lower at 190 °C than at 115 °C (Table 1). As shown in Table 1, there was no significant difference in CO content, which indicates that there was no degradation of CA into CO due to oxidation, as it is suggested in published studies [18,22].

By considering a confidence level of 95%, the linear effects of the temperature (A) as well as all quadratic effects (A^2, B^2, and C^2) were significant (Table 2) with *p*-value below 0.05. There were also interactions between variables A and B in significant scale, with a *p*-value lower than 0.05 (Table 2). Empirical relationships allowed linking responses studied and key variables involved in the model. From ANOVA, the coefficient of determination R^2 was determined to be higher than 80% for the three considered responses.

Table 2. Summary of the ANOVA for the central composite design.

Variables	Responses					
	Mass Yield		RA Content		CA Content	
	F-Ratio	*p-Value*	*F-Ratio*	*p-Value*	*F-Ratio*	*p-Value*
Temperature (A)	445.98	0	0.31	0.5871	106.25	0
Pressure (B)	0.26	0.6225	0.12	0.7389	2.12	0.1756
Extraction time (C)	0.86	0.3743	0.04	0.8487	0.89	0.3685
A^2	28.41	0.0003	32.84	0.0002	9.79	0.0107
A × B	0.38	0.5508	5.17	0.0463	0.07	0.7911
A × C	0.55	0.4758	0.57	0.4661	4.74	0.0546
B^2	1.81	0.208	0.19	0.6738	21.88	0.0009
B × C	0.46	0.5125	0.29	0.6002	1.18	0.3020
C^2	1.97	0.1904	3.50	0.0907	49.02	0
Error R^2 (%)	97.9518		82.1394		94.8633	
R^2 adjusted for d.f (%)	96.1084		66.0649		90.2402	
Optimal conditions predicted	A = 190 °C B = 40 bar C = 33 min		A = 100 °C B = 40 bar C = 3 min		A = 190 °C B = 130 bar C = 4 min	

Three-dimensional surface responses of a multiple nonlinear regression model (Figure 5) illustrate the linear and quadratic effects together with the interaction effects on the responses given in Table 1. Figure 5 highlights the behavior of the three responses as a function of two variables: Temperature (A) and pressure (B). In each plot, the extraction time (C) was fixed at the central value ("0"). The most influential effect was the linear terms of temperature (A) as can be seen in Table 2, with low *p*-values: 0, 0.5871, 0 for mass yield, RA content, and CA content, respectively.

As expected, the model confirmed that mass extraction yield increased with temperature (B). Influence of quadratic terms given in Table 2 is illustrated in Figure 5 by observation of the surface curvatures of the plots. Optimal settings for the maximization of each response are presented in Table 2. An optimization of the desirability was carried out to obtain optimal factors' settings for the multi-responses' maximization. The settings which simultaneously maximized mass yield, RA content, and CA content were: Temperature A = 183 °C, pressure B = 130 bar, extraction time C = 3 × 1 min.

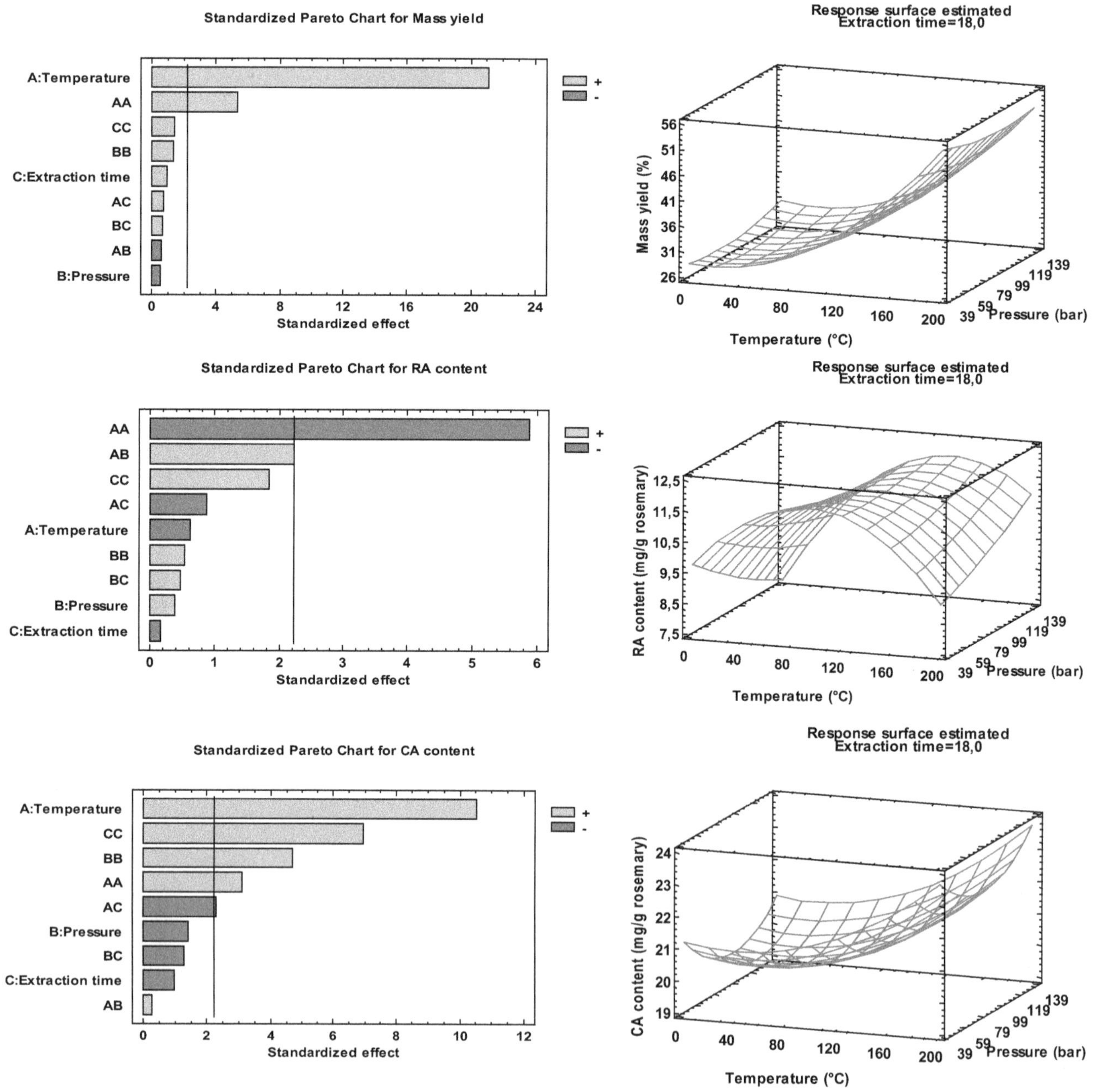

Figure 5. Standardized Pareto charts and response surfaces estimated for the optimization of PLE parameters.

3.3. *Statistical Comparison of CSE and PLE*

In order to assess the reliability of the PLE extraction technique to replace CSE for raw material active content determination, statistical analysis was performed on repeated trials. For this purpose, 16 experiments were run, 8 with the conventional Soxhlet technique and 8 with the optimized PLE extraction (Table 3). The RA, CA, and CO contents (mg/g rosemary) were analyzed and reported.

The mean value of RA content obtained with PLE was similar to the value obtained using CSE (10 ± 1 and 9.9 ± 0.5 mg/g rosemary, respectively) and the mean value of CA content obtained with PLE was higher than the value obtained using CSE (21 ± 1 and 17.7 ± 0.9 mg/g rosemary, respectively). The relative standard deviations were 10% for PLE and CSE, which means that the results were little dispersed around the mean value. This was confirmed using the Fisher–Snedecor's test, which gave no significant difference in the dispersion of the results for the two different processes ($\alpha = 0.05$) for both RA and CA content. The student test was then applied in order to check if there was any significant difference between the mean values of RA and CA content obtained by the two processes. The tabular value obtained by the student test table for $\alpha = 0.05$ was 2.10 and the calculated value was 0.35 for RA content and 1.83×10^{-6} for CA content, which meant that there was no significant difference between the mean values from a statistical point of view. The statistical tests validated the PLE technique as a good alternative to the conventional one for the determination of active compounds' content in rosemary.

Extraction performance of CSE and PLE are compared in Table 3 in terms of mass extraction yield and active contents. Mass yield of optimized PLE extraction (47.6 ± 0.5%) was higher than mass yield obtained with CSE (26 ± 1%). High pressure and high temperature generated during PLE enabled extracting more compounds from the plant material [16]. Extraction of active compounds such as RA and CA was improved with PLE (Table 3). The use of drastic extraction conditions during PLE did not lead to the degradation of compounds, even the most thermosensitive such as CA. This absence of degradation could be explained by the absence of oxygen during PLE due to nitrogen flushing and a short contacting duration between the solvent and the matrix (around 45 min against 8 h with CSE). A higher CO content was obtained with CSE than with PLE, respectively 4.4 ± 0.8 mg/g rosemary and 1.9 ± 0.3 mg/g rosemary, suggesting the degradation of CA into CO during CSE (Table 3). Due to higher extraction yields achieved with PLE, purity of the active compounds in extracts was lower, this extraction technique was less selective regarding these compounds. However, the goal of this analytical technique was not to reach high purities of RA and CA in extracts, but to drain completely the plant material. High purity in RA and CA in the final extract was not considered as a response to maximize. PLE with optimized conditions seems to be a good technique to quantitatively extract RA, CA, and CO from ground rosemary leaves, with better performance than CSE in term of active content yields.

Table 3. Reproducibility of extraction and statistical comparison test between Pressurized Liquid Extraction (PLE) and Conventional Soxhlet Extraction (CSE).

Experiments		1	2	3	4	5	6	7	8	Mean (%)	SD (%)	RSD (%)	Variance S^2	F_{CAL}	F_{TAB}	t_{CAL}	t_{TAB}
Extraction yield (%)	CSE	26.5	26.2	28.9	26.3	28.2	28.3	28.7	30.2	27.91	1.45	5.21	1.85	0.31	3.79	2.55×10^{-14}	2.1
	PLE	47.8	44.7	46.7	46.2	46.5	45.5	45.3	46.7	46.20	0.97	2.11	0.83				
RA content (mg/g)	CSE	9.71	9.23	9.98	9.14	9.70	10.39	10.23	10.49	9.86	0.51	5.13	0.22	0.11		0.35	
	PLE	11.74	10.88	9.15	9.95	8.59	9.96	9.87	9.92	10.01	0.97	9.65	0.82				
CA content (mg/g)	CSE	18.18	17.08	17.38	17.56	17.61	16.16	18.67	18.78	17.68	0.86	4.88	0.65	0.69		1.83×10^{-6}	
	PLE	22.46	20.18	22.38	20.62	21.26	19.99	20.25	21.88	21.13	1.01	4.78	0.89				
CO content (mg/g)	CSE	3.51	4.29	4.34	5.64	4.25	5.36	3.62	4.14	4.39	0.75	17.14	0.50	0.01		1.67×10^{-7}	
	PLE	2.20	1.72	1.69	2.00	1.68	1.96	1.50	2.16	1.86	0.25	13.58	0.06				

SD, standard deviation; RSD, relative standard deviation; F_{CAL}, F value calculated using Fisher-Snedecor's test; F_{TAB}, F value tabulated for $\alpha = 0.05$ and 7 of freedom; $t_{CAL} = t$ value calculated using student's test; $t_{TAB} = t$ value tabulated for $\alpha = 0.05$ and 14 degrees of freedom.

3.4. *Eco-Footprint: CSE vs. PLE Processes*

The two extraction techniques were evaluated according to the six principles of green extraction developed by Chemat et al. [24]. The six parameters considered were calculated as follows:

- Raw material (Principle 1): Mass of plant material required for an analysis (in g).
- Solvent (Principle 2): Mass of solvent required for an analysis (in g).
- Energy (Principle 3): Energy consumption for the analysis considering extraction and evaporation steps based on the energy transfer equation [25–27] (in kWh).
- By-products (Principle 4): Amount of waste generated by an analysis (solvent and plant material) (in g).
- Process (Principle 5): Time of an analysis including steps of preparation, extraction, evaporation, and cleaning (in h).
- Product recovery (Principle 6): (Mass of final product recovered) / (mass of available product in the plant material) (in %).

In Figure 6, it is important to notice that for each principle, a value close to the center is a positive result whereas a value far from the center corresponds to a negative result. Thus, for "Product recovery", the center corresponds to a maximum of actives extracted.

Compared to CSE, PLE enabled reducing extraction time by a factor 8, from 9 h 40 min to 1 h 10 min. As well, PLE required less solvent, around 50 mL against 300 mL for CSE. Waste of solvent is a real problem in analytical laboratories because usually solvents are not recycled. It is all the more important to minimize the amount of solvent required for an analysis. Thus, during PLE less waste is produced by an analysis in terms of solvent and spent residue. In PLE extraction, less raw material was needed (3 g against 10 g for CSE).

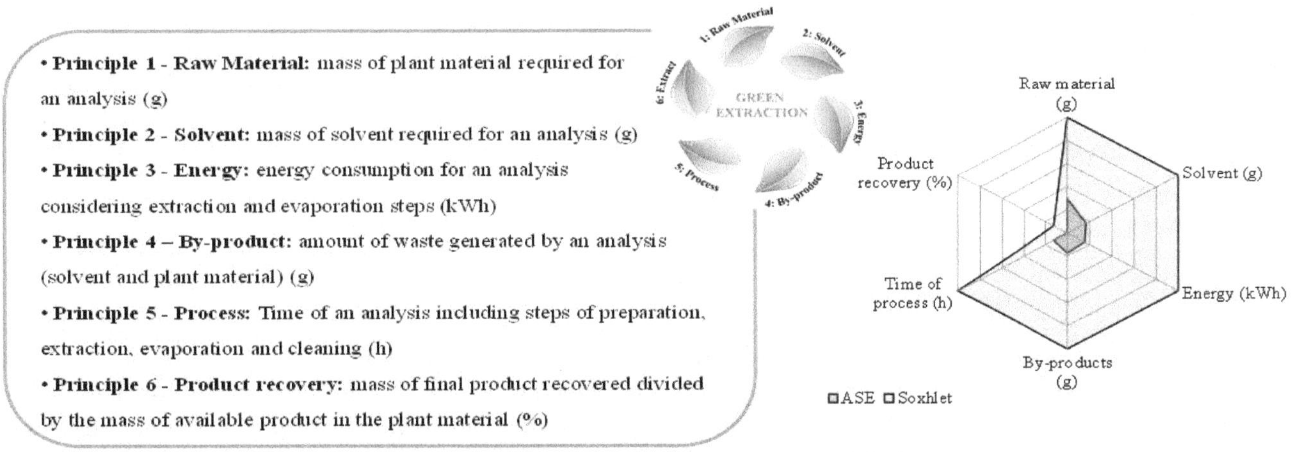

Figure 6. Eco-footprint of PLE vs. CSE processes.

More than the economical aspect, it can be very practical in a sourcing demarche where, regularly, only few quantities of raw material are available. Energy consumption was lower for PLE extraction, because even if the extraction temperature was higher (183 °C instead of 78 °C for CSE), less solvent had to be heated (50 mL against 300 mL), and there were only 3 heating cycles during PLE against 20 cycles during CSE. Another positive aspect of PLE compared to CSE was the percentage of product recovery (100% for PLE and 87.7% for CSE). Higher pressure and temperature during PLE allowed extracting more actives, and their degradation was avoided thanks to the absence of oxygen in the system and the short time of contact between the matrix and the solvent. Finally, the reduced cost of extraction was advantageous for the PLE method in terms of time, amount of raw material and solvent, product recovery, and waste generated. The eco-footprint of PLE was 33 times lower than CSE, with 2.96 area units for PLE against 100 area units for CSE, represented in Figure 6. The implementation of this technique in industrial quality control laboratories could be advantageous compared to CSE in terms of capital expenses and economic savings.

4. Conclusions

Optimization of PLE was carried out using a central composite design methodology. Maximization of extraction was obtained combining three PLE parameters: Temperature (183 °C), pressure (130 bar), and static duration of extraction (3 min). Given the high temperatures tested, carnosol was monitored to follow degradation of carnosic acid and no increase of carnosol was evidenced. To evaluate if PLE could replace CSE, a statistical comparison of extraction performances of the two processes was performed. Ultimately, the eco-footprint of PLE and CSE were determined considering consumption of raw material, solvent, energy, and time. PLE proved to be a rapid, clean, and environmentally friendly technique for determination of active content in plant matrices.

Author Contributions: Conceptualization, M.H., N.R., A.B., A.S.F.-T. and F.C.; methodology, M.H., N.R., A.B., A.S.F.-T. and F.C.; software, M.H., N.R., A.B., A.S.F.-T. and F.C.; validation, M.H., N.R., A.B., A.S.F.-T. and F.C.; formal analysis, M.H., N.R., A.B., A.S.F.-T. and F.C.; investigation M.H., N.R., A.B., A.S.F.-T. and F.C.; resources M.H., N.R., A.B., A.S.F.-T. and F.C.; data curation, M.H., N.R., A.B., A.S.F.-T. and F.C.; writing—original draft preparation, M.H., N.R., A.B., A.S.F.-T. and F.C.; writing—review and editing, M.H., N.R., A.B., A.S.F.-T. and F.C.; visualization, M.H., N.R., A.B., A.S.F.-T. and F.C.; supervision, M.H., N.R., A.B., A.S.F.-T. and F.C.; project administration, M.H., N.R., A.B., A.S.F.-T. and F.C.; funding acquisition, A.B. and F.C. All authors have read and agreed to the published version of the manuscript.

Acknowledgments: Authors acknowledge Anthony Aldebert and Cindy Gonzalez for their help on the analytical aspects of this work.

References

1. Luque de Castro, M.; Ayuso, L. Soxhlet extraction. In *Encyclopedia of Separation Science*; Wilson, I., Ed.; Academic Press: San Diego, CA, USA, 2000; pp. 2701–2709.
2. Rodríguez-Solana, R.; Salgado, J.; Domínguez, J.; Cortés-Diéguez, S. Characterization of fennel extracts and quantification of estragole: Optimization and comparison of accelerated solvent extraction and soxhlet techniques. *Ind. Crops Prod.* **2014**, *52*, 528–536. [CrossRef]
3. Wang, L.; Weller, C. Recent advances in extraction of nutraceuticals from plants. *Trends Food Sci. Technol.* **2006**, *17*, 300–312. [CrossRef]
4. Luque de Castro, M.; Priego-Capote, F. 2.05-soxhlet extraction versus accelerated solvent extraction a2. In *Comprehensive Sampling and Sample Preparation*; Pawliszyn, J., Ed.; Academic Press: Oxford, UK, 2012; pp. 83–103.
5. Steyaningsih, W.; Saputro, I.E.; Palma, M.; Barroso, C.G. Pressurized liquid extraction of phenolic compounds from rice (Oryza sativa) grains. *Food Chem.* **2016**, *192*, 452–459. [CrossRef] [PubMed]
6. Feuereisen, M.M.; Barraza, M.G.; Zimmermann, B.F.; Schieber, A.; Schulze-Kaysers, N. Pressurized liquid extraction of anthocyanins and biflavonoids from *Schinus terebinthifolius* Raddi: A multivariate optimization. *Food Chem.* **2017**, *214*, 564–571. [CrossRef] [PubMed]
7. Andreu, V.; Pico, Y. Pressurized liquid extraction of organic contaminants in environmental and food samples. *Trends Anal. Chem.* **2019**, *118*, 709–721. [CrossRef]
8. Saldaña, M.; Valdivieso-Ramírez, C. Pressurized fluid systems: Phytochemical production from biomass. *J. Supercrit. Fluids* **2015**, *96*, 228–244. [CrossRef]
9. Liang, X.; Nielsen, N.J.; Christensen, J.H. Selective pressurized liquid extraction of plant secondary matabolites; *Convallaria majalis* L. as a case. *Anal. Chim. Acta* **2020**, *X4*, 100040.
10. Pereira, D.T.V.; Tarone, A.G.; Cazarin, C.B.B.; Barbero, G.F.; Martinez, J. Pressurized liquid extraction of bioactive compounds from grape marc. *J. Food Eng.* **2019**, *240*, 105–113. [CrossRef]
11. De Oliveira, G.A.; de Oliveira, A.; da Conceição, E.; Leles, M. Multiresponse optimization of an extraction procedure of carnosol and rosmarinic and carnosic acids from rosemary. *Food Chem.* **2016**, *211*, 465–473. [CrossRef]
12. Ribeiro-Santos, R.; Carvalho-Costa, D.; Cavaleiro, C.; Costa, H.; Albuquerque, T.; Castilho, M.; Ramos, F.;

Melo, N.; Sanches-Silva, A. A novel insight on an ancient aromatic plant: The rosemary (*rosmarinus officinalis* L.). *Trends Food Sci. Technol.* **2015**, *45*, 355–368. [CrossRef]

13. Borrás, L.; Arráez-Román, D.; Herrero, M.; Ibáñez, E.; Segura-Carretero, A.; Fernández-Gutiérrez, A. Comparison of different extraction procedures for the comprehensive characterization of bioactive phenolic compounds in *rosmarinus officinalis* by reversed-phase high-performance liquid chromatography with diode array detection coupled to electrospray time-of-flight mass spectrometry. *J. Chromatogr. A.* **2011**, *1218*, 7682–7690.
14. Herrero, M.; Plaza, M.; Cifuentes, A.; Ibanez, E. Green processes for the extraction of bioactives from rosemary: Chemical and functional characterization via ultra-performance liquid chromatography-tandem mass spectrometry and in-vitro assays. *J. Chromatogr. A* **2010**, *1217*, 2512–2520. [CrossRef]
15. Hossain, B.; Barry-Ryan, C.; Martin-Diana, A.; Brunton, N. Optimisation of accelerated solvent extraction of antioxidant compounds from rosemary (*rosmarinus officinalis* L.), marjoram (*origanum majorana* L.) and oregano (*origanum vulgare* L.) using response surface methodology. *Food Chem.* **2011**, *126*, 339–346. [CrossRef]
16. Rodríguez-Solana, R.; Salgado, J.; Domínguez, J.; Cortés-Diéguez, S. Comparison of soxhlet, accelerated solvent and supercritical fluid extraction techniques for volatile (gc–ms and gc/fid) and phenolic compounds (hplc–esi/ms/ms) from lamiaceae species. *Phytochem. Anal.* **2015**, *26*, 61–71. [CrossRef] [PubMed]
17. Herrero, M.; Arráez-Román, D.; Segura, A.; Kenndler, E.; Gius, B.; Raggi, M.; Ibáñez, E.; Cifuentes, A. Pressurized liquid extraction–capillary electrophoresis–mass spectrometry for the analysis of polar antioxidants in rosemary extracts. *J. Chromatogr. A* **2005**, *1084*, 54–62. [CrossRef] [PubMed]
18. Jacotet-Navarro, M.; Rombaut, N.; Fabiano-Tixier, A.; Danguien, M.; Bily, A.; Chemat, F. Ultrasound versus microwave as green processes for extraction of rosmarinic, carnosic and ursolic acids from rosemary. *Ultrason. Sonochem.* **2015**, *27*, 102–109. [CrossRef]
19. Petersen, M.; Abdullah, Y.; Benner, J.; Eberle, D.; Gehlen, K.; Hücherig, S.; Janiak, V.; Kim, K.; Sander, M.; Weitzel, C.; et al. Evolution of rosmarinic acid biosynthesis. *Phytochemistry* **2009**, *70*, 1663–1679. [CrossRef]
20. Bernatoniene, J.; Cizauskaite, U.; Ivanauskas, L.; Jakstas, V.; Kalveniene, Z.; Kopustinskiene, D. Novel approaches to optimize extraction processes of ursolic, oleanolic and rosmarinic acids from *rosmarinus officinalis* leaves. *Ind. Crops Prod.* **2016**, *84*, 72–79. [CrossRef]
21. Birtić, S.; Dussort, P.; Pierre, F.; Bily, A.; Roller, M. Carnosic acid. *Phytochemistry* **2015**, *115*, 9–19. [CrossRef]
22. Ho, C.; Osawa, T.; Huang, M.; Rosen, R. *Food Phytochemicals for Cancer Prevention II; Teas, Spices, and Herbs*; American Chemical Society: Washington, DC, USA, 1994; Volume 547, p. 388.
23. Rodríguez-Rojo, S.; Visentin, A.; Maestri, D.; Cocero, M. Assisted extraction of rosemary antioxidants with green solvents. *J. Food Eng.* **2012**, *109*, 98–103. [CrossRef]
24. Chemat, F.; Abert-Vian, M.; Cravotto, G. Green extraction of natural products: Concept and principles. *Int. J. Mol. Sci.* **2012**, *13*, 8615–8627. [CrossRef] [PubMed]
25. Larkin, J. Thermodynamic properties of aqueous non-electrolyte mixtures i. Excess enthalpy for water + ethanol at 298.15 to 383.15 k. *J. Chem. Thermodyn.* **1975**, *7*, 137–148. [CrossRef]
26. Newsham, D.; Mendez-Lecanda, E. Isobaric enthalpies of vaporization of water, methanol, ethanol, propan-2-ol, and their mixtures. *J. Chem. Thermodyn.* **1982**, *14*, 291–301. [CrossRef]
27. Pingret, D.; Fabiano-Tixier, A.; Le Bourvellec, C.; Renard, C.; Chemat, F. Lab and pilot-scale ultrasound-assisted water extraction of polyphenols from apple pomace. *J. Food Eng.* **2012**, *111*, 73–81. [CrossRef]

4

Pressurized Solvent Extraction with Ethyl Acetate and Liquid Chromatography—Tandem Mass Spectrometry for the Analysis of Selected

Renata Raina-Fulton * and Aisha A. Mohamad

Department of Chemistry & Biochemistry, Trace Analysis Facility, University of Regina; 3737 Wascana Parkway, Regina, SK S4S 0A2, Canada; aam837@uregina.ca
* Correspondence: renata.raina@uregina.ca

Abstract: The extraction of powdered nutraceuticals is challenging due to the low water content and high concentration of matrix components that can lead to significant matrix effects in liquid chromatography-positive ion electrospray ionization-tandem mass spectrometry (LC-ESI$^+$-MS/MS). In this study we assess the feasibility of using pressurized solvent extraction with ethyl acetate to reduce the co-extraction of polar matrix components. Pigment attributed to chlorophyll was removed with in-cell clean-up utilizing Anasorb 747, Florisil®, and C18. Visible inspection of the extracts showed that pigment was removed from matcha, a powdered green tea sample. Pressurized solvent extraction with in-cell clean-up can be utilized to remove pigments from powdered samples such as nutraceuticals. Average matrix effect of the 32 target analytes that observed mass spectrometric signal suppression or soft MS signal enhancement was $-41 \pm 19\%$ with the majority of analytes having a protonated molecular ion with m/z of 250 to 412. As generally moderate signal suppression was observed for conazole fungicides and structurally related compounds analyzed by LC-ESI$^+$-MS/MS, it is recommended that matrix matched or standard addition calibration is used for quantitation. Catachins, other polyphenols, and caffeine are expected to contribute to the matrix effects observed in LC-ESI$^+$-MS/MS. Diniconazole, fenbuconazole, and tebufenozide were the only target analytes with severe MS signal enhancement. Low levels (0.002–0.004 mg/kg) of prothioconazole-desthio and flusilazole were detected, along with trace levels of tebuthiuron in matcha.

Keywords: matcha; conazole fungicides; pressurized solvent extraction; pesticide residue analysis

1. Introduction

Conazole fungicides are a critical group of fungicides used on a wide variety of crops including tea products. They are predominated de-methylation inhibitors and include triazoles and imidazole fungicides. Triazoles are most widely used for blister blight in teas produced in Asia. There has been a rise in commercially available powder nutraceutical products which may be consumed in multiple ways including in drinks (tea or mixed in with other drinks) or food products (e.g., baked goods and chocolate). This necessitates the need for analytical methods capable of dealing with finely powdered samples with high pigment levels.

We have previously reviewed methods for analysis of fungicides in nutraceutical products and presented some of the analytical challenges to a wide variety of fungicides currently in use [1,2]. The analysis of conazole fungicides has been accomplished by LC-MS/MS, GC-MS/MS or GC-MS methods, and is more prone than other pesticides to a variety of issues including stability issues of target analytes, pH sensitivity, carry-over problems largely attributed to strong adsorption of conazole fungicides on surfaces including tubing and other components of LC-MS/MS or GC-MS systems, and isobaric interferences in MS detection even when using tandem MS methods [1,3–9]. Tandem mass spectrometry provides good sensitivity and selectivity for the analysis of conazoles fungicides. Chromatographic resolution is still required due to the structural similarity and presence of isotopes for many conazoles such that isobaric interferences are common. The number of individual conazole fungicides registered for use worldwide is also increasing with small modifications to the structure such that there is a higher potential for isobaric interferences in MS detection. In addition until recently few deuterated or C-13 isotopes of conazole fungicides were commercially available such that conazoles used for medical or veterinary applications were previously utilized as internal standards [1,4]. Due to the ability of LC-MS/MS to provide short analysis times and good MS sensitivity and selectivity it is often preferred when a wide range of conazole fungicdes are analyzed. Selected conazoles may have better or similar sensitivity when GC-MS methods are used for their analysis [1,7].

Methods for the extraction and analyses of pesticides in dry nutraceutical products are more limited than other food products [2]. In addition to the very complex matrix of nutraceutical products, these products are predominantly sold in dry powder form or as tablets or capsules which is less compatible to extraction methods. There are high concentrations of pigments. Tea products such as matcha have high levels of chlorophyll, caffeine, and polyphenols [10–12]. Most extraction and clean-up methods for fruits and vegetables which have higher water content are based on quick, easy, cheap, effective, rugged, and safe (QuEChERS) or modified QuEChERS approaches, and when dry powders are extracted a wetting step is required for the acetonitrile salt-out extraction typically followed by filtration [7–9]. Sample size must be controlled for dispersive solid phase extraction (dSPE) clean-up to avoid saturation of sorbent materials. A variety of sorbents have been used for clean-up of tea matrices in dSPE or SPE including graphitized carbon black (GCB), graphite carbon/aminopropylsilanized silica gel (carbon-NH_2), primary secondary amine (PSA) or silica, and are efficient for the removal of chlorophyll, catechins and caffeine in infused teas and other sample matrices [9,12–15]. Some methods have been able to successfully use graphitized carbon black in sample cleanup for analysis of selected conazoles, while others have found lower recoveries [8]. Planar pesticides including selected conazole fungicides are known to strongly bind to graphitized carbon black which is the most commonly used sorbent to remove pigments in extracts.

Pesticides in tea products have been predominately analyzed after infusion or a wetting step. The Japan Official Method and modified versions of this method for pesticide residues utilize a wetting step with 20 mL of water per 5 g sample for a 30-min period followed by homogenization with acetonitrile and subsequent filtration [13,14]. The salt-out acetonitrile extraction step is followed by a portion of the extract removed for subsequent clean-up. These approaches rely on good transfer rates of the fungicides from the solid matrix into water such that the fungicides must have high water solubility. Poor recoveries of <20% have been reported for conazole fungicides even with a wetting step in the extraction procedure in a variety of sample matrices including teas [14]. Desired recoveries for pesticide residue analyses methods are 85–110%. Only selected methods have completed the extraction of some pesticides or catechins in tea directly into methanol or 50/50 *v*/*v*% ethyl acetate/hexane [3,16]. Analysis of strobilurin fungicides that were extracted with ethyl acetate using pressurized solvent extraction (PSE) has been accomplished for particles collected on filters and matcha [4,17]. PSE is an alternative extraction approach to QuEChERS that allows for direct extraction of target analytes into an organic solvent without the need for the powdered sample of low water content to undergo a time-consuming wetting step or for the extract obtained to be filtered prior to clean-up. Using PSE with ethyl acetate as the extraction solvent recoveries of conazole fungicides and deuterated internal standards (diazinond_{10}) are 85–110% with within batch recoveries of $\pm$10% [4]. The objective of this study was to evaluate an adapted pressurized solvent extraction method with in-cell clean-up of matcha green tea powders targeting initial removal of pigment. We focused on the evaluation of recoveries and the matrix effects of the remaining matrix in the extract in the quantitation of a large range of conazole fungicides by LC-ESI^+-MS/MS.

2. Materials and Methods

2.1. Materials

Ethyl acetate, acetonitrile, and methanol were of pesticide grade and supplied by Fisher Scientific. Deionized water (18 MΩ resistivity) was from a Nanopure Diamond system (Barnstead International, Dubuque, IA, USA). Aqueous solvents were passed through a 0.45 μm membrane filter from Nuclepore (Watman, Florham Park, NJ, USA). Formic acid (>88.0%) was obtained from VWR Scientific (West Chester, PA, USA). Solids or stock solutions at 100 μg/mL of pesticide standards of all test analytes were purchased from Chem Service, Inc. (West Chester, PA, USA). Solid of propiconazole-phenyld_3 was purchased from Sigma-Aldrich (Oakville, ON, Canada). Solids of individual pesticides (~1 mg) were dissolved in 1mL of methanol with further dilution to prepare individual stock solutions at 100 μg/mL in methanol and stored at −4 °C. A further standard stock solution at 1 μg/mL of all target analytes was prepared in methanol for use in preparation of calibration standards.

Filters used for weighing the matcha solid were LABX Berkshire Engineering Clean, 10 cm diameter, and filters used in the 66 mL ASE extraction were glass fiber 934-AH™, 3.0 cm diameter (VWR Scientific). Matcha is a powdered green tea product and samples were obtained from three different manufacturers that distributed product within Canada and were labeled as organic products. Anasorb 747 (40/80 mesh) was obtained from SKC Inc. (Eighty Four, PA, USA). C18 and Florisil® were obtained from Sigma-Aldrich (Oakville, ON, Canada).

2.2. Sample Preparation

Pressurized solvent extraction with in-cell clean-up was used to extract the target analytes from the matcha samples (see Figure 1). An ASE 100 pressurized solvent extraction system (Dionex, Sunnyvale, USA) was used for extraction with the following extraction parameters: temperature 100 °C; static mode time of 30 min at 1500 psi; four static cycles; 60% flush volume; purge time with nitrogen (UHP) at the end of 600 s. The 66 mL extraction cell was loaded from bottom to top as follows: 934-AH™ filter; 2 g of C18; filter; 4 g Florisil®; filter; 30 g Anasorb 747; two filters; 1 g of Matcha weighed in folded cleanroom grade LABX filter paper (10 cm). The LABX filter paper holding the pre-weighed matcha sample was folded to a diameter of 3 cm to fit into the extraction cell.

The extraction solvent was ethyl acetate with total volume of extraction after the four static cycles of ~130 mL. A 1 mL volume of 2-propanol was added to the extract as a keeper for the drying step. The sample extract was then dried to 4–5 mL in a SPE apparatus under slight vacuum, transferred to a 15 mL vial with methanol used to rinse the extraction bottle three times, and dried to approximately 1 mL at 0.5 mL/hr. The extract was diluted by a factor of 0.43 (65/150) with addition of internal standard (propiconazole-phenyld_3 at 50 ng/mL) for final analysis. Standard addition calibration with internal standard was used for final analysis of the three brands of organic matcha samples. For standard addition calibrations the standard amounts added were from MDL to 80 ng/mL.

For recovery evaluation 1 g pre-weighed matcha powdered samples were spiked with 100 μL of 1 μg/mL conazole standard mix (equivalent to 0.1 mg/kg) and allowed to dry prior to loading into the extraction cell. Calibration standards (solvent-only) and matrix matched standards were prepared for evaluation of matrix effects from 0.6 to 100 ng/mL with matrix added at 1/10 dilution. The matrix was obtained from the extraction of a matcha sample with no detectable fungicides.

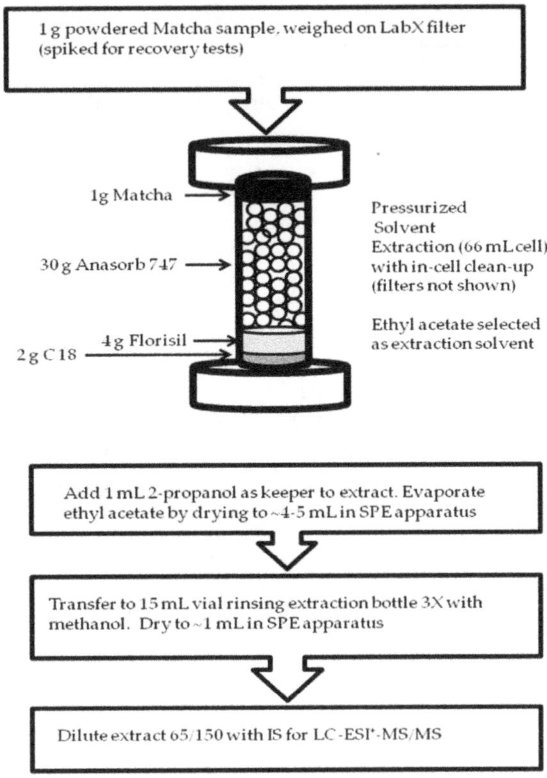
1 g powdered Matcha sample, weighed on LabX filter (spiked for recovery tests)
1g Matcha
30 g Anasorb 747
4 g Florisil
2 g C18
Pressurized Solvent Extraction (66 mL cell) with in-cell clean-up (filters not shown)
Ethyl acetate selected as extraction solvent
Add 1 mL 2-propanol as keeper to extract. Evaporate ethyl acetate by drying to ~4-5 mL in SPE apparatus
Transfer to 15 mL vial rinsing extraction bottle 3X with methanol. Dry to ~1 mL in SPE apparatus
Dilute extract 65/150 with IS for LC-ESI+-MS/MS

Figure 1. Sample preparation method for conazole fungicide analysis.

2.3. LC-ESI$^+$-MS/MS Analysis

LC analysis was performed with a Waters (Milford, MA, USA) LC system consisting of a 1525 μm binary pump and a column heater at 21 °C. A LEAP Technologies autosampler (Carrboro, NC, USA) was used for 5 μL injections at 100 μL/s and 1 pre- and post-cleans with ethyl acetate followed by methanol to minimize carry-over. A guard column (4 × 2.0 mm id, Gemini) was connected to the analytical column, Synergi Polar-RP, 550 × 2.00 mm id, 2.5 μm particle size (Phenomenex, Torrance, CA, USA). A pre-injection of 5 μL of 2-propanol was completed prior to each sample run at the initial conditions (3 v% methanol in 0.05 v% formic acid aqueous mobile phase) with a flow rate of 0.15 mL/min held for 10 min. The pre-injection of 2-propanol was completed at the initial mobile phase conditions to reduce carry-over issues. A mobile phase gradient was used for the separation of target analytes with initial conditions at 3 v% methanol in 0.05 v% formic acid aqueous mobile phase. The gradient of 0.1 v% formic acid in acetonitrile was changed linearly as follows: 0 to 1.5 min 0%; 2.5 min at 20%; 3 min at 35%; 10 min at 45%; 16 min at 50%; 18 min at 60%; 20 min at 75%; 25 min at 80% and held to 30 min. The mobile phase is re-equilibrated to 0% acetonitrile in 15 min and held for an additional 5 min prior to the pre-injection of 2-propanol. All analytes eluted in the first 25 min.

The Waters LC system was connected to a Waters Quattro Premier triple quadrupole mass spectrometer, operated in electrospray positive ion mode. The temperature of the source was set to 120 °C, desolvation temperature to 350 °C, desolvation gas flow was 750 L/h, and cone gas flow was 150 L/h. The optimized settings for ESI$^+$ were capillary voltage of 3.1 kV; and Rf lens, 0.1 V. The collision gas used for SRM was argon (UHP) at 0.15 mL/min or 4×10^{-4} mbar. Cone voltage and collision energy were set up in the MS method as previously report [4], and shown in Table 1 for target analytes. Infusion experiments were conducted for each individual target analyte to determine the SRM conditions with a syringe pump flow rate of 50 μL/min. Table S1 (see Supplementary Material) shows the regression coefficient of the matrix matched calibration curve for each analyte at the quantitative SRM. The ratio of response for SRM1/SRM2 are within relative standard deviation criteria of <20% for all target analytes.

Table 1. Selected Reaction Monitoring Transitions (SRMs) for Target Analytes Analyzed by LC-ESI$^+$-MS/MS.

Target Analyte	Retention Time (min)	Quantitative SRM, Confirmation SRM (Cone Voltage, Collision Energy)	Method Detection Limits with Matrix Matched Standards (mg/kg)
Benzotriazole	8.24	120→65 (40,17), 120→92 (40,17)	0.0006
Sulfathiazole	8.46	256→155 (20,15), 256→92 (20,25)	0.002
Imazamox	9.09	306→261 (40,20), 306→217 (40,20)	0.010
Sulfamethizole	9.34	271→156 (20,15), 271→92 (20,25)	0.002
Tebuthiuron (thiadiazolylurea herbicide)	10.48	229→172 (25,15), 229→116 (25,25)	0.001
Tricyclazole (benzothiazole fungicide)	12.06	190→163 (35,20), 190→136 (35,25)	0.002
Sulfentrazone	15.36	387→307 (35,20), 389→309 (35,20)	0.010
Imazalil	15.8	297→159 (20,25), 297→201 (20,20)	0.010
Thioconazole	15.93	391→130 (20,20), 391→360 (20,10)	0.010
Azaconazole	16.71	300→159 (30,25), 300→231 (30,15)	0.001
Triadimenol	18.19	296→70 (15,15), 298→70 (15,15), 296→99 (15,10)	0.002
Paclobutrazol (plant growth regulator with triazole moiety)	18.58	294→70 (30,20), 295→70 (25,20), 296→70 (15,15)	0.010
Triticonazole	19.39	318→70 (20,15), 320→70 (20,20)	0.010
Cyproconazole	19.76	292→70 (30,15), 294→70(30,20)	0.002
Hexaconazole	20.58	314→70 (25,20), 316→70(25,20)	0.010
Uniconazole (uniconazole-P)	20.94	292→70 (30,15), 294→70 (30,20)	0.010
Etaconazole	21.58	328→159 (30,25), 330→161 (30,25), 328→187 (30,30)	0.001
Prochloraz	21.61	376→70 (15,25),378→70 (15,25), 376→308 (15,15)	0.010
Myclobutanil	21.73	289→70 (25,15), 291→70 (25,15)	0.010
Triadimefon	21.73	295→70 (25,20), 297→70 (25,20)	0.002
Prothioconazole (analyzed as prothioconazole-desthio)	21.75	314→70 (25,20), 312→70 (25,20), 312→125 (25,20)	0.005
Tebuconazole	21.94	308.5→70 (35,20), 310.5→70 (35,20), 308.5→125 (35,20)	0.001
Bromuconazole	22.01	376→159 (30,25), 378→159 (30,25)	0.010
Penconazole	22.12	284→70 (25,15), 284→159 (25,15)	0.010
Metconazole	22.15	321→70 (30,20), 323→70 (30,20)	0.010
Diniconazole	22.46	326→70(35,25), 328→70(35,25), 326→159 (35,20)	0.0006
Epoxiconazole	22.46	330→121 (25,20), 332→121 (25,20), 330→123 (25,20)	0.010
Tetraconazole	22.46	372→159 (30,25), 372→70 (30,25)	0.010
Biteranol	22.73	338→99 (15,15), 338→269 (15,15)	0.002
Propiconazole	22.73	342→159 (30,25), 342→69 (30,25)	0.010
Flusilazole	22.94	316→165 (30,25), 316→248 (30,15)	0.0006
Fenbuconazole	23.12	337→70 (30,20), 337→125 (30,20)	0.001
Tebufenozide (insecticide)	23.12	353→133 (12,17), 353→297 (12,17)	0.002
Difenoconazole	23.64	406→251 (30,25), 408→253 (30,25)	0.010
Etoxazole	25.02	360→57 (35,25), 360→141 (35,30), 360→177.5 (35,20)	0.010
Propiconazole-phenyld$_3$ (internal standard)	22.80	347→164 (50,25), 349→166 (50,25), 347→69 (50,25), 349→69 (50,25)	NA

3. Results

3.1. Cell Design for Pressurized Solvent Extraction of Matcha

Matcha was used as a model case for analysis of conazole fungicides and structurally related pesticides in powdered samples with high levels of pigments in the sample matrix. Potential interferences in MS detection from the matrix of matcha samples includes chlorophyll, caffeine, catechins and other polyphenols which if present in samples injected would co-elute in a reversed-phase LC separation [10,11,13,14,16]. These matrix components can lead to interferences in MS detection and signal suppression or enhancement in MS detection. Powdered samples may require additional contact time with solvents to ensure complete extraction based on our prior work on extraction of fungicides from particles collected on glass fiber filters [17]. The pressurized solvent extraction procedure with in-cell clean-up was designed to firstly remove the pigment attributed to chlorophyll from the sample. The sample was loaded at the top of the extraction cell held in LABX filter paper for easier removal of the solid after extraction. Two glass fiber filters (3 cm diameter) were placed between the sample and Anasorb 747 sorbent to remove residue solids that pass through the LABX filter paper holding the matcha powdered sample. Ethyl acetate was selected as it provides good recoveries for conazole fungicides from other solid materials and the use of a less polar solvent than acetonitrile, acetonitrile/water or methanol reduces co-extraction of more polar matrix components that can lead to suppression of MS signal and interferences in MS detection. Pressured solvent extraction with ethyl acetate has been used in the extraction of catechins from teas including matcha with lower levels of caffeine found in the extracts [16]. PSE with ethyl acetate or liquid–liquid extractions have also been used to extract selected conazole and strobilurin fungicides from other teas, and solid sorbents used in air sampling [2–4]. Ethyl acetate has also been used with QuEChERS rather than acetonitrile for extraction of fungicides from fruit and vegetable matrices [9]. To reduce co-extraction of less polar matrix components such as fat soluble co-extracts more polar solvents such as acetonitrile or acetone are selected for extraction and have been used to extract fungicides from soil, plant and animal based foods [18,19].

The extraction cell size (66 mL) was selected to enable 30 g of Anasorb 747 to be loaded in the cell (directly below the sample) and this amount of sorbent was required to remove the green pigment coloration of the extracts. Without the use of Anasorb 747 extracts were too high in chlorophyll content such that after preconcentrated (drying) step there was precipitation. To remove the residual color in the ethyl acetate extracts 4 g of Florisil® was required which was placed in the extraction cell below the Anasorb 747. We choose to also add 2 g C18 to aid in removal of some potential matrix components including pigment. Conazole fungicides recoveries have been shown to be good with both C18 and Florisil® SPE clean-up [3,4,13–15]. Filters were placed in the extraction cell before and after each sorbent layer to allow the sorbents to be removed separately. Visible inspection of the sorbent materials after extraction showed that Florisil® and to a lesser extent C18 were removing the residue pigment after the solvent had passed through the Anasorb 747. Graphitized carbon black and primary secondary amine (PSA) were also tested but were not used as the extracts obtained still had green coloration after the pressurized solvent extraction step. PSA has been found to work less effectively for clean-up in the presence of high pigment levels [15]. The use of Anasorb 747 over graphitized carbon black for pressurized solvent extraction was preferred for improved flow characteristics (larger particle size (20/40 mesh) and was the most efficient sorbent tested at removing most of the color of the extracts. Other carbon based sorbents that have been used in the literature include Carbon-X in a SPE format for green tea supplements [15]. Based on our prior procedure where we extracted fungicides from particles on filters we selected four static stages followed by a 60% flush after each static stage to ensure adequate contact time of the solvent with the finely powdered matcha sample. Fungicides are known to strong bind to many solid sorbent materials. An addition extraction with 4 static stages showed no visible color in the extract and also no presence of pesticides.

3.2. *Modifications to the LC-ESI$^+$-MS/MS Analysis*

We modified our existing LC-ESI$^+$-MS/MS method to include some new fungicides or additional structurally related pesticides that have become commercially available since our initial development (see Table 1) [4]. Sulfentrazone, which is a triazolone herbicide, has few existing methods [5,6]. We also re-optimized the separation for shorter total analysis time from ~45 min to 25 min with the addition of a small percentage of methanol rather than 5% 2-propanol in the aqueous mobile phase. In the prior method 2-propanol was added in the aqueous mobile phase to reduce carry-over issues, however, additional of methanol to the aqueous mobile phase rather than 2-propanol results in better chromatographic resolution for isobaric compounds. To avoid issues with carry-over from sorption of matrix or conazole fungicides in the LC-MS/MS system we completed a pre-injection of 2-propanol prior to each run for 10 min at initial mobile phase conditions.

Good chromatographic resolution was obtained for all compounds with the same or similar SRMs. The conazole fungicides shown in Figure 2A,B have similar molecular weight range are chlorinated and produce the m/z = 70 fragment that is attributed to the triazole moiety. Good separation of these compounds and retention time stability must be obtained to avoid false positive identification as particularly cyproconazole and uniconazole-P give response at SRM of 292>70 and 294>70 (quantification and confirmation SRMs). Co-elution of these fungicides with isobaric interferences occurs and worsens with matrix issues [1,4,17,20]. Similarly hexaconazole and prothionconazole-desthio elute in a similar retention time range and give response at 314>70 (Figure 2C) as well as 312>70 and 316>70. Prothioconazole-desthio has a significantly higher abundance than hexaconazole at 314>70. A small amount of 2-propanol could be added to the aqueous mobile phase to further reduce potential sorption of matrix or conazoles in the LC-MS/MS system with a recommendation to kept the percentage below 2% if methanol is also used in the aqueous mobile phase to ensure adequate chromatographic resolution of isobaric fungicides without the need for long analysis times. When matrix or conazole sorption occurs in the system this is usually evident by a shift to longer retention times and broadening of peak shapes.

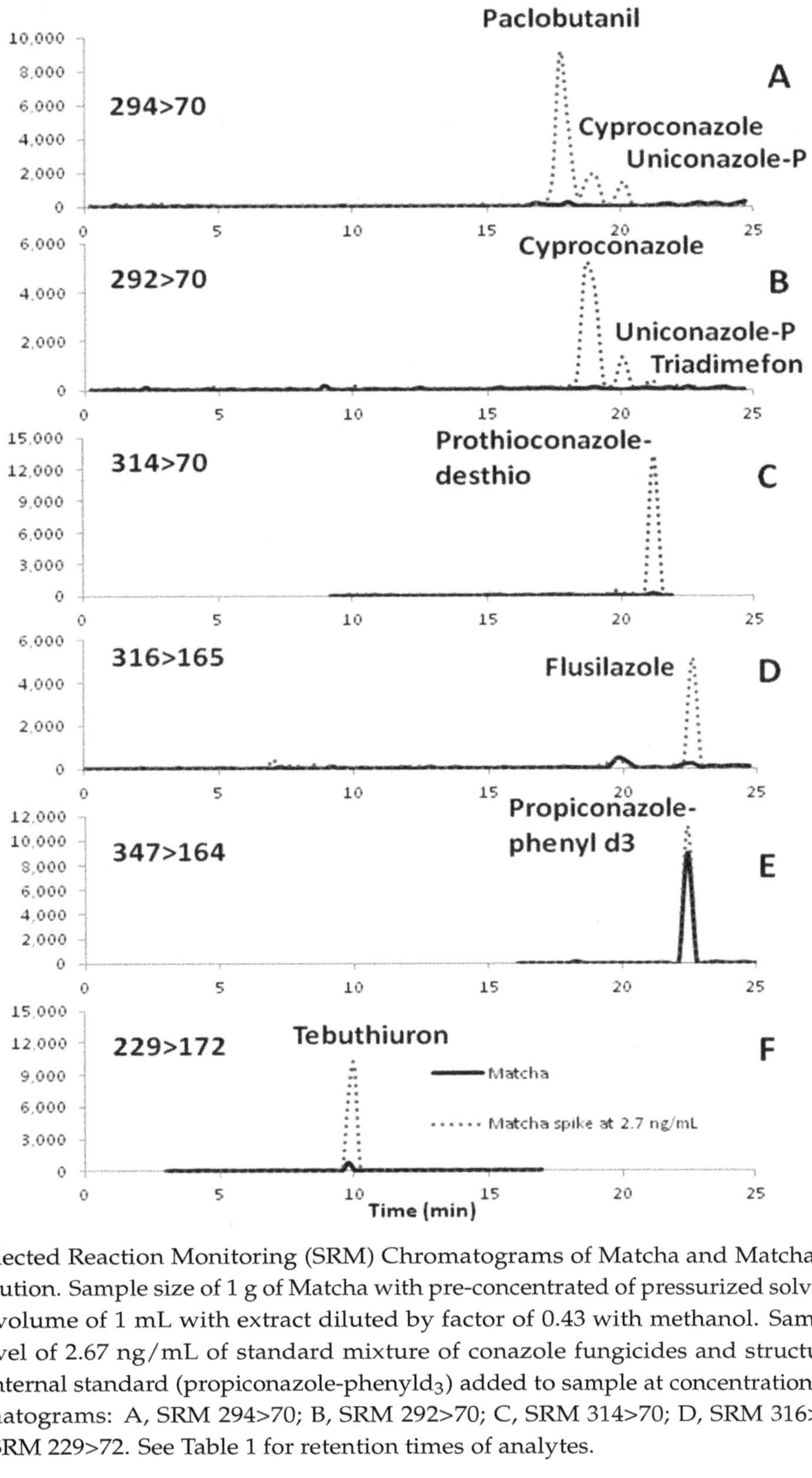

Figure 2. Selected Reaction Monitoring (SRM) Chromatograms of Matcha and Matcha Spiked with Standard Solution. Sample size of 1 g of Matcha with pre-concentrated of pressurized solvent extraction extract to a volume of 1 mL with extract diluted by factor of 0.43 with methanol. Sample extract is spiked at level of 2.67 ng/mL of standard mixture of conazole fungicides and structurally similar pesticides. Internal standard (propiconazole-phenyld$_3$) added to sample at concentration of 50 ng/mL. SRM Chromatograms: A, SRM 294>70; B, SRM 292>70; C, SRM 314>70; D, SRM 316>165; E, SRM 347>164; F, SRM 229>72. See Table 1 for retention times of analytes.

3.3. Method Detection Limits and Calibration

Table 1 shows the quantitative and confirmation SRM for all target analytes. All new conazole or structurally similar analytes (imazamox, tricyclazole, sulfentrazone, etaconazole, etoxazole) were separated from each other. Etaconazole at higher concentrations can give a response at 328→70, and 326→70, however the response of diniconazole is much stronger at these SRMs and the two conazoles are separated. Table 1 shows the method detection limits (MDLs) as determined by the lowest concentrations of standard (matrix matched) that deviate from the regression line by <25%. The method detection limits for 29 of the 35 analytes were slightly higher than we previously reported with solvent-based standards [4], but still in a desirable range of 0.0006–0.010 mg/kg. The new target analytes (imazamox, sulfentrazone, and etacoanzole) had detection limits of 0.010 mg/kg. Method detection limit for etaconazole and tricyclazone were lower at 0.001 and 0.002 mg/kg, respectively. The ratio of SRM1/SRM2 were determined on the same day of analysis from matrix matched standards, with relative standard deviation <20% (see Supplementary Material Table S1). Our working calibration range was MDL to 0.080 mg/kg. Correlation coefficient of calibration curves obtained from matrix matched standards was >0.94 (see Table S1). Tetraconazole observed a lower r^2 (0.91) which was attributed to severe matrix effects.

3.4. Matrix Effects and Recoveries of Pressurized Solvent Extraction

To evaluate the extent of matrix suppression or enhancement on MS signal in LC-ESI$^+$-MS/MS the slope of best fit line obtained with calibration standards of the same concentration in calibration standards with matrix added (m_{matrix}) and those obtained with solvent-only ($m_{solvent}$) were compared. The matrix effect (ME) was calculated using the following formula:

$$\text{ME} = ((\text{mmatrix} - \text{msolvent}) - 1) * 100\% \quad (1)$$

The matrix was obtained from an extract of a matcha powder sample which was shown to have no target analytes detectable (signal/noise ratio < 3). Most fungicides observed soft (+20 to −20%) or moderate matrix effects (±20–50%). Signal suppression rather than enhancement in MS detection was more commonly observed after removal of the chlorophyll from the sample extract. Average matrix effect of 32 target analytes with suppression or soft signal enhancement was −41 ± 19%. The majority of these target analytes have a mass to charge of 250 to 412 for the protonated molecular ion. Diniconazole, fenbuconazole, and tebufenozide were the only analytes with severe MS signal enhancement (102%, 81%, 225%, respectively). A larger number of analytes (6 of 8) with severe signal suppression (range −55% to −75%) eluted between 20–25 min in the separation. Febuconazole, diniconazole, and difenconazole also observed shifts in the baseline (see Figure S1) and had low recoveries (see Table 2). For conazole fungicides matrix matched standards or standard addition calibration and use of an internal standard are necessary to provide reliable quantitation. Addition of higher amounts of matrix than 50% of solvent composition of injected sample lead to instability of retention times such that the samples for subsequent analysis were diluted generally by a factor of 0.43 (65/150) with methanol.

Table 2 shows that acceptable recoveries (70–120%) were obtained for selected conazole fungicides including tebuthiuron, triadimenol, myclobutanil, and triadimenfon. The recovery of prothioconazole-desthio which was detected in one of the matcha samples was 69% with a relative standard deviation >20%. Cyproconazole and paclobutanil also had recoveries in the 50–70% range (see Table 2). While these recoveries are not ideal they are still an improvement over methods reported for this difficult sample matrix [2]. Other less commonly analyzed conazoles observed low recoveries (not listed in Table 2 due to recoveries <20%), although MS signal suppression was moderate indicating that either these fungicides were strongly adsorbed on Anasorb 747 or not adequately removed from the powdered matcha sample. Additional extraction with ethyl acetate did not show any detectable levels of fungicides.

Table 2. Matrix Effect and Detected Concentration of Selected Fungicides in Matcha.

Target Analyte (SRM)	Recovery, Spiked at 0.01 mg/kg [1] (Average ±SD, n = 4)	% Matrix Effect	Detected Concentration in Matcha (mg/kg) [2]
Selected Analytes with MS Signal Suppression or Soft Enhancement			
Tebuthiuron (229→172)	80.7 ± 4.70	−19%	ND
Sulfentrazone (387→307)	64.0 ± 18.3	−35%	ND
Triadimenol (296→70)	109.5 ± 11.7	−32%	ND
Paclobutanil (295→70)	51.8 ± 14.0	−38%	ND
Cyproconazole (292→70)	69.3 ± 12.9	−37%	ND
Uniconazole (292→70)	23.9 ± 12.9	−29%	ND
Myclobutanil (291→70)	84.9 ± 38.3	−53%	ND
Triadimenfon (295→70)	96.1 ± 34.6	−44%	ND
Hexaconazole (314→70)	12.1 ± 19.8	−75%	ND
Prothioconazole-desthio (314→70)	69.2 ± 29.2	−54%	0.0035
Flusilazole (316→165)	40.4 ± 28.9	6%	0.0024
Propiconazole (342→159)	41.9 ± 25.0	−54%	ND
Etaconazole (330→161)	49.1 ± 11.3	−5%	ND
Azaconazole (300→159)	32.3 ± 11.0	−33%	ND
Difenconazole (406→251)	20.4 ± 87.9	−47%	ND
Analytes with Severe Signal Enhancement			
Diniconazole (326→70)	21.5 ± 5.40	102%	ND
Fenbuconazole (337→70)	29.9 ± 8.10	81%	ND

[1] Matrix Matched Standards; [2] Standard addition calibration; ND (not detected) < MDL.

For subsequent analysis of matcha product samples (labeled as organic) from three different manufacturers standard addition calibration was utilized with propiconazole-phenyld$_3$ for the internal standard. Target analytes that were at or near our method detection limits included tebuthiuron, prothioconazole-dethio, and flusilazole. Figure 2C–E shows the SRM chromatograms for sample injection with and without standard addition for prothioconazole-desthio, flusilazole, and tebuthiuron. Two of the three commercial products did not contain any target analytes above the method detection limit. Tebuthiuron levels were just below our MDL for the injected sample at a dilution factor of 0.43 (see Figure 2). With improved extraction and clean-up the analysis of these target analytes would be more reliable and based on the current method detectable amounts are in the range of 2–4 μg/kg which is comparable to other reports of conazole fungicides in green tea leaves [2,21]. There have been no prior reports of detectable levels of conazole fungicides or tebuthiuron in matcha.

4. Discussion

An additional clean-up step following the pressurized solvent extraction with in-cell clean-up was not used. SPE clean-up following extraction and pre-concentration of the extract is a commonly used approach to further remove matrix. Given the sorbents used in the in-cell clean-up the most common alternative sorbent for SPE clean of extracts would be silica to remove residual caffeine in the ethyl acetate extracts, however it has already been reported that a large number of conazole fungicides have poor recoveries with this sorbent material [13,15]. Filtration with polyvinyldifluoride membrane filters has been shown to reduce epicatechin and epigallocatechin gallate concentrations in acetonitrile extracts [10], and this could be further evaluated for recoveries of conazole fungicides in different solvents used for pressurized solvent extraction.

Ethyl acetate is a less polar solvent than typically used in modified QuEChERS methods, Official Japan method, or modification of this method that involve a wetting step of the solid followed by an acetonitrile salt-out extraction. It has been shown that ethyl acetate extracts obtained with pressurized solvent extraction have lower caffeine content but high concentrations of catechins and polyphenols [16]. Other options for further removal of polyphenols without removal of conazole fungicides includes the use of other non-carbon based sorbents such as polymeric resins or green extraction resins such as chotisan which still need to be evaluated for application to extraction of nutraceuticals [21]. Including a more nonpolar solvent such as hexane or cyclohexane with ethyl acetate as an extraction solvent mixture for pressurized solvent extraction may also further reduce the presence of catechins and other polyphenols in the extract solvent [3,19]. Difenoconazole has been shown to be efficiently extracted with both ethyl acetate and 50/50 *v*/*v* mixtures of ethyl acetate and hexane from chrysanthemum flower tea [3]. Poor recoveries of difenconazole and matrix effects in our study indicate that both MS interferences and strong sorption onto sorbents used for clean-up are the most likely sources of low recoveries of conazole fungicides. Poor extraction recoveries <20% were observed for all target analytes not reported in Table 2 with the potential source including strong adsorption onto the Anasorb 747 (a carbon based material) or poor contact of the solvent with the powdered sample which may also result in larger (>20%) relative standard deviation of recoveries for some conazole fungicides. MS signal suppression was moderate for most analytes, and matrix matched calibration curves in the desired calibration range from MDL-0.080 mg/kg had correlation coefficients >0.94 (see Table S1).

The standard deviation of the recoveries of conazole fungicides from matcha were also larger for some fungicides (>20%) with recoveries in the range of 60–70% or lower. There were also noticeable differences in matrix interferences in repeated extractions of the same manufacturer's sample of matcha as seen in baseline shifts in some SRM chromatograms. Lower extraction recoveries for biteranol, cyproconazole, simeconazole, and triadimenol for the matcha samples relative to other tea samples have also been reported [14]. To obtain relative standard deviation for recoveries in the desired range (<20%) for all target analytes it may be necessary to increase the contact of the solvent with the fine powdered sample by mixing the matcha with a sorbent such as a polymeric resin or Florisil®. This may also allow us to reduce the amount or need for Anasorb 747 for removal of chlorophyll in the extracted solvent.

5. Conclusions

The clean-up of extracts from powdered nutraceutical samples is a challenge. Extracts of matcha infused in water have inheritably better relative standard deviation of recoveries of selected conazoles as only the water soluble components are extracted, however other conazoles still show poor recoveries [14]. Evaluation of a wider range of conazole fungicides and other structurally similar pesticides showed that there are similar issues when using PSE with poor extraction recoveries for the less commonly analyzed conazole fungicides. In general matrix suppression of the MS signal was moderate indicating a need for matrix matched or standard addition calibration. PSE with in-cell clean-up can be used to remove pigments from powdered nutraceutical products such as matcha. There are more commercially available products with high pigment levels such that new extraction and clean-up methods need to be developed. This analysis shows that conazole fungicides strongly adsorb to many of the sorbent materials used for pigment removal such that alternate sorbents from carbon based materials need to be evaluated in the future. Matcha is a very difficult sample matrix and selected conazole fungicides can be extracted directly into ethyl acetate with subsequent in-cell clean-up to remove chlorophyll. The first detection of flusilazole, prothioconazole-desthio and trace levels of tebuthiuron are presented. The MS signal suppression that is still present is attributed to the presence of polyphenols including catechins that future extraction and clean-up methods need to address. This work shows that issues with MS signal suppression are also caused by other components of matcha than pigment. Further reducing the polarity of the extraction solvent while maintaining good recoveries for conazole fungicides, and clean-up with Florisil®, polymeric or alternative green sorbent materials should be evaluated. Of the carbon based sorbents and other sorbents evaluated Anasorb 747 was best able to remove high amounts of the green pigment from ethyl acetate extracts such that filtration was not required after the pre-concentration of extracts. Anasorb 747 also has potential for re-use after clean-up to reduce sample analysis costs. To improve the method development strategy utilizing green analytical approaches, improvements in PSE with in-cell clean-up are needed to deal with difficult sample matrices without the need for subsequent off-line clean-up steps [22,23]. To further reduce matrix effects caused by more polar matrix components, a membrane filter could be tested for use with PSE in-cell clean-up. Some membrane filters strongly absorb polyphenols, and this may aid in reducing matrix issues. Pressurized solvent extraction is well suited to address filtration in the extraction cell as extracts are filtered as the solvent is pushed out of the extraction cell with nitrogen gas.

Supplementary Materials:
Figure S1: Selected Reaction Monitoring Chromatograms of Diniconazole With and Without Matrix Added, Table S1: Method Detection Limit, Regression Coefficient of Matrix Matched Calibration Curve, and Percentage Matrix Effects for Target Analytes Analyzed by LC-ESI$^+$-MS/MS.

Author Contributions: Conceptualization, methodology, data curation, writing-review, R.R.-F.; methodology and data curation, A.A.M.

References

1. Raina-Fulton, R.; Behdarvandan, A.; Mohamad, A.A. The challenges of fungicide analyses using gas chromatography and liquid chromatography-mass spectrometry methods. *Austin Environ. Sci.* **2018**, *3*, 1031.
2. Raina-Fulton, R.; Aborkhees, G.; Behdarvandan, A. Analysis of herbicide and/or pesticide residues in dietary botanical supplements. *Encycl. Anal. Chem.* **2018**, *A9603*, 1–126. [CrossRef]
3. Xue, J.; Li, H.; Liu, F.; Xue, J.; Chen, X.; Zhan, J. Transfer of difenoconazole and azoxystrobin residues from chrysanthemum flow tea to its infusion. *Food Addit. Contam. Part A* **2014**, *31*, 666–674. [CrossRef] [PubMed]
4. Raina, R.; Smith, E. Detection of azole fungicides in atmospheric samples collected in the Canadian prairies. *J. AOAC Int.* **2012**, *95*, 1350–1356. [CrossRef] [PubMed]
5. Pihlström, T.; Blomkvist, G.; Friman, P.; Pagard, U.; Österdahl, B. Analysis of pesticide residues in fruit and vegetables with ethyl acetate extraction using gas and liquid chromatography with tandem mass spectrometric detection. *Anal. Bioanal. Chem.* **2007**, *389*, 1773–1789. [CrossRef] [PubMed]
6. Akiyama, Y.; Matsuoka, T.; Mitsuhashi, T. Multi-residue screening method of acidic pesticides in agricultural products by liquid chromatography/time of flight mass spectrometry. *J. Pestic. Sci.* **2009**, *34*, 265–272. [CrossRef]
7. Zayats, M.F.; Leschev, S.M.; Zayats, M.A. An improved extraction method of rapeseed oil sample preparation for the subsequent determination in it of azole class fungicides by gas chromatography. *Anal. Chem. Res.* **2015**, *3*, 37–45. [CrossRef]
8. Grimalt, S.; DeHouck, P. Review of analytical methods for the determination of pesticide residues in grapes. *J. Chromatogr. A* **2016**, *1433*, 1–23. [CrossRef] [PubMed]
9. Jadhav, M.R.; Oulkar, D.P.; Shabeer, A.T.P.; Banerjee, K. Quantitative screening of agrochemical residues in fruits and vegetables by buffered ethyl acetate extraction and LC-MS/MS. *J. Agric. Food Chem.* **2015**, *63*, 4449–4456. [CrossRef] [PubMed]
10. Goto, T.; Yoshida, Y.; Kiso, M.; Nagashima, H. Simultaneous analysis of individual catechins and caffeine in green tea. *J. Chromatogr. A* **1996**, *749*, 295–299. [CrossRef]
11. Weiss, D.J.; Anderton, C.R. Determination of catechins in matcha green tea by micellar electrokinetic chromatography. *J. Chromatogr. A* **2003**, *1011*, 173–180. [CrossRef]
12. El-Aty, A.M.A.; Choi, J.; Rahman, M.; Kim, S.; Tosun, A.; Shim, J. Residues and contaminants in tea and tea infusions: A review. *Food Addit. Contam. Part A* **2014**, *31*, 1794–1804. [CrossRef] [PubMed]
13. Saito, S.; Nemoto, S.; Matsuda, R. Simultaneous determination of pesticide residues in tea by LC-MS/MS-Modification of Japanese official multiresidue method. *Jpn. J. Food Chem. Saf.* **2014**, *21*, 27–36. [CrossRef]
14. Saito, S.; Nemoto, S.; Teshima, R. Multiresidue determination of pesticides in tea by gas chromatography-tandem mass spectrometry. *J. Environ. Sci. Health Part B Pestic. Food Contam. Agric. Wastes* **2015**, *50*, 760–776. [CrossRef] [PubMed]
15. Chen, Y.; Lopez, S.; Hayward, D.G.; Park, H.Y.; Wong, J.W.; Kim, S.S.; Wan, J.; Reddy, R.M.; Quinn, D.J.; Steinger, D. Determination of multiresidue pesticides in botanical dietary supplements using gas chromatography-triple-quadrupole mass spectrometry (GC-MS/MS). *J. Agric. Food Chem.* **2016**, *64*, 6125–6132. [CrossRef] [PubMed]
16. Villanueva Bermejo, D.; Ibáñez, E.; Reglero, G.; Turner, C.; Fornari, T.; Rodriguez-Meizoso, I. High catechins/low caffeine powder from green tea leaves by pressurized liquid extraction and supercritical antisolvent precipitation. *Sep. Purif. Technol.* **2015**, *148*, 49–56. [CrossRef]
17. Raina-Fulton, R. Determination of neonicotinoid insecticides and strobilurin fungicides in particle

phase atmospheric samples by liquid chromatography-tandem mass spectrometry. *J. Agric. Food Chem.* **2015**, *63*, 5152–5162. [CrossRef] [PubMed]
18. Chitescu, C.L.; Oosterink, E.; de Jong, J.; Stolker, A.A.M. Ultrasonic or accelerated solvent extraction followed by U-HPLC-high mass accuracy MS for screening of pharmaceuticals and fungicides in soil and plant samples. *Talanta* **2012**, *88*, 653–662. [CrossRef] [PubMed]
19. Wu, G.; Bao, X.; Zhao, S.; Wu, J.; Han, A.; Ye, Q. Analysis of multi-pesticide residues in the foods of animal origin by GC-MS coupled with accelerated solvent extraction and gel permeation chromatography cleanup. *Food Chem.* **2011**, *126*, 646–654. [CrossRef]
20. Bakirci, G.T.; Hişil, Y. Fast and simple extraction of pesticide residues in selected fruits and vegetables using tetrafluoroethane and toluene followed by ultrahigh-performance liquid chromatography/tandem mass spectrometry. *Food Chem.* **2012**, *135*, 1901–1913. [CrossRef] [PubMed]
21. Karcher, S.; Kornmüller, A.; Jekel, M. Screening of commercial sorbents for the removal of reactive dyes. *Dyes Pigments* **2001**, *51*, 111–125. [CrossRef]
22. Gałuszka, A.; Migaszewski, Z.M.; Konieczka, P.; Namieśnik, J. Analytical Eco-Scale for assessing the greenness of analytical procedures. *Trends Anal. Chem.* **2012**, *37*, 61–72. [CrossRef]
23. Płotka-Wasylka, J. A new tool for the evaluation of the analytical procedure: Green Analytical Procedure Index. *Talanta* **2018**, *181*, 204–209. [CrossRef] [PubMed]

Mycotoxins and Mycotoxin Producing Fungi in Pollen

Aleksandar Ž. Kostić [1,*], Danijel D. Milinčić [1], Tanja S. Petrović [2], Vesna S. Krnjaja [3], Sladjana P. Stanojević [1], Miroljub B. Barać [1], Živoslav Lj. Tešić [4] and Mirjana B. Pešić [1]

1 Chemistry and Biochemistry, Faculty of Agriculture, University of Belgrade, Nemanjina 6, 11080 Belgrade, Serbia; danijel.milincic@agrif.bg.ac.rs (D.D.M.); sladjas@agrif.bg.ac.rs (S.P.S.); baracm@agrif.bg.ac.rs (M.B.B.); mpesic@agrif.bg.ac.rs (M.B.P.)
2 Preservation and Fermentation, Faculty of Agriculture, University of Belgrade, Nemanjina 6, 11080 Belgrade, Serbia; tpetrovic@agrif.bg.ac.rs
3 Institute for Animal Husbandry, Autoput 16, 11080 Belgrade, Serbia; vesnakrnjaja.izs@gmail.com
4 Analytical Chemistry, Faculty of Chemistry, University of Belgrade, Studentski Trg 12-16, 11158 Belgrade, Serbia; ztesic@chem.bg.ac.rs
* Correspondence: akostic@agrif.bg.ac.rs

Abstract: Due to its divergent chemical composition and good nutritional properties, pollen is not only important as a potential food supplement but also as a good substrate for the development of different microorganisms. Among such microorganisms, toxigenic fungi are extremely dangerous as they can synthesize mycotoxins as a part of their metabolic pathways. Furthermore, favorable conditions that enable the synthesis of mycotoxins (adequate temperature, relative humidity, pH, and a_W values) are found frequently during pollen collection and/or production process. Internationally, several different mycotoxins have been identified in pollen samples, with a noted predominance of aflatoxins, ochratoxins, fumonisins, zearalenone, deoxynivalenol, and T-2 toxin. Mycotoxins are, generally speaking, extremely harmful for humans and other mammals. Current EU legislation contains guidelines on the permissible content of this group of compounds, but without information pertaining to the content of mycotoxins in pollen. Currently only aflatoxins have been researched and discussed in the literature in regard to proposed limits. Therefore, the aim of this review is to give information about the presence of different mycotoxins in pollen samples collected all around the world, to propose possible aflatoxin contamination pathways, and to emphasize the importance of a regular mycotoxicological analysis of pollen. Furthermore, a suggestion is made regarding the legal regulation of pollen as a food supplement and the proposed tolerable limits for other mycotoxins.

Keywords: pollen; fungi; mycotoxins; aflatoxins; ochratoxins; fumonisins; T-2 toxin; zearalenone; deoxynivalenol

Key Contribution: This review gives an overview of scientific data about pollen contamination with different mycotoxins and mycotoxin producing fungi. Also; importance of standard mycotoxicological pollen analysis is emphasized. Inclusion of pollen in the legal regulation; as potential food supplement; is suggested.

---

1. Introduction

Pollen grain, as a male gametophyte of flowering plants, is produced and released from anthers during pollination [1]. Two of the most important pollinators are insects (in the case of entomophilous plants it is, above all, the honey bee (*Apis mellifera* L.)) and, in the case of anemophilous plants, wind. Pollen is prime food for bees due to its amazing diversity of nutritionally important constituents-proteins, lipids, carbohydrates, vitamins, and minerals [2,3]. For the same reasons, floral or bee-collected pollen is potentially a good food supplement for human nutrition [4–8]. Because of great its sensitivity, pollen grain contains a significant quantity of secondary plant metabolites, as part of the plant's defense mechanism, such as different phenolic compounds [9–16] or carotenoids [17,18] and possesses substantial antioxidant properties, which is important for its application as a food supplement [19,20]. Besides the nutritionally important and desirable components, pollen can contain some contaminants such as toxic elements [2,21–23]. Due to optimal water (moisture) content, water activity (a_W), and pH-value, pollen often presents an ideal medium for the development of different microorganisms—bacteria, mold, and yeast. As a result of the presence of mold and yeast, the production of mycotoxins can occur. Mycotoxins are secondary metabolites of different fungi species which are toxic to vertebrates and can lead to some disorders and diseases, or, at worst, death in humans and other animals [24]. The scientific "history" of mycotoxins started in 1962 during a great veterinary crisis when about 100,000 turkeys died in England due to being fed with contaminated peanuts that contained secondary metabolites of *Aspergillus flavus* [24]. The occurrence of mycotoxins in different types of feed and food has been recorded [25–31] and it was found to be strongly dependent on several factors such as climatic conditions (including geographical position of growing region, temperature, and relative humidity) before, during, or after feed/food production [32]. The European Commission (EC Commission Regulation No 1881/2006) sets maximum tolerable levels for several types of mycotoxins (aflatoxins B, G, and/or M, ochratoxin A (OTA), patulin, fumonisins B_1 and B_2, deoxynivalenol, and zearalenone) in different types of foods (nuts, cereals, dried fruits, juices, milk, etc.) [33] but without information pertaining to bee products such as honey, pollen, or bee bread.

The aim of this review is to make a cross-check of current data about contamination of pollen with different types of mycotoxins as well as mycotoxin producing fungi. Also, the effort to emphasize the importance of mycotoxin estimation of pollen samples as obligatory part of their microbiological analysis will be made.

2. Mycotoxins in Pollen

More than a hundred mycotoxins are known, and most of them are produced by some of the species belonging to one of three fungi genera: *Aspergillus*, *Penicillium* and/or *Fusarium* [34]. According to the available literature [35–50] the presence of the following mycotoxins in pollen has been investigated or proved with appropriate analytical methods and analysis: Aflatoxins (AFs), ochratoxins (OTs), fumonisins (FBs), zearalenone (ZEN), deoxynivalenol (DON), and its acetoxy derivative, T-2 toxin (T-2), HT-2 toxin, fusarenon-X, diacetoxyscirpenol, nivalenol, neosolaniol, roridin A, verrucarrin A, α-β-dehydrocurvularin, phomalactone,6-(1-propenyl)-3,4,5,6-tetrahydro-5-hydroxy-4H-pyran-2-one, 5-[1-(1hydroxibut-2-enyl)]-dihydrofuran-2-one and 5-[1-(1-hydroxibut-2-enyl)]-furan-2-one.

2.1. *Aflatoxins*

Aflatoxins are the product of the metabolism of different fungi species which belong to *Aspergillus* genus with *A. flavus* and *A. parasiticus* strains as the main producers [24]. They can be synthetized in fungi's spores and mycelium or secreted as exotoxins [25]. The most toxic and dangerous aflatoxins are aflatoxin B_1 and B_2 (Figure 1) [34]. Both aflatoxin B_1 and B_2 are carcinogenic for humans and animals, and are listed in Group 1 of carcinogenic substances according to International Agency for Research on Cancer (IARC) [51]. The liver is the organ that suffers most from the effects of aflatoxins [52]. Ingestion of these toxins can lead to aflatoxicosis, as an acute form of poisoning, or, in the case of long-term exposure, to the development of liver cancer [52]. Hydroxylated AFB-forms presented in milk are aflatoxin M1 and M2 [24] which are possibly carcinogenic for humans (IARC Group 2A of carcinogenic substances) [34,51]. Furthermore, two other forms of AF exist: Aflatoxin G1 and G2 (Figure 1).

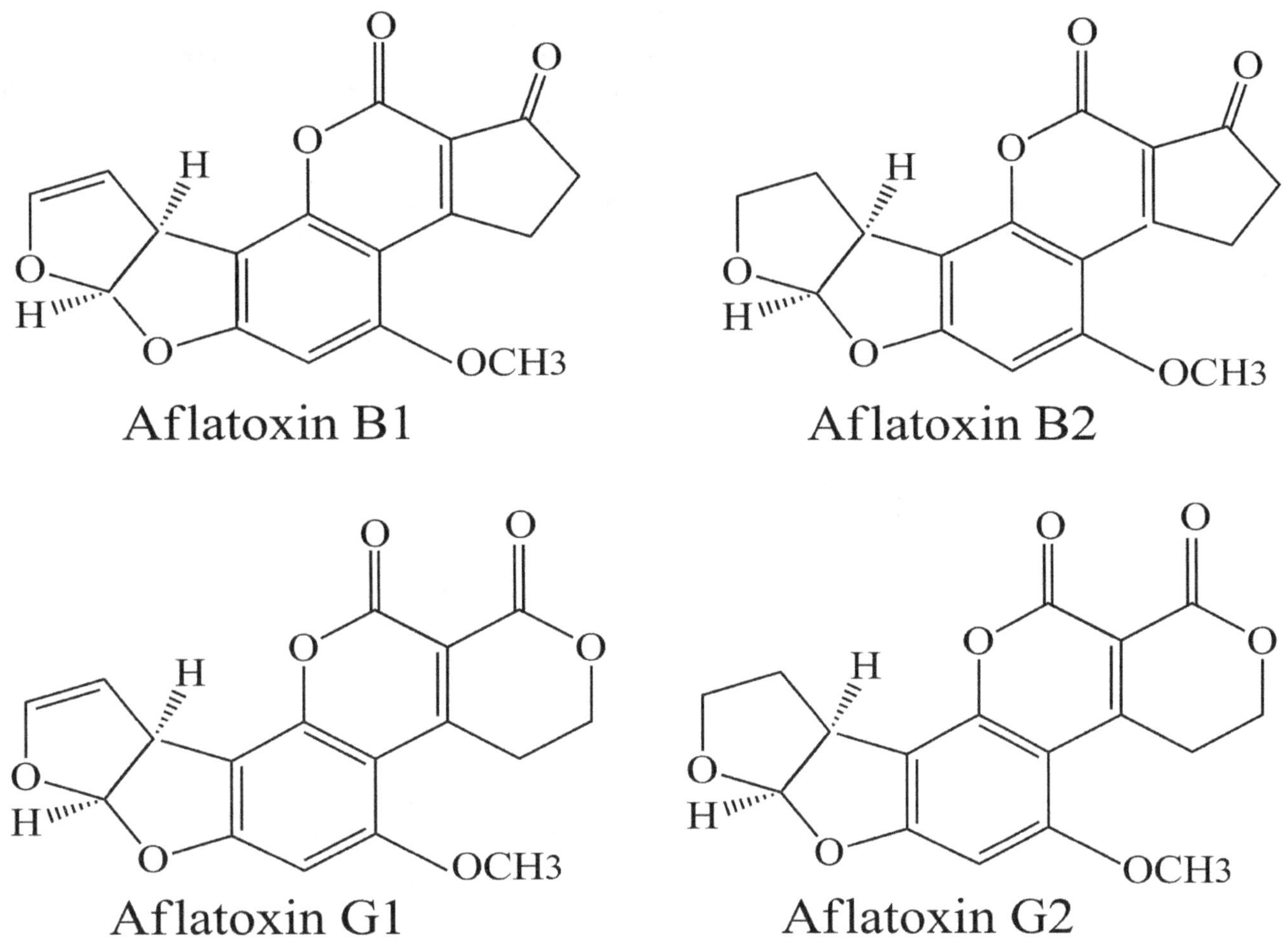

Figure 1. Chemical structures of aflatoxin B_1, B_2, G_1, and G_2.

2.1.1. Contamination of Pollen with Aflatoxins—Possible Ways

Pollen often presents a suitable substrate for the proliferation of various microorganisms due to its favorable moisture content, water activity (a_w), and pH-value. External conditions such as relative humidity and temperature, different stages of pollen production, and storage conditions have been shown to lead to microbiological contamination of pollen [35]. According to data found in the literature, pH-value ranging between 4.0 and 6.5 have been shown to be suitable for the development of bacteria, mold, and yeast while the minimal a_w-values sufficient for the growth of *Aspergillus* and *Penicillium* spp. have been shown to be 0.71 to 0.96 [53] i.e., 0.55 in the case of pollen [54]. Microbiological contamination is strongly pH and temperature dependent and is also conditioned by the type of microorganism [53]. If proper conditions have been achieved in any phase of pollen production, the growth of microbes will occur which can cause aflatoxin production and the contamination of pollen. In addition to production process and human hygiene practices, which are the most important sources of aflatoxin contamination, sometimes microbe growth can be triggered by infected flowering plants [25,48]. Namely, during the flowering and the pollination process, *Aspergillus* spp. spores can germinate on female flower parts. Following this, the toxigenic fungal spores placed in the pollen tubes will grow and further infect the egg-cells [25]. If bees visit these flowers, the contaminated pollen grains will be transferred into the hives. Since there is intensive contact between bees when in the hive (due to highly organized bee societies) their "home" is the third possible source of aflatoxin pollen contamination [48]. As aflatoxins show detrimental effects on bee health, the incidence of these compounds in hives is undesirable. It is for this reason that the occurrence and production of propolis in hives is an effective way for bees to deal with AFs toxicity [55,56] which could indicate that this source of pollen contamination with aflatoxins is at least probable. In the past, aflatoxin occurrence in feed and food was a characteristic of tropic or sub-tropic regions due to favorable climatic conditions. Recently, with climatic changes, which extensively influences weather conditions in temperate areas (such as the majority of Europe), the presence of aflatoxins in these areas is becoming more frequent. The detection of aflatoxins in samples of pollen from the most diverse parts of the world (Table 1) is in accordance with this fact and is becoming a growing problem. Interestingly, in our previous investigation [48] the majority of examined pollen samples were sterile but all were contaminated with AFB_1. This situation confirms three hypotheses:

- There are different ways of pollen contamination with aflatoxin(s).
- These toxins remain in samples with or without presence of appropriate fungi.
- It is extremely important to always perform mycotoxicological analysis together with microbiological characterization of pollen.

Table 1. Toxigenic fungi and concentration level reported for aflatoxins in pollen samples from different countries.

No. of Examined Pollen Samples	Geographical Origin	Analytical Methods	Isolated Mycotoxins Producing Fungi Species	AF Types and Concentration Range(s)	Reference
20	Spain	ELISA test	/	Total AFs: below 5 μg/kg	[35]
20	Spain	HPLC (with fluorescent detection)	/	AFB_1 and AFB_2: below limit detection (BLD)	[37]
87 + 3	Spain + Argentina	HPLC (with fluorescent detection)	*A. flavus* *A. parasiticus*	AFB_1, AFB_2, AFG_1 and AFG_2: not determined.	[38]
5	China	Cyclic voltametry	/	AFB_1: 0.00–0.52 μg/kg	[39,42]
1	Epirus (Western Greece)	HPLC (with fluorescent detection)	not detected	AFB_1: not detected	[40]
45	Slovakia	ELISA test	*A. flavus*, *A. parasiticus.*	Total AFs: 13.60–16.20 μg/kg (in poppy pollen) 3.15–5.40 μg/kg (in rape pollen) 1.20–3.40 μg/kg (in sunflower pollen)	[41]
33	Serbia	ELISA test	*A. flavus*	AFB_1: 3.49–14.02 μg/kg	[44]
20	China	LC-MS/MS	/	AFB_1, AFB_2, AFG_1 and AFG_2: below limit detection (BLD)	[45]
27	Brazil	Qualitative analysis	*A. flavus*	AFB_1 and AFB_2: not determined	[47]
26	Serbia	ELISA test	*A. flavus*	AFB_1: 3.15–17.32 μg/kg	[48]
30	Egypt	Thin-layer chromatography	*A. flavus*	AFB_1 AFB_2, AFG_1 and AFG_2 were not determined.	[49]
9	Portugal	ELISA test	Not detected	Not detected AFB1	[50]

ELISA—enzyme linked immunosorbent assays; AFs—aflatoxins; AFB_1—aflatoxin B_1; AFB_2—aflatoxin B_2; AFG_1—aflatoxin G_1; AFG_2—aflatoxin G_2.

2.1.2. Quantification of Aflatoxins in Pollen Samples

Results of different studies about the determination of aflatoxin content in pollen samples with diverse palynological (botanical) and geographical origins are given in Table 1.

2.2. Ochratoxins

Ochratoxins (OTs) are a group of chemical compounds (Figure 2) derived from shikimic acid metabolic pathway with ochratoxin A (OTA) as a major food contaminant [57]. The main OTs-producers are different *Aspergillus* species with a special emphasis on *Aspergillus niger* strains since they are industrially important due to their applications for enzyme and citric acid production. Furthermore, one species (*P. verrucosum*) belonging to *Penicillium* genus can be the source of ochratoxins [24]. OTA belongs to the IARC 2B group which means that it is a possible carcinogen for humans [51]. The kidneys are the most vulnerable organs effected by OTA. OTA has been noted as having a strong influence on the endemic disease 'Balkan nephropathy', as well as porcine nephropathy, which has been documented in several Scandinavian countries [24].

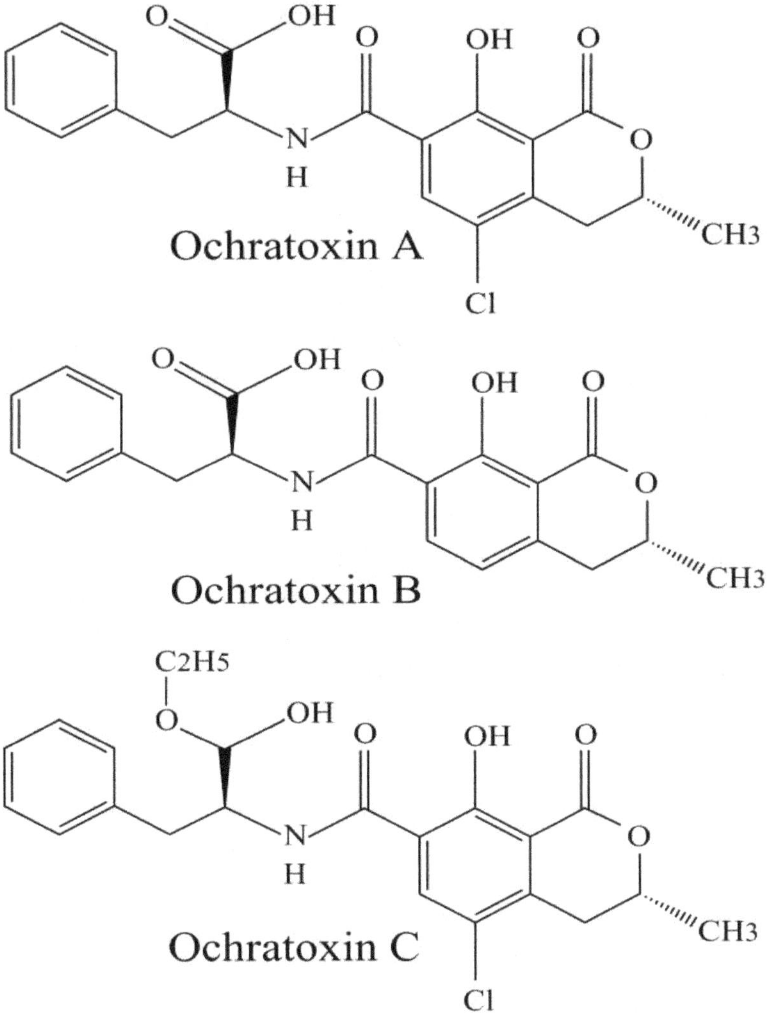

Figure 2. Chemical structures of ochratoxins A, B, and C.

Ochratoxins in Pollen

Besides many types of food (nuts, meat products, barley, oats, rye, wheat, wine, dried fruits, coffee, and coffee products) where the presence of OTA has been recorded [24,57], in some herbs, bottled water [57], and pollen samples, this mycotoxin has also been observed. Xue et al. [45] conducted an examination of 20 bee pollen samples from North China for the presence of OTA p by LC-MS/MS analysis. The obtained results showed that none of the studied pollen samples were contaminated with OTA. These results can be associated with the dry weather conditions during the collection period. The same situation was observed in the case of 20 bee pollen samples that originated from Spain [37]. However, HPLC analysis of 90 Spanish and Argentinian bee pollen samples in [38] confirmed the presence of several *Aspergillus* (*A. carbonarius*, *A. ochraceus* and *A. niger*), and *Penicillium* (*P. verrucosum*) species with the ability to produce OTA. Significant contamination of bee pollen was determined in a case of Slovakian samples [41]. In total, 45 samples were divided in three groups of 15 samples originating from poppy, rape, and sunflower plants. Determined OTA concentration ranges in poppy, rape, and sunflower pollen samples were 6.12 to 10.98 μg/kg, 3.24 to 9.87 μg/kg, and 0.23 to 6.93 μg/kg, respectively. In Spain, by analyzing the toxigenic potential of *A. ochraceus* in various substrates (bee pollen, maize, wheat, and rice) Medina et al. [36] found that OTA production in bee pollen was statistically significantly higher than that found in the production of tested cereals, regardless of the incubation time (7, 14, 21, 28 days). Likewise, positive correlations have been found between the proportion of bee pollen added to the yeast extract sucrose broth inoculated with spores of *A. ochraceus* and OTA level [36]. Based on all of the above, it can be assumed that bee pollen may represent a significant risk factor for the occurrence of OTA in the food chain.

2.3. The Other Mycotoxins Examined in Pollen

2.3.1. Fumonisins

Fumonisins (FBs) are a group of mycotoxins predominantly connected with maize (grown as endophyte in both vegetative or reproductive tissues) and maize products but can be found in many cereals and products made from these plants [24,58]. Although maize is an anemophilic plant due to its high pollen production [7] it is not a rare that bees collect its pollen during the pollen collection season [4]. In that sense, it is possible to find pollen samples contaminated with FBs. The first report about FB food contamination dates back to 1988. The main representative of this mycotoxin group is fumonisin B_1 (FB_1) [24,58]. It is sorted in IARC 2B group of carcinogenic substances [51]. Moreover, fumonisins B_2, B_3, and B_4 also exist (Figure 3) [57]. Fungi belonging to *Fusarium* genus are the most important FBs producers, especially two species: *F. proliferatum* and *F. verticillioides* as well as *A. alternata* from *Alternaria* spp. It is important to point out that the presence of these microbes does not mean that FBs contamination is guaranteed [24]. In an investigation by Kačaniová et al. [41] the presence of both, *F. proliferatum* and *F. verticillioides* was confirmed in thirty i.e., forty-five bee pollen samples, respectively but FBs were quantified only in the samples originating from sunflower (fifteen samples). This observation confirms the previously mentioned hypothesis, that despite the presence of *Fusarium* spp. in some material, appropriate weather conditions or insect damage are necessary for FBs production [24]. The range of FBs concentrations in these samples is given in Table 2.

Table 2. Concentration level reported for mycotoxins other than aflatoxins in pollen samples from different countries.

No. of Contaminated/ Examined Pollen Samples	Geographical Origin	Analytical Methods	Isolated Mycotoxin Producing Fungi Specie(s)	Mycotoxin Types and Concentration Range(s)	Reference
15/45 were contaminated	Slovakia	ELISA test	*F. proliferatum*, *A. alternata* Keissl.	Total FBs: 6.30–12.60 μg/kg	[41]
45	Slovakia	ELISA test	*F. graminearum*	ZEN: 311.00–361.30 μg/kg (in poppy pollen) 137.10–181.60 μg/kg (in rape pollen) 115.60–147.40 μg/kg (in sunflower pollen)	[41]
45	Slovakia	ELISA test	*F. graminearum*, *F. oxysporum*, *F. proliferatum*, *F. sporotrichioides*, *F. verticillioides*	T-2 toxin: 113.90–299.60 μg/kg (in poppy pollen) 197.10-265.70 μg/kg (in rape pollen) 173.60–364.90 μg/kg (in sunflower pollen)	[41]
45	Slovakia	ELISA test	*F. graminearum*, *F. oxysporum*, *F. proliferatum*, *F. sporotrichioides*, *F. verticillioides*	DON: 183.10–273.90 μg/kg (in poppy pollen) 189.60–244.70 μg/kg (in rape pollen) 133.30–203.50 μg/kg (in sunflower pollen)	[41]
2/15	Spain	GC/MS	/	neosolaniol: 22 i.e., 30 μg/kg nivalenol: 1 μg/kg	[43]

ELISA—enzyme linked immunosorbent assays; FBs—fumonisins; ZEN—zearalenone; DON—deoxynivalenol.

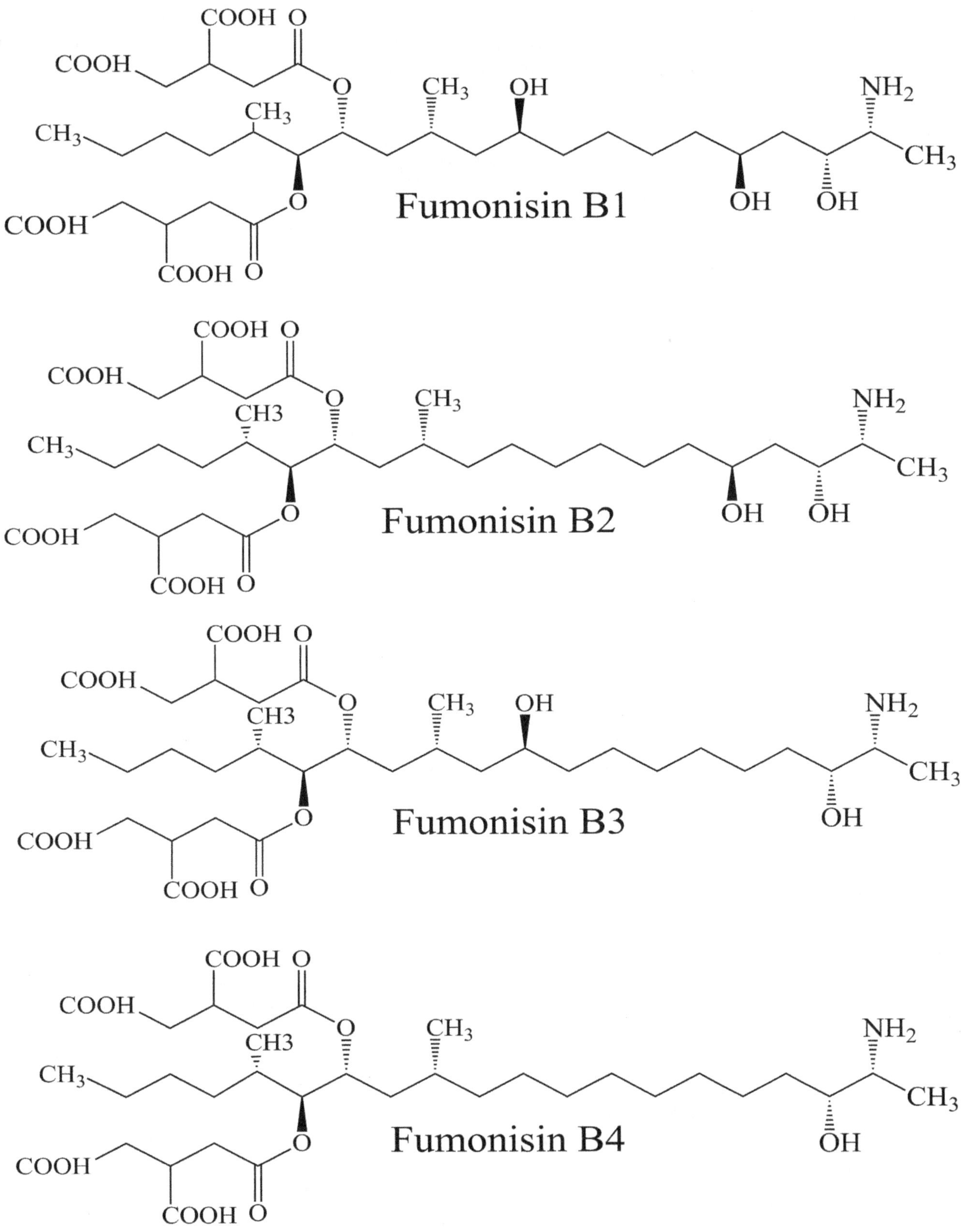

Figure 3. Chemical structures of fumonisins B_1, B_2, B_3, and B_4.

2.3.2. Zearalenone

Zearalenone (ZEN) (Figure 4) is mycoestrogen with limited toxicity that is produced by several *Fusarium* species: *F. graminearum*, *F. culmorum*, *F. crookwellense*, and *F. equiseti*. It is regularly present in crops and crop products [24]. According to IARC this macrocyclic lactone is classified in group 3 which means that it is not classifiable as to its carcinogenicity to humans [51]. In the case of pollen, the significant contamination with ZEN was recorded in Slovakian bee samples [41] (Table 2).

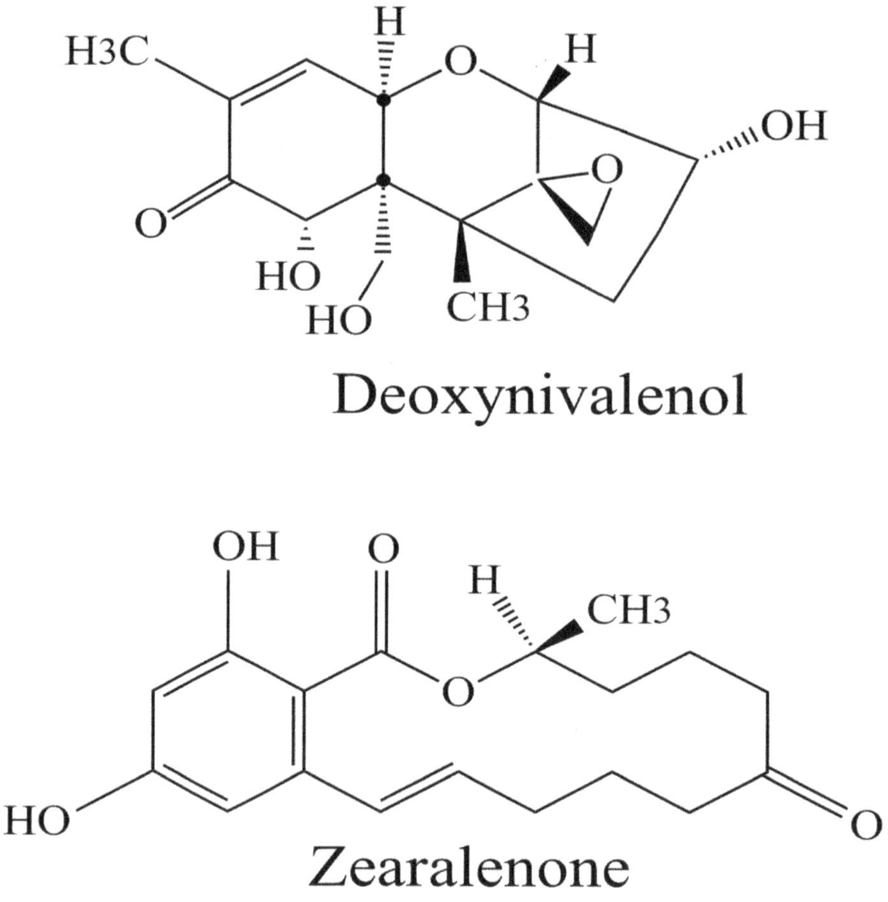

Figure 4. Chemical structures of deoxynivalenol and zearalenone.

2.3.3. Trichothecenes Group of Mycotoxins

In a study from Slovakia [41], the authors also reported the contamination of all examined bee pollen samples with T-2 toxin and deoxynivalenol (Figure 4). Both toxins belong to trichothecene compounds, the sesquiterpenoid metabolites obtained after microbiological activity of several fungi from the following genera: *Fusarium* (primary source), *Trichoderma*, *Myrothecium*, *Phomopsis*, etc., [24]. Together with ZEN, they were the most dominant quantified mycotoxins in the pollen samples. Additionally, the presence of DON and T-2 toxin was checked in fifteen pollen samples from Spain, but the content of these mycotoxins was below limit detection of applied GC/MS method [43]. In the same study, the authors examined the presence of several other *Fusarium* spp. producing mycotoxins: 3-acetyl-deoxynivalenol, fusarenon-X, diacetoxiscirpenol, nivalenol, neosolaniol, and HT-2 toxin. All the above-mentioned compounds belong to trichothecene terpenoid's derivatives. It was determined that some of the samples were contaminated with neosolaniol and nivalenol (Table 2), while all other examined toxins were below limit detection. A report made by Cirigiliano et al. [46] should also be mentioned as their study was the first to detect seven specific mycotoxins (roridin A, verrucarrin A, α-β-dehydrocurvularin, phomalactone,6-(1-propenyl)-3-,4,5,6-tetrahydro-5-hydroxy-4H-pirane-2-one, 5-[1-(1-hydroxibut-2-enyl)]-dihydrofuran-2-one and 5-[1-(1-hydroxibut-2-enyl)]-furan-2-one) in beehives from Argentina with pronounced antifungal effect. Roridin A, verrucarin A, and α-β-dehydrocurvularin were isolated from strains of fungi *Myrothecium verrucaria* while other mycotoxins were obtained as result of *Nigrospora sphaerica* strains activity. Their structures were confirmed by 1D and 2D-NMR spectroscopy.

3. Mycotoxin Producing Fungi in Pollen

The microbiological quality of pollen is equally important as its chemical composition due to its safety use. Although the examination of mycotoxins in pollen began mostly in the last decade, the determination of different microbes (bacteria, mold, and yeast) present in pollen samples started much earlier—at the end of 1970s with studies by Gilliam [59,60]. Considering that a long period of time usually passes between collection of pollen samples and its application as food supplement (or as medicament), there is a great chance for the development of some toxigenic fungi [41]. Their presence may indicate mycotoxin production in pollen with or without their quantification. In that sense, this review also gives information on pollen investigations concerning the presence of mycotoxin producing fungi [41] made without further mycotoxicological analysis. The results of a cross-check of the available literature data, with appropriate comments and information, are given in Table 3.

Table 3. Toxigenic fungi and yeast reported in pollen samples from different countries.

No. of Examined Pollen Samples	Geographical Origin	Detected Microbial Class	Microbial Species or/and Total Microbial	Microbial Count	Observations	Reference
Unknown number of samples of floral and bee-collected almond pollen	unknown	Mold		No. of fungal isolates:	*Mucor* spp. was the dominant mold in floral pollen but not identified in bee-collected pollen. *Aureobasidium pullulans*, *P. corylophilum*, *P. crustosum* and *Rhizopus nigricans* were identified only in bee-collected pollen.	[61]
			Alternaria spp.	6		
			Cladosporium spp.	5		
			Penicillium spp.	5		
			Aspergillus spp.	3		
			Mucor spp.	19		
90 samples of bee pollen	Spain (87 samples) Argentina (3 samples)	Mold	*Aspergillus* section *Nigri*	$1.4 \times 10–2.3 \times 10^2$ cfu/g	The results show the occurrence of different mold species in pollen samples. *Penicillium*, *Alternaria*, and *Aspergillus* spp. were present in 90%, 86.6%, and 80% of samples, respectively. Predominant *Aspergillus* species was *A. niger*. The species of the genus *Fusarium* were isolated in 53.3%.	[38]
			A.flavus +*A. parasiticus*	$1.7 \times 10–2.5 \times 10$ cfu/g		
			Other *Aspergillus* spp.	2×10 cfu/g		
			P. verrucosum	1.4×10^2 cfu/g		
			Other *Penicillium* spp.	$1.3 \times 10^2–4.3 \times 10^3$ cfu/g		
			Fusarium spp.	$16–9.5 \times 10^1$ cfu/g		
			Cladosporium spp.	$6 \times 10–1.4 \times 10^3$ cfu/g		
			Alternaria spp.	$6 \times 10–5.2 \times 10^2$ cfu/g		
			Rhizopus spp.	$2 \times 10–9 \times 10$ cfu/g		
			Mucor spp.	$8–2.2 \times 10^2$ cfu/g		
			Botrytis spp.	$8–3 \times 10$ cfu/g		
			Epicoccum spp.	5–10 cfu/g		
		Yeast	Not specified	$3.6 \times 10^2–7.3 \times 10^3$ cfu/g		
42 samples of dehydrated bee pollen	Brazil	Mold/Yeast	Not specified	Total mold and yeast count: $10^2–1.3 \times 10^4$ cfu/g	About 12% of pollen samples were contaminated with mold and yeast above the limit (1×10^4) for a total mold and yeast proposed by Brazilian legislation.	[62]
30 samples of bee pollen	Slovakia	Microscopic fungi (mold)	*Alternaria* spp. *Cladosporium* spp. *Penicillium* spp. *Fusarium* spp. *Aspergillus* spp. (*A. flavus*, *A. ochraceus*) *Mucor* spp. *Trichoderma* spp. *Acremonium* spp. *Scopulariopsis* spp. *Rhizopus* spp. *Botrytis* spp.	Total mold and yeast count: $1.1 \times 10^2–4.57 \times 10^5$ cfu/g	The dominant fungi isolated from pollen samples were colonies of *A. alternata*, *Cladosporium cladosporoides*, and *Penicillium* spp. Also, the presence of well-known mycotoxicogenic species such as *A. flavus* and *A. ochraceus* were detected.	[63]

Table 3. *Cont.*

No. of Examined Pollen Samples	Geographical Origin	Detected Microbial Class	Microbial Species or/and Total Microbial	Microbial Count	Observations	Reference
19 samples of bee pollen	Mexico	Fungi (mold)	*A. flavus*	Incidence of mold genus (%):	Fungi contamination was generally low. The highest contamination was in three samples handled without packages.	[64]
			Alternaria spp.	3.6%		
			Penicillium spp.	2.9%		
			Fusarium spp.	2.9%		
			Aspergillus spp.	3.6%		
			Mucor spp.	3.1%		
			Rhizopus spp.	0.7%		
8 samples of bee pollen	Slovakia	Mold	*Alternaria* spp. *Cladosporium* spp. *Penicillium* spp. *Aspergillus* spp. *Mucor* spp. *Aureobasidium* spp. *Humicola* spp. *Monodictys* spp. *Paecilomyces* spp. *Rhizopus* spp. *Mortierella* spp. *Trichosporiella* spp. *Harpografium* spp. *Mortierella* spp.	Total mold and yeast count: 107–4688 cfu/g	The results show that in all analyzed samples of pollen 21 fungal species of 13 genera of microscopic fungi were detected. The dominant identified species, over 62% of the isolates belonged to following genera: *Mucor, Rhizopus, Aspergillus, Alternaria,* and *Paecilomyces.*	[65]
28 samples (fresh and dried bee pollen)	Cuba	Mold/Yeast	Not specified	Total mold and yeast count: 10^4–1.5×10^5 cfu/g	All samples had quantified number of mold and yeast above proposed limits (10^4 cfu/g for the fresh and 10^2 cfu/g for dried pollen). Nevertheless, in the dry pollen, a smaller number of high contaminated samples were recorded. Drying could not be used as reliable method for obtaining pollen with acceptable microbiological quality.	[66]
8 samples of commercial bee pollen	Portugal (4 samples) Spain (3 samples) Unknown origin (1 sample)	Mold Yeast	Not specified Individually identified yeast	Total mold and yeast count: <10 to 9.4×10^2 cfu/g	All samples were contaminated with yeast and mold. Further, yeast species were identified, and results indicated the presence of five different genus of yeast which can influence the risk of food-borne illness and spoilage or can serve as an indicator of a lack of hygiene standards.	[67]

Table 3. *Cont.*

No. of Examined Pollen Samples	Geographical Origin	Detected Microbial Class	Microbial Species or/and Total Microbial	Microbial Count	Observations	Reference
Unknown	Portugal	Mold/Yeast	Not specified	Total mold and yeast count: $<10^4$ cfu/g	Generally, yeast and mold were identified in 60% of all examined samples. pH and a_w values had a strong impact on the total microbe number in pollen.	[54]
22 samples of organic bee pollen	Portugal	Mold/Yeast	Not specified	Total mold and yeast count: <10–3560 cfu/g	In all samples of organic bee pollen, the presence of mold and yeast was detected, but their individual species were not identified.	[68]
3 samples of pollen	Algeria	Mold/Yeast	Not specified	Total mold and yeast count: 5×10^4–4×10^5 cfu/g	/	[69]
33 samples of bee pollen	Serbia	Mold	*Alternaria* spp. *Mucor* spp. *Rhizopus* spp. *Cladosporium* spp. *Epicoccum* spp. *Acremonium* spp.	Total mold count: 1×10^3–1×10^5 cfu/g	See Table 1.	[44]
27 samples of dried bee pollen	Brazil	Mold		Total mold count: 1×10^2–5×10^2 cfu/g Incidence of mold genus (%):	Total mold count depends on growing media.	[47]
			Aspergillus spp. (*A. flavus; A. fumigatus; A. versicolor; A. ochraceus; A. carbonarius; A. terreus; A. oryzae*)	85%		
			Cladosporium spp.	63%		
			Penicillium spp. (*P. citrinum; P. citreonigrum; P. glabrum; P. oxalicum*)	41%		
			Alternaria spp.	19%		
			Wallemia spp. and *Eurotium* spp.	11%		
			Mucor spp.	7%		
			Curvularia spp., *Paecilomyces* spp. and *Fusarium* spp. (*F. camptoceras*)	4%		
45 samples of dehydrated bee pollen	Brazil	Mold Yeast	Not specified Identified different species	Total mold and yeast count: <10–7.67×10^3 cfu/g	/	[70]
21 samples of bee pollen (*Melipona* bees)	Brazil	Mold/Yeast	Not specified	/	All samples were sterile without presence of any mold or yeast species.	[71]

Table 3. *Cont.*

No. of Examined Pollen Samples	Geographical Origin	Detected Microbial Class	Microbial Species or/and Total Microbial	Microbial Count	Observations	Reference
40 samples of bee pollen	Italy	Mold	*Cladosporium* spp. *Alternaria* spp. *Humicola* spp. *Mucoraceae* *Acremonium* spp. *Penicillium* spp. (*P. chrysogenum; P. brevicompacticum*) *Aspergillus* spp. (*A. flavus; A. nidulans; A.miger; A. terreus*)	Total mold count: 4–568 cfu/g	In all pollen samples at least one fungal isolate was detected. *Cladosporium* spp. was the most frequently detected mold. *Aspergillus* spp. and *Penicillium* spp., as a potentially mycotoxicogenic mold, were also identified in 8 i.e., 22 pollen samples.	[72]
Dehydrated (electric oven, EO) or lyophilized (L) bee pollen samples	Brazil	Mold/Yeast	Not specified	Total mold and yeast count: 99–242 cfu/g (EO) 16–935 cfu/g (L)	Number of quantified mold and yeast depended on time (April or September) of collection.	[73]
26 samples of bee pollen	Serbia	Mold		Total mold count:	See Table 1	[48]
			Alternaria spp.	1×10^3 cfu/g		
			Mucor spp.	1×10^3 cfu/g		
			Rhizopus spp.	1×10^3 cfu/g		
			Trichoderma spp.	1×10^4 cfu/g		
1 sample of bee pollen	Not known	Mold/Yeast	Not specified	Total mold and yeast count: >21 cfu/g	Presence of yeast and mold can be responsible for the potential presence of toxins in the samples.	[74]
18 samples of commercial bee pollen	Argentina	Mold/Yeast	Not specified	Total mold and yeast count: $<10^2$ cfu/g	The total fungi number is specified for 28% of the samples.	[75]
62 samples of dehydrated bee pollen	Brazil	Mold/Yeast	Not specified	Total mold and yeast count: 1.9×10^2–7.62×10^2 cfu/g	The microbial contamination is dependent on geographical origin of samples.	[76]
8 samples of commercial bee pollen	Algeria	Mold/Yeast	Not specified	Total mold and yeast count: 10^4–2.8×10^5 cfu/g	/	[77]

Table 3. *Cont.*

No. of Examined Pollen Samples	Geographical Origin	Detected Microbial Class	Microbial Species or/and Total Microbial	Microbial Count	Observations	Reference
32 (13 fresh (F) and 19 dried (D) samples of bee pollen)	Bulgaria	Mold	Identified mold: *Aspergillus* spp. *Fusarium* spp. *Penicillium* spp. (*P. brevicompactum*) *Alternaria* spp. *Cladosporium* spp. Other species	Total mold count: 5.6×10^2 –3.7×10^4 cfu/g (F) 150–1.1×10^4 cfu/g (D)	The results show that the values for fungal colony count were significantly lower in the dried pollen samples. 136 fungal isolates were identified. Among detected isolates, genus *Penicillium* was dominant while the genus *Fusarium* was the least fungal contaminant. Dominant species isolated from 14 different samples was *P. brevicompactum*.	[78]
19 samples of stored pollen of five stingless bee species	Brazil	Mold/Yeast	Not specified	Total mold and yeast count: 4.2 $\times 10^1$ cfu/g (1 sample only)	The results show that only for the stored pollen of the stingless bee specie *Frieseomellite varies* it was possible to enumerate mold and yeast.	[79]
bee pollen samples	Colombia	Mold/Yeast	Not specified	Total mold and yeast count: 3×10^2–2×10^5 cfu/g	Number of quantified microbes is strongly dependent on applied temperature for drying of samples.	[80]

4. Legislations of Mycotoxins Level in Food and Pollen

In order to prevent undesirable consequences and to protect consumers health, the European Commission, as well as some other international agencies, have proposed maximum permissible concentrations (MPC) for several mycotoxins in different types of food [33,81]. Maximum permissible concentrations vary due to differences in food origin and greater/less possibility of contamination with mycotoxins, as well as because of smaller or larger intake in meals. For instance, the MPC for AFB_1 alters from 0 to 8 μg/kg [33]. Zero tolerance is established for milk and dairy products due to regular daily consumption while the maximal value has been proposed for groundnut-based food. Furthermore, for sensitive groups (such as infants and children), special lower limits have been usually established. The proposed limits are subject to corrections as a result of the development of new, more precise, and sensitive analytical methods for determining the content of mycotoxins [81]. In Table 4 current EU MPC values for some food types are given.

Table 4. Examples for the current maximum permissible concentrations (MPC) for some mycotoxins in different types of food/food supplements.

Food/Food Supplements	Mycotoxin(s)	MPC Value(s)	Reference
Groundnuts used as components for food production	AFB_1	8 μg/kg	[33]
	Sum of AFB_1, AFB_2, AFG_1 and AFG_2	15 μg/kg	
Groundnuts for direct human consumption	AFB_1	2 μg/kg	[33]
	Sum of AFB_1, AFB_2, AFG_1 and AFG_2	4 μg/kg	
Dried fruits used as components for food production	AFB_1	5 μg/kg	[33]
	Sum of AFB_1, AFB_2, AFG_1 and AFG_2	10 μg/kg	
Dried fruits for direct human consumption	AFB_1	2 μg/kg	[33]
	Sum of AFB_1, AFB_2, AFG_1 and AFG_2	4 μg/kg	
Raw milk used for consumption and dairy productions, infant formulae and infant-milk	AFB_1	0 μg/kg	[33]
	Sum of AFB_1, AFB_2, AFG_1 and AFG_2	0 μg/kg	
Unprocessed cereals	OTA	5 μg/kg	[33]
Cereals based products	OTA	3 μg/kg	[33]
Instant coffee	OTA	10 μg/kg	[33]
Roasted coffee	OTA	5 μg/kg	[33]

The Scientific Committee of Food requested and obtained from the European Food Safety Authority (EFSA) current data for Tolerable Weekly Intake (TWI) for OTA—0.12 μg/kg of body weight (bw) [82]. Recently, EFSA published new information about the potential increase of maximum allowable level (from 4 to 10 μg/kg) for total AFs in peanuts and processed products, requested by EU Commission [83]. The CONTAM panel (EFSA Panel on Contaminants in the Food Chain) strongly opposed this request due to the significant increase of cancer risk (factor value = 1.6–1.8). For other mycotoxins proposed Tolerable Daily Intake (TDI) values are: 2 μg/kg bw for nivalenol, 0.25 μg/kg bw for ZEN [84], 2 μg/kg (provisional maximum TDI) for FBs [85], 1 μg/kg bw for DON [86], 0.1 μg/kg bw for the sum of T-2 and HT-2 toxins [87], 0.06 μg/kg for combined trichothecenes mycotoxins group [33]. In these legislations, there is no information about proposed limits for mycotoxins in pollen. In 2008 Campos et al. [2] proposed that in the case of AFB_1 occurrence in pollen the MPC value should be set at 2 μg/kg i.e., 4.2 μg/kg for total AFs. To the best of our knowledge, this is the only proposal which defines the level of some mycotoxins in pollen. Since this paper gives an overview about the presence of different mycotoxins in pollen samples originating from various locations around the world, it will be of great importance to define some tolerable levels for other fungi-produced toxins in pollen, especially for OTA. Moreover, current values for AFB_1 and AFs should be reconsidered and checked due to an increasingly frequent aflatoxin contamination caused by climatic changes. Special concerns exist due to mixed (cross) contamination of pollen samples as confirmed by the presented data. Previously, several authors [32,88,89] confirmed that some combined mycotoxins have a more distinct detrimental effect on human health. Furthermore, Manafi et al. [90] have shown that AFs and T-2 toxin synergistically influenced the decrease of total serum protein and albumin levels in broiler chickens as well as decreased antibody titers. It is therefore of the utmost importance to evaluate the toxicological impact of mycotoxin combinations on animal and human health risks.

5. Conclusions and Future Perspectives

Pollen could be used as a food supplement which can be attributed to its appropriate chemical composition. The microbiological quality of pollen is equally important as its nutritional characteristics. The fungal contamination of different feed/food, including pollen will be more frequent as a result of intensive climatic changes. The quality of pollen can be significantly influenced by the presence of toxigenic fungi. Since it has been proved that the absence of microbial contamination in pollen does not exclude the presence of mycotoxins, mycotoxicological analyses should also be included as a regular control measure together with microbiological tests. Since aflatoxins and ochratoxins are proven as carcinogenic substances, their presence in pollen is extremely undesirable. Therefore, it is important to monitor mold and mycotoxin levels in feed/food in order to avoid adverse health effects. The incorporation of pollen as a food supplement in current legislation will be useful. Proposed quality parameters need to cover tolerable daily/weekly intake for different mycotoxins as well as their sum. In order to obtain reliable and accurate recommendations for pollen quality control, further studies on the toxicological impact of mycotoxin combinations should be conducted.

Author Contributions: All authors participated in the creation and conceptualization of the article. A.Ž.K. and D.D.M. conducted the literature search. A.Ž.K., D.D.M., and M.B.P. wrote the manuscript. T.S.P., V.S.K., S.P.S., M.B.B., Ž.L.T. and M.B.P. controlled and critically reviewed language and manuscript content during preparation. All authors read and approved final manuscript.

Acknowledgments: Authors would like to thank to Vladimir Kostić for technical support in Graphical apstract preparation.

References

1. Borg, M.; Brownfield, L.; Twell, D. Male gametophyte development: A molecular perspective. *J. Exp. Bot.* **2009**, *60*, 1465–1478. [CrossRef]
2. Campos, G.R.M.; Bogdanov, S.; Almeida-Muradian, L.B.; Szczesna, T.; Mancebo, Y.; Frigerio, C.; Ferreira, F. Pollen composition and standardization of analytical methods. *J. Apic. Res.* **2008**, *47*, 154–161. [CrossRef]
3. Bogdanov, S. Pollen: Collection, harvest, composition, quality. In *Bee Product Science (The Pollen Book)*; 2012; Chapter 1; Available online: http://www.bee-hexagon.net/pollen/collection-harvest-composition-quality/ (accessed on 23 January 2019).
4. Kostić, A.Ž.; Barać, M.B.; Stanojević, S.P.; Milojković-Opsenica, D.M.; Tešić, Ž.L.; Šikoparija, B.; Radišić, P.; Prentović, M.; Pešić, M.B. Physicochemical properties and techno-functional properties of bee pollen collected in Serbia. *LWT Food Sci. Technol.* **2015**, *62*, 301–309. [CrossRef]
5. Kostić, A.Ž.; Kalušerović, L.M.; Dojčinović, B.P.; Barać, M.B.; Babić, V.B.; Mačukanović-Jocić, M.P. Preliminary investigation of mineral content of pollen collected from different Serbian maize hybrids—Is there any potential nutritional value? *J. Sci. Food Agric.* **2017**, *97*, 2803–2809. [CrossRef] [PubMed]

6. Kostić, A.Ž.; Pešić, M.B.; Trbović, D.; Petronijević, R.; Dramićanin, A.; Milojković-Opsenica, D.M.; Tešić, Ž.L. Fatty acid's profile of Serbian bee-collected pollen—Chemotaxonomic and nutritional approach. *J. Apic. Res.* **2017**, *56*, 533–542. [CrossRef]
7. Kostić, A.Ž.; Mačukanović-Jocić, M.P.; Špirović Trifunović, B.D.; Vukašinović, I.Ž.; Pavlović, V.B.; Pešić, M.B. Fatty acids of maize pollen-quantification, nutritional and morphological evaluation. *J. Cereal Sci.* **2017**, *77*, 180–185. [CrossRef]
8. Conte, P.; Del Caro, A.; Balestra, F.; Piga, A.; Fadda, C. Bee pollen as a functional ingredient in gluten-free bread: A physical-chemical, technological and sensory approach. *LWT Food Sci. Technol.* **2018**, *90*, 1–7. [CrossRef]
9. Campos, M.; Markham, M.R.; Mitchell, K.A.; da Cuhna, A.P. An approach to the characterization of bee pollens via their flavonoid/phenolic profiles. *Phytochem Anal.* **1997**, *8*, 181–185. [CrossRef]
10. Serra-Bonvehí, J.; Torrentó, S.M.; Lorente, C.E. Evaluation of polyphenolic and flavonoid compounds in honeybee-collected pollen produced in Spain. *J. Agric. Food Chem.* **2001**, *49*, 1848–1853. [CrossRef]
11. Campos, M.G.; Webby, F.B.; Markham, M.R.; Mitchell, K.A.; da Cuhna, A.P. Age-induced diminution of free radical scavenging capacity in bee pollens and the contribution of constituent flavonoids. *J. Agric. Food Chem.* **2003**, *51*, 742–745. [CrossRef]
12. Di Paola-Naranjo, R.D.; Sánchez, S.J.; Paramás, A.M.G.; Gonzalo, J.C.R. Liquid chromatographic-mass spectrometric analysis of anthocyanin composition of dark blue bee pollen from *Echium Plantagineum*. *J. Chromatogr. A* **2004**, *1054*, 205–210. [CrossRef] [PubMed]
13. Almaraz Abarca, N.; Campos da Graça, M.; Ávila-Reyes, J.A.; Naranjo-Jiménez, N.; Corral, J.H.; González-Valdez, L.S. Antioxidant activity of polyphenolic extract of monofloral honeybee-collected pollen from mesquite (*Prosopis juliflora, Leguminosae*). *J. Food Compos. Anal.* **2007**, *20*, 119–124. [CrossRef]
14. Ferreres, F.; Pereira, D.M.; Valentão, P.; Andrade, P.B. First report of noncoloured flavonoids in *Echium plantagineum* bee pollen: Differentation of ismomers by liquid chromatography/ion trap mass spectometry. *Rapid Commun. Mass Spectrom.* **2010**, *24*, 801–806. [CrossRef]
15. Ares, A.M.; Valverde, S.; Bernal, J.L.; Nozal, M.J.; Bernal, J. Extraction and determination of bioactive compounds from bee pollen. *J. Pharm. Biomed. Anal.* **2018**, *147*, 110–124. [CrossRef] [PubMed]
16. De-Melo, A.A.M.; Estevinho, L.M.; Moreira, M.M.; Delerue-Matos, C.; da Silva de Freitas, A.; Barth, O.M.; de Almeida-Muradian, L.B. Phenolic profile by HPLC-MS, biological potential, and nutritonal value of a promising food: Monofloral bee pollen. *J. Food Biochem.* **2018**, *42*, e12536. [CrossRef]
17. Almeida-Muradian, L.B.; Pamplona, L.C.; Coimbra, S.; Ortrud, M.B. Chemical composition and botanical evaluation of dried bee pollen pellets. *J. Food Compos. Anal.* **2005**, *18*, 105–111. [CrossRef]
18. Mărgăoan, R.; Mărghitaş, L.A.; Dezmirean, D.S.; Dulf, F.V.; Bunea, A.; Socaci, S.A.; Bobiş, O. Predominant and secondary pollen botanical origins influence the carotenoid and fatty acid profile in fresh honeybee-collected pollen. *J. Agric. Food Chem.* **2014**, *62*, 6306–6316. [CrossRef]
19. Krystyjan, M.; Gumul, D.; Ziobro, R.; Korus, A. The fortification of biscuits with bee pollen and its effect on physicochemical and antioxidant properties in biscuits. *LWT Food Sci. Technol.* **2015**, *63*, 640–646. [CrossRef]
20. De Florio Almeida, J.; Soares dos Reis, A.; Serafini Heldt, L.F.; Pereira, D.; Bianchin, M.; de Moura, C.; Plata-Oviedo, M.V.; Haminiuk, C.W.I.; Ribeiro, I.S.; Fernades Pinto da Luz, C.; et al. Lyophilized bee pollen extract: A natural antioxidant source to prevent lipid oxidation in refrigerated sausages. *LWT Food Sci. Technol.* **2017**, *76*, 299–305. [CrossRef]
21. Kostić, A.Ž.; Pešić, M.B.; Mosić, M.D.; Dojčinović, B.P.; Natić, M.N.; Trifković, J.Đ. Mineral content of some bee-collected pollen from Serbia. *Arch. Ind. Hyg. Toxicol.* **2015**, *66*, 251–258. [CrossRef]
22. Sattler, J.A.G.; de Melo Machado, A.A.; do Nascimento, K.S.; de Melo Pereira, I.L.; Mancini-Filho, J.; Sattler, A.; de Almeida-Muradian, L.B. Essential minerals and inorganic contaminants (barium, cadmium, lithium, lead and vanadium) in dried bee pollen produced in Rio Grande do Sul State, Brazil. *Food Sci. Technol.* **2016**, *36*, 505–509. [CrossRef]
23. Altunaltmaz, S.S.; Tarhan, D.; Aksu, F.; Barutçu, U.B.; Or, M.E. Mineral element and heavy metal (cadmium, lead and arsenic) levels of bee pollen in Turkey. *Food Sci. Technol.* **2017**, *37* (Suppl. S1), 136–141. [CrossRef]
24. Bennet, J.W.; Klich, M. Mycotoxins. *Clin. Microbiol. Rev.* **2003**, *16*, 497–516. [CrossRef]
25. Hanssen, E.; Jung, M. Control of aflatoxins in the food industry. *Pure Appl. Chem.* **1973**, *35*, 239–250. [CrossRef]

26. Bosco, F.; Mollea, C. Mycotoxins in food. In *Food Industrial Processes—Methods and Equipment*; Valdez, B., Ed.; Intech Open Limited: London, UK, 2012; Chapter 10; pp. 169–200. ISBN 978-953-307-905-9.
27. Stanković, S.; Lević, J.; Ivanović, D.; Krnjaja, V.; Stanković, G.; Tančić, S. Fumonisin B1 and its co-occurrence with other fusariotoxins in naturally-contaminated wheat grain. *Food Control* **2012**, *23*, 384–388. [CrossRef]
28. Krnjaja, V.; Mandić, V.; Lević, J.; Stanković, S.; Petrović, T.; Vasić, T.; Obradović, A. Influence of N-fertilization on Fusarium head blight and mycotoxin levels in winter wheat. *Crop Prot.* **2015**, *67*, 251–256. [CrossRef]
29. Abrunhosa, L.; Morales, H.; Soares, C.; Calado, T.; Vila-Cha, A.S.; Pereira, M.; Venâncio, A. A review of mycotoxins in food and feed products in Portugal and estimation of probable daily intake. *Crit. Rev. Food Sci. Nutr.* **2016**, *56*, 249–265. [CrossRef]
30. Bijelić, Z.; Krnjaja, V.; Stanković, S.; Muslić-Ružić, D.; Mandić, V.; Škrbić, Z.; Lukić, M. Occurrence of moulds and mycotoxins in grass-legume silages influenced by nitrogen fertilization and phenological phase at harvest. *Rom. Biotech. Lett.* **2017**, *22*, 12907–12914.
31. Krnjaja, V.; Stanković, S.; Obradović, A.; Petrović, T.; Mandić, V.; Bijelić, Z.; Božić, M. Trichothecene genotypes of *Fusarium graminearum* populations isolated from winter wheat crops in Serbia. *Toxins* **2018**, *10*, 460. [CrossRef]
32. Smith, M.-C.; Madec, S.; Coton, E.; Hymery, N. Natural co-occurrence of mycotoxins in foods and feeds and their in vitro combined toxicological effects. *Toxins* **2016**, *8*, 94. [CrossRef] [PubMed]
33. EC Commission. Setting of maximum levels for certain contaminants in foodstuffs—Regulation No. 1881/2006. *Official J. of the EU.* **2006**, *L364*, 5–24.
34. Van Egmond, H.P. Mycotoxins: Risks, regulations and European co-operation. *J. Nat. Sci. Matica Srpska Novi Sad.* **2013**, *125*, 7–20. [CrossRef]
35. Serra-Bonvehi, J.; Escolà Jordà, R. Nutrient composition and microbiological quality of honey bee-collected pollen in Spain. *J. Agric. Food Chem.* **1997**, *45*, 725–732. [CrossRef]
36. Medina, Á.; González, G.; Sáez, J.M.; Mateo, R.; Jiménez, M. Bee pollen, a substrate that stimulates ochratoxin A production by *Aspergillus ochraceus* Wilh. *Syst. Appl. Microbiol.* **2004**, *27*, 261–267. [CrossRef] [PubMed]
37. Garcia-Villanova, R.J.; Cordón, C.; González-Paramás, A.M.; Aparicio, P.; Garcia Rosales, M.E. Simultaneous immunoaffinity column cleanup and hplc analysis of aflatoxins and ochratoxin a in spanish bee pollen. *J. Agric. Food Chem.* **2004**, *52*, 7235–7239. [CrossRef] [PubMed]
38. González, G.; Hinojo, M.J.; Mateo, R.; Medina, A.; Jiménez, M. Occurrence of mycotoxin producing fungi in bee pollen. *Int. J. Food Microbiol.* **2005**, *105*, 1–9. [CrossRef] [PubMed]
39. Zaijun, L.; Zhongyun, W.; Xiulan, S.; Yinjun, F.; Peipei, C. A sensitive and highly stable electrochemical impendace immunosensor based on the formation of silica gel-ionic liquid biocompatible film on the glassy carbon electrode for the determination of aflatoxin B1 in bee pollen. *Talanta* **2010**, *80*, 1632–1637. [CrossRef]
40. Pitta, M.; Markaki, P. Study of aflatoxin B1 production by *Aspergillus parasiticus* in bee pollen of Greek origin. *Mycotoxin Res.* **2010**, *26*, 229–234. [CrossRef]
41. Kačaniová, M.; Juráček, M.; Chlebo, R.; Kňazovická, V.; Kadasi-Horáková, M.; Kunová, S.; Lejková, J.; Haščík, P.; Mareček, J.; Šimko, M. Mycobiota and mycotoxins in bee pollen collected from different areas of Slovakia. *J. Environ. Sci. Health Part B* **2011**, *46*, 623–629. [CrossRef]
42. Vidal, J.C.; Bonel, L.; Ezquerra, A.; Hernández, S.; Bertolín, J.R.; Cubel, C.; Castillo, J.R. Electrochemical affinity biosensors for detection of mycotoxins: A review. *Biosens. Bioelectron.* **2013**, *49*, 146–158. [CrossRef]
43. Rodríguez-Carasco, Y.; Font, G.; Mañes, J.; Berrada, H. Determination of mycotoxins in bee pollen by gas chromatography−tandem mass spectrometry. *J. Agric. Food Chem.* **2013**, *61*, 1999–2005. [CrossRef] [PubMed]
44. Petrović, T.; Nedić, N.; Paunović, D.; Rajić, J.; Matović, K.; Radulović, Z.; Krnjaja, V. Natural mycobiota and aflatoxin B1 presence in bee pollen collected in Serbia. *Biotechnol. Anim. Husb.* **2014**, *30*, 731–741. [CrossRef]
45. Xue, X.; Selvaraj, J.N.; Zhao, L.; Dong, H.; Liu, F.; Liu, Y.; Li, Y. Simultaneous determination of aflatoxins and ochratoxin a in bee pollen by low-temperature fat precipitation and immunoaffinity column cleanup coupled with LC-MS/MS. *Food Anal. Methods* **2014**, *7*, 690–696. [CrossRef]
46. Cirigliano, A.M.; Rodríguez, M.A.; Godeas, A.M.; Cabrera, G.M. Mycotoxins from beehive pollen mycoflora. *J. Sci. Res. Rep.* **2014**, *3*, 966–972. [CrossRef]
47. Valadares Deveza, M.; Keller, K.M.; Affonso Lorenzon, M.C.; Teixeira Nunes, L.M.; Oliveira Sales, E.; Barth, O.M. Mycotoxicological and palynological profiles of commercial brands of dried bee pollen. *Braz. J. Microbiol.* **2015**, *46*, 1171–1176. [CrossRef]

48. Kostić, A.Ž.; Petrović, T.S.; Krnjaja, V.S.; Nedić, N.M.; Tešić, Ž.L.; Milojković-Opsenica, D.M.; Barać, M.B.; Stanojević, S.P.; Pešić, M.B. Mold/aflatoxin contamination of honey bee collected pollen from different Serbian regions. *J. Apic. Res.* **2017**, *56*, 13–20. [CrossRef]
49. Hosny, A.S.; Sabbah, F.M.; El-Bazza, Z.E. Studies on microbial decontamination of Egyptian bee pollen by γ-irradiation. *Egypt Pharm. J.* **2018**, *17*, 190–200. [CrossRef]
50. Estevinho, L.M.; Dias, T.; Anjos, O. Influence of the storage conditions (frozen vs dried) in health-related lipid indexes and antioxidants of bee pollen. *Eur. J. Lipid Sci. Technol.* **2018**, *2018*, 1800393. [CrossRef]
51. Vidal, A.; Mengelers, M.; Yang, S.; De Saeger, S.; De Boevre, M. Mycotoxin biomarkers of exposure: A comprehensive review. *Compr. Rev. Food Sci. Food Saf.* **2018**, *17*, 1127–1155. [CrossRef]
52. Neal, G.E. Genetic implications in the metabolism and toxicity of mycotoxins. *Toxicol. Lett.* **1995**, *82/83*, 861–867. [CrossRef]
53. Magan, N.; Lacey, J. Effect of temperature and pH on water realtions of field and storage fungi. *Trans. Br. Mycol. Soc.* **1984**, *82*, 71–81. [CrossRef]
54. Estevinho, L.M.; Rodrigues, S.; Pereira, A.P.; Feás, X. Portugese bee pollen: Palynological study, nutritional and microbiological evaluation. *Int. J. Food Sci. Technol.* **2012**, *47*, 429–435. [CrossRef]
55. Niu, G.; Johnson, R.M.; Berenbaum, M.R. Toxicity of mycotoxins to honeybees and its amelioration by propolis. *Apidologie* **2011**, *42*, 79–87. [CrossRef]
56. Temiz, A.; Şener Mumcu, A.; Özkök Tüylü, A.; Sorkun, K.; Salih, B. Antifungal activity of propolis samples collected from different geographical regions of Turkey against two food-related molds, *Aspergillus versicolor* and *Penicillium aurantiogriseum*. *Gida* **2013**, *38*, 135–142. [CrossRef]
57. Tao, Y.; Xie, S.; Xu, F.; Liu, A.; Wang, Y.; Chen, D.; Pan, Y.; Huang, L.; Peng, D.; Wang, X.; et al. Ochratoxin A: Toxicity, oxidative stress and metabolism (Review). *Food Chem. Toxicol.* **2018**, *112*, 320–331. [CrossRef] [PubMed]
58. Cendoya, E.; Chiotta, M.L.; Zachetti, V.; Chulze, S.N.; Ramirez, M.L. Fumonisins and fumonisin-producing *Fusarium* occurrence in wheat and wheat by products: A review. *J. Cereal Sci.* **2018**, *80*, 158–166. [CrossRef]
59. Gilliam, M. Microbiology of pollen and bee bread: The yeasts. *Apidologie* **1979**, *10*, 43–53. [CrossRef]
60. Gilliam, M. Microbiology of pollen and bee bread: The genus *Bacillus*. *Apidologie* **1979**, *10*, 269–274. [CrossRef]
61. Gilliam, M.; Prest, D.B.; Lorenz, B.J. Microbiology of pollen and bee bread: Taxonomy and enzimology of molds. *Apidologie* **1989**, *20*, 53–68. [CrossRef]
62. Carelli Barreto, L.M.R.; Cunha Funari, S.R.; de Oliveira Rosi, R. Composição e qualidade do pólen apícola proveniente de sete estados Brasileiros e do distrito federal. *Bol. Ind. Anim.* **2005**, *62*, 167–175.
63. Kačániová, M.; Pavličová, S.; Haščík, P.; Kociubinski, G.; Kňazovická, V.; Sudzina, M.; Sudzinova, J.; Fikselová, M. Microbial communities in bees, pollen and honey from Slovakia. *Acta Microbiol. Immunol. Hung.* **2009**, *56*, 285–295. [CrossRef] [PubMed]
64. Bucio Villalobos, C.M.; López Preciado, G.; Martínez Jaime, O.A.; Torres Morales, J.J. Micoflora asociada a granos de polen recolectados por abejas domésticas (*Apis mellifera* L.). *Rev. Electron. Nova Sci.* **2010**, *4*, 93–103. [CrossRef]
65. Brindza, J.; Gróf, J.; Bacigálová, K.; Ferianc, P.; Tóth, D. Pollen microbial colonization and food safety. *Acta Chim. Slov.* **2010**, *3*, 95–102.
66. Puig-Peña, Y.; del-Risco-Ríos, C.A.; Álvarez-Rivera, V.P.; Leiva-Castillo, V.; García-Neninger, R. Comparación de la calidad microbiológica del polen apícola fresco y después de un proceso de secado. *Rev. CENIC. Cienc. Biol.* **2012**, *43*, 23–27.
67. Nogueira, C.; Iglesias, A.; Feás, X.; Estevinho, M.L. Commercial bee pollen with different geographical origins: A comprehensive approach. *Int. J. Mol. Sci.* **2012**, *13*, 11173–11187. [CrossRef] [PubMed]
68. Feás, X.; Pilar Vázquez-Tato, M.; Estevinho, L.; Seijas, J.A.; Iglesias, A. Organic bee pollen: Botanical origin, nutritional value, bioactive compounds, antioxidant activity and microbiological quality. *Molecules* **2012**, *17*, 8359–8377. [CrossRef] [PubMed]
69. Hani, B.; Dalila, B.; Saliha, D.; Daoud, H.; Mouloud, G.; Seddik, K. Microbiological sanitary aspects of pollen. *Adv. Environ. Biol.* **2012**, *6*, 1415–1420.
70. De-Melo Machado, A.A.; Estevinho, M.L.M.F.; Almeida-Muradian, L.B. A diagnosis of the microbiological quality of dehydrated bee-pollen produced in Brazil. *Lett. Appl. Microbiol.* **2015**, *61*, 477–483. [CrossRef]
71. Santa Bárbara, M.; Machado, C.S.; da Silva Sodré, G.; Dias, L.G.; Estevinho, L.M.; Lopes de Carvalho, C.A. Microbiological assessment, nutritional characterization and phenolic compounds of bee pollen from *Mellipona mandacaia* Smith, 1983. *Molecules* **2015**, *20*, 12525–12544. [CrossRef]

72. Nardoni, S.; D'Ascenzi, C.; Rocchigiani, G.; Moretti, V.; Mancianti, F. Occurrence of molds from bee pollen in Central Italy—A preliminary study. *Ann. Agric. Environ. Med.* **2016**, *23*, 103–105. [CrossRef]
73. De-Melo Machado, A.A.; Fernandes Estevinho, M.L.M.; Gasparotto Sattler, J.A.; Rodrigues Souza, B.; da Silva Freitas, A.; Barth, O.M.; Bicudo Almeida-Muradian, L. Effect of processing conditions on characteristics of dehydrated bee-pollen and correlation between quality parameters. *LWT Food Sci. Technol.* **2016**, *65*, 808–815. [CrossRef]
74. Grabowski, N.T.; Klein, G. Microbiology of processed edible insect products—Results of a preliminary survey. *Int. J. Food Microbiol.* **2017**, *243*, 103–107. [CrossRef] [PubMed]
75. Libonatti, C.; Andersen-Puchuri, L.; Tabera, A.; Varela, S.; Passucci, J.; Basualdo, M. Caracterización microbiológica de polen comercial. Reporte preliminar. *Rev. Electron. Vet.* **2017**, *18*, 1–5.
76. Aparecida Soares de Arruda, V.; Vieria dos Santos, A.; Figueiredo Sampaio, D.; da Silva Araújo, E.; de Castro Peixoto, A.L.; Fernandes Estevinho, L.M.; de Almeida-Muradian, B.L. Microbiological quality and physicochemical characterization of Brazilian bee pollen. *J. Apic. Res.* **2017**, *56*, 231–238. [CrossRef]
77. Adjlane, N.; Hadj Ali, L.M.; Benamara, M.; Bounadi, O.; Haddad, N. Qualite microbiologique du pollen produit par les apiculteurs et commercialise en Algerie. *Rev. Microbiol. Ind. San et Environ.* **2017**, *11*, 31–39.
78. Beev, G.; Stratev, D.; Vashin, I.; Pavlov, D.; Dinkov, D. Quality assessment of bee pollen: A cross sectional survey in Bulgaria. *J. Food Qual. Hazards Control* **2018**, *5*, 11–16. [CrossRef]
79. Figueredo Santa Bárbara, M.; Santiago Machado, C.; da Silva Sodré, G.; de Lima Silva, F.; Alfredo Lopes de Carvalho, C. Microbiological and physicochemical characterization of the pollen stored by stingless bees. *Braz. J. Food Technol.* **2018**, *21*, e2017180. [CrossRef]
80. Zuluaga-Domínguez, C.; Serrato-Bermudez, J.; Quicazán, M. Influence of drying-related operations on microbiological, structural and physicochemical aspects for processing of bee-pollen. *Eng. Agric. Environ. Food.* **2018**, *11*, 57–64. [CrossRef]
81. Arroyo-Manzanares, N.; Huertas-Pérez, J.F.; García-Campaña, A.M.; Gámiz-Gracia, L. Mycotoxin analysis: New proposals for sample treatment. *Adv. Chem.* **2014**, *2014*, 547506. [CrossRef]
82. European Food Safety Authority (EFSA). Opinion of the Scientific Panel on contaminants in the food chain of the EFSA on a request from the Commission related to ochratoxin A in food. *EFSA J.* **2006**, *4*, 365. [CrossRef]
83. EPSA Panel on Contaminants in the Food Chain (Contam); Knutsen, H.K.; Barregard, L.J.A.; Bingami, M.; Bruschweiler, B.; Ceccatelli, S.; Cottrill, B.; Dinovi, M.; Edler, L.; Grasl-Kraup, B.; et al. Effect on public health of a possible increase of the maximum level for 'aflatoxin total' from 4 to 10 μg/kg in peanuts and processed products thereof, intended for direct human consumption or use as an ingredient in foodstuffs-statement. *EFSA J.* **2018**, *16*, 5175. [CrossRef]
84. EPSA Panel on Contaminants in the Food Chain (Contam). Scientific Opinion on the risks for public health related to the presence of zearalenone in food. *EFSA J.* **2011**, *9*, 2197. [CrossRef]
85. EPSA Panel on Contaminants in the Food Chain (Contam); Knutsen, H.K.; Alexander, J.; Barregard, L.J.A.; Bingami, M.; Bruschweiler, B.; Ceccatelli, S.; Cottrill, B.; Dinovi, M.; Grasl-Kraup, B.; et al. Scientific Opinion on the risks for human and animal health related to the presence of modified forms of certain mycotoxins in food and feed. *EFSA J.* **2014**, *12*, 3916. [CrossRef]
86. EPSA Panel on Contaminants in the Food Chain (Contam). Risks to human and animal health related to the presence of deoxynivalenol and its acetylated and modfied forms in food and feed. *EFSA J.* **2017**, *15*, 4718. [CrossRef]
87. EPSA Panel on Contaminants in the Food Chain (Contam). Scientific Opinion on the risks for animal and public health related to the presence of T-2 and HT-2 toxin in food and feed. *EFSA J.* **2011**, *9*, 2481. [CrossRef]
88. Šegvić Klarić, M. Adverse effects of combined mycotoxins. *Arh. Ind. Hyg. Toxikol.* **2012**, *63*, 519–530. [CrossRef]
89. Šegvić Klarić, M.; Rašić, D.; Peraica, M. Deleterious effects of mycotoxin combinations involving Ochratoxin, A. *Toxins* **2013**, *5*, 1965–1987. [CrossRef] [PubMed]
90. Manafi, M.; Umakantha, B.; Mohan, K.; Narayana Swamy, H.D. Synergistic effects of two commonly contaminating mycotoxins (Aflatoxin and T-2 toxin) on biochemical parameters and immune status of broiler chickens. *World Appl. Sci. J.* **2012**, *17*, 364–367.

6

Potential Antagonistic Effects of Acrylamide Mitigation during Coffee Roasting on Furfuryl Alcohol, Furan and 5-Hydroxymethylfurfural

Dirk W. Lachenmeier [1,*], Steffen Schwarz [2], Jan Teipel [1], Maren Hegmanns [1], Thomas Kuballa [1], Stephan G. Walch [1] and Carmen M. Breitling-Utzmann [3]

1 Chemisches und Veterinäruntersuchungsamt (CVUA) Karlsruhe, Weissenburger Strasse 3, 76187 Karlsruhe, Germany; jan.teipel@cvuaka.bwl.de (J.T.); maren.hegmanns@cvuaka.bwl.de (M.H.); thomas.kuballa@cvuaka.bwl.de (T.K.); stephan.walch@cvuaka.bwl.de (S.G.W.)

2 Coffee Consulate, Hans-Thoma-Strasse 20, 68163 Mannheim, Germany; schwarz@coffee-consulate.com

3 Chemisches und Veterinäruntersuchungsamt Stuttgart, Schaflandstr. 3/2, 70736 Fellbach, Germany; Carmen.Breitling-Utzmann@cvuas.bwl.de

* Correspondence: Lachenmeier@web.de

Abstract: The four heat-induced coffee contaminants—acrylamide, furfuryl alcohol (FA), furan and 5-hydroxymethylfurfural (HMF)—were analyzed in a collective of commercial samples as well as in *Coffea arabica* seeds roasted under controlled conditions from very light Scandinavian style to very dark Neapolitan style profiles. Regarding acrylamide, average contents in commercial samples were lower than in a previous study in 2002 (195 compared to 303 μg/kg). The roasting experiment confirmed the inverse relationship between roasting degree and acrylamide content, i.e., the lighter the coffee, the higher the acrylamide content. However, FA, furan and HMF were inversely related to acrylamide and found in higher contents in darker roasts. Therefore, mitigation measures must consider all contaminants and not be focused isolatedly on acrylamide, specifically since FA and HMF are contained in much higher contents with lower margins of exposure compared to acrylamide.

Keywords: coffee; acrylamide; furfuryl alcohol; furan; 5-hydroxymethylfurfural; risk assessment

1. Introduction

Acrylamide is a heat-induced contaminant with frequent occurrence in foods and beverages [1–4]. It has been classified by the International Agency for Research on Cancer (IARC) as probably carcinogenic to humans (group 2A) [5]. The EFSA suggested that its margin of exposure indicates a concern for neoplastic effects based on animal evidence [6]. Coffee is an important topic in reduction of acrylamide, because its consumption may lead to 20–30% of total daily intake [7].

Following the first findings of acrylamide in foods and research into its formation mechanism [8,9], it was quickly discovered that coffee behaves differently from all other foods. While typically, the acrylamide content rises with color or browning degree due to its origin as a Maillard reaction product, for coffee, its content decreases from light to very dark roasts [10]. The maximum of acrylamide is formed very early in the roast and then decreases until the desired roasting degree is reached. Experimental studies have shown that the final acrylamide content purely depends on the roasting degree but not on the profile by which this degree is achieved (i.e., neither very slow nor very quick roasting methods have any influence) [10]. Currently, literature offers only speculation into the breakdown product of acrylamide during roasting or the reaction leading to its degradation [11].

Acrylamide is a product formed during coffee roasting by the Maillard reaction, a major pathway comprising the reaction between asparagine and reducing sugars [12,13]. The formation capacity is limited by the amount of asparagine [14], which is the reason for higher acrylamide contents found in *Coffea canephora* ("robusta") coffee due to its higher asparagine content.

Mitigation options may start with agronomy (e.g., species and variety selection, fertilization etc.) and roasting, but have also included strategies during processing such as asparaginase addition or lactic acid bacteria, none of which left the feasibility stage [15]. Careful removal of defective coffee beans is recommended, because these contain significantly higher amounts of asparagine (>2 fold), which is a major precursor of acrylamide formation [7,16]. Storage of coffee may lead to considerable reduction, but the final brew preparation is believed to have little influence due to the excellent water-solubility of acrylamide [15]. Some authors suggested that the variation detected in commercial samples may predominantly reflect differences in storage time [17]. Supercritical fluid extraction can be applied to reduce acrylamide by up to 79% [18]. Vacuum processing was suggested as a measure to reduce acrylamide in medium roasted coffee by 50% [19].

From all these factors, roasting was the predominant focus of previous research, and consistent findings hint that an increased roasting degree leads to a decrease in acrylamide formation [10,14,20–24].

Following several years of voluntary industry action with minimization concept [25], mitigation measures and benchmark levels for the reduction of the presence of acrylamide in food were recently implemented in an EU regulation [26]. The producers need to identify the critical roast conditions to ensure minimal acrylamide formation. They also need to ensure that the level of acrylamide in coffee is lower than the benchmark level of 400 μg/kg.

Besides acrylamide, coffee may contain further heat-induced contaminants that were also classified by IARC. Namely, furan [27], and furfuryl alcohol (FA) [28,29] are possibly carcinogenic to humans (group 2B). For 5-hydroxymethylfurfural (HMF), some evidence of carcinogenic activity was found in animal experiments [30,31], but the compound has not yet been classified by IARC. Out of these, furan is the compound in coffee studied most intensely, including large surveys [32–34], while less research is available on furfuryl alcohol [35–38] and HMF [39].

2. Materials and Methods

2.1. Analytical Methodology

The analysis of acrylamide was conducted according to the standard method EN 16618:2015 using liquid chromatography in combination with tandem mass spectrometry (LC/MS/MS) [40]. In deviation to this standard, samples were defatted with a mixture of isohexane and butyl methyl ether. Furthermore solid-phase extraction (SPE) was only used for clean-up, not for concentrating the acrylamide [11]. With this method, a limit of detection (LOD) of 10 μg/kg, and a limit of quantification (LOQ) of 30 μg/kg can be achieved. A repeatability relative standard deviation (RSDr) of 6% was determined within our laboratory. The method was applied successfully in several proficiency tests.

Analysis of furan was conducted using headspace-GC-MS and quantification with internal standard (furan-d_4) as previously described [41]. A multipoint calibration (0.65–12.94 mg/kg) was used for quantification in SIM-Mode on a GC 7890B with MSD 5977B (Agilent Technologies, Waldbronn, BW, Germany) instead of the previously used standard addition. With this method, a LOD of 0.36 mg/kg and a LOQ of 1.2 mg/kg was achieved (0.5 g coffee sample weight). A RSDr of 3.5% was determined within our laboratory.

Analysis of furfuryl alcohol (FA) and 5-hydroxymethylfurfural (HMF) was accomplished using nuclear magnetic resonance (NMR) spectroscopy as previously described [42]. The within-laboratory RSDr was 6% for FA and 8% for HMF. LOD and LOQ were 12 and 39 mg/kg for FA and 6 and 23 mg/kg for HMF, respectively.

2.2. *Samples and Roasting Experiments*

Samples were obtained from official sampling for food control purposes in the German federal state Baden-Württemberg from all stages of trade, mainly supermarkets and artisanal roasters. For roasting experiments, two directly imported single estate terrace coffees (*Coffea arabica* and *canephora*) were supplied by Amarella Trading (Mannheim, BW, Germany).

Twelve separate 2.4 kg batches of coffee beans were roasted using an FZ-94 Laboratory Roaster (CoffeeTech, Tel Aviv, Israel). Roasting was conducted using either pure *Coffea arabica* or pure *Coffea canephora* samples. The roasting profiles (e.g., regarding temperature endpoints) were based on expert roasters' experience as best suitable for the intended coffee roast type. The systematically different roast profiles were recorded and controlled using Artisan v1.5.0 (Artisan-Scope.org, Poing, BY, Germany, 2018, https://artisan-scope.org).

2.3. *Risk Assessment Methodology and Statistics*

Risk assessment was conducted using the margin of exposure (MOE) methodology according to the method for comparative risk assessment previously published for alcoholic beverages [3]. Statistical correlations were assessed using linear regression analysis calculated with OriginPro V7.5 (OriginLab Corporation, Northampton, MA, USA) with R being the correlation coefficient and p being the significance of Pearson's test for linear relation. p values below 0.05 are assumed as being significant.

3. Results

3.1. *Results of Roasting Experiments*

Two green coffee samples (*Coffea arabica and canephora*) were subjected to roasting using six different profiles, namely coffee roasting (quick and slow drying), espresso roasting (quick and slow drying) as well as Scandinavian roasting (very light roasting) and Neapolitan roasting (very black roasting). The roasting profiles for the *C. canephora* roasting are shown in Figure 1. Profiles for *C. arabica* roasting were similar (data not shown).

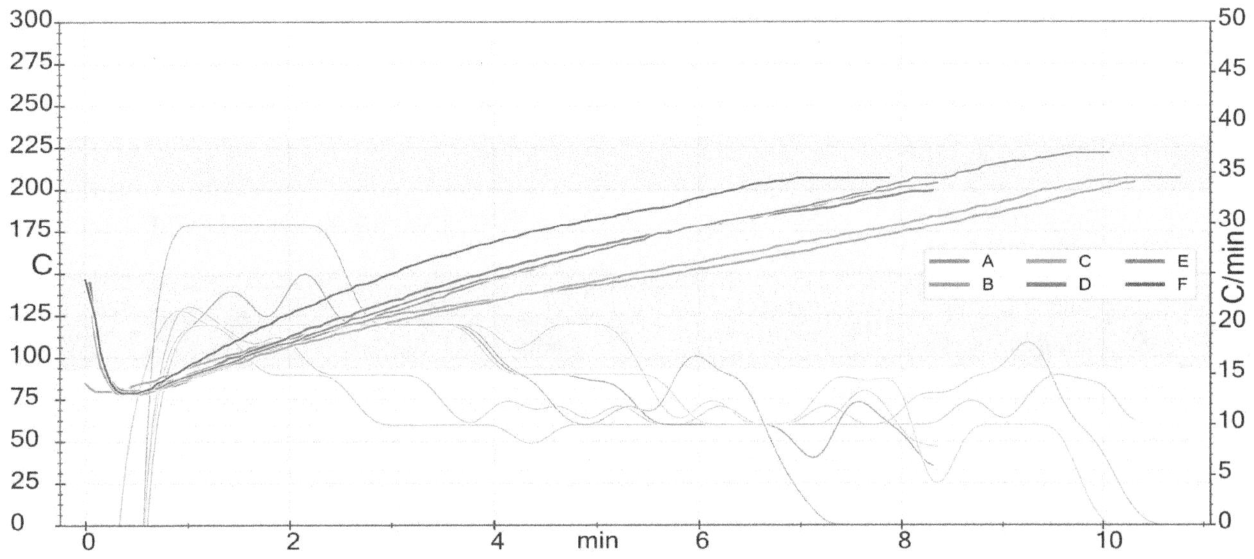

Figure 1. Profiles of experimental coffee roasting (A: Coffee quick drying; B: Coffee slow drying; C: Espresso slow drying; D: Scandinavian coffee; E: Espresso Neapolitan; F: Espresso quick drying).

Some numerical descriptors of the roasting profiles are provided in Table 1 as well as the analytical results for the samples. The individual roasting profile had a significant influence on the contents of the process contaminants. The area under the curve (AUC) is inversely related to acrylamide content ($R = -0.59$; $p = 0.045$; $n = 12$), while the contents of furfuryl alcohol ($R = 0.78$; $p = 0.003$; $n = 12$) and furan ($R = 0.63$; $p = 0.027$; $n = 12$) are positively correlated to this roasting parameter, independent of the coffee species. Furan ($R = 0.65$; $p = 0.021$; $n = 12$) and furfuryl alcohol ($R = 0.82$; $p = 0.001$; $n = 12$) are significantly positively correlated to drop temperature. The other parameters were not significantly correlated with any analyte.

Table 1. Indicators of roasting (data for *C. canephora* roast; *C. arabica* data similar) and analytical results of roasted coffee (*C. arabica*/*C. canephora*).

Profile	Charge [1] [°C]	Drop [2] [min]	Drop [2] [°C]	AUC [3] [°C·min]	Acrylamide [μg/kg]	Furfuryl Alcohol [mg/kg]	Furan [mg/kg]	HMF [mg/kg]
Scandinavian coffee	145	08:22	200	555	470/480	70/93	<1.2/2.5	40/47
Coffee quick drying	140	08:24	204	566	200/390	124/94	1.7/2.7	74/49
Coffee slow drying	85	10:21	205	673	210/420	128/92	1.5/2.6	62/43
Espresso quick drying	147	07:55	203	625	170/300	170/117	2.5/4.9	66/47
Espresso slow drying	140	10:48	207	762	150/290	173/133	2.6/5.0	78/42
Neapolitan espresso	145	10:06	222	796	130/250	223/189	3.6/7.6	84/32

[1] Temperature at charge of roaster. [2] Drop = end of roast. [3] Area under the curve (indicator how much total energy the beans have received during roasting).

There is an inverse linear statistically significant relationship between acrylamide and furfuryl alcohol ($R = -0.85$; $p < 0.001$; $n = 12$), and between acrylamide and HMF ($R = -0.73$; $p = 0.007$; $n = 12$). None of the other pairs for contaminants were significantly correlated; however, in tendency, acrylamide and furan were also inversely correlated, while furfuryl alcohol is positively correlated with HMF and furan.

3.2. *Results of Commercial Sample Analyzes*

The full results of analysis are provided in Appendix A, Table 1. The results are summarized in Table 2. From the sub-group of samples analyzed for both acrylamide and furfuryl alcohol, an inverse linear relationship was detected ($R = -0.59$; $p = 0.008$; $n = 19$). However, no correlation between HMF and acrylamide was detected, while HMF and furfuryl alcohol were positively correlated ($R = 0.50$; $p = 0.007$; $n = 28$). The data set of furan analysis was too small for meaningful statistical analysis.

Despite the low number of samples, the comparison of results in Table 2 suggests that the acrylamide content in roasted coffee and in instant coffee may have decreased over the years. None of the samples has exceeded the new EU benchmark levels.

Table 2. Comparison of acrylamide analysis results from 2002 with current results (summary from Annex A, Table S1).

Category according to EU Regulation 2017/2158	Year of Analysis	Number of Samples	Average [μg/kg]	Median [μg/kg]	90th Percentile [μg/kg]
Roast coffee	2002 (data from [11])	5	303	313	461
Roast coffee	2015	4	118	130	138
Roast coffee	2018	22	195	165	306
Instant (soluble coffee)	2013	6	642	686	831
Instant (soluble coffee)	2015	7	483	356	805
Instant (soluble coffee)	2016	5	379	269	664
Instant (soluble coffee)	2018	13	555	600	842
Coffee substitutes exclusively from cereals	2013–2018	6	401	418	563
Coffee substitutes from a mixture of cereals and chicory	2012–2018	16	587	525	805

3.3. *Comparative Risk Assesment of Heat-Induced Contaminants in Coffee*

Finally, the results of comparative risk assessment using the margin of exposure methodology are shown in Table 3. The risk assessment uses survey data from the literature due to the restricted, non-representative sampling in the current study.

While the contents of acrylamide and furan are much lower than the contents of furfuryl alcohol and HMF, the toxicity thresholds of both compounds are also much lower, with acrylamide being the compound showing effects at the lowest concentration of all four compounds. Nevertheless, due to the higher exposure, HMF and furfuryl alcohol have the lowest margins of exposure. Three of the compounds, acrylamide, furfuryl alcohol and HMF, have MOEs below 10,000. Furan falls below this threshold only in worst-case scenarios (P95 exposure) and can be seen as a compound with lower risk. However, HMF is believed to operate by a non-genotoxic mechanism and hence an uncertainty factor of 100 (instead of 10,000 for genotoxic compounds) may be sufficient to exclude public health concerns.

Table 3. Risk assessment of several roasting contaminants in coffee.

Contaminant	Average/P95 Content in Roasted Coffee	Average/P95 Exposure for Drinking 1 Cup of Coffee [1]	Toxicological Threshold [2]	Average/P95 Margin of Exposure (MOE) [3]
Acrylamide	249/543 μg/kg [6]	0.05/0.10 μg/kg bw/day	0.18 mg/kg bw/day (BDML10) [43]	3800/1700
Furfuryl alcohol	251/392 mg/kg [35]	0.05/0.07 mg/kg bw/day	53 mg/kg bw/day (NOEL) [44]	1110/710
HMF	689/1688 mg/kg [39]	0.13/0.32 mg/kg bw/day	79 mg/kg bw/day (BMDL10) [30]	600/250
Furan	38/107 μg/L [33]	0.12/0.14 μg/kg bw/day [33]	1.28 mg/kg bw/day (BMDL10) [45]	42,134/3113 [33]

[1] Calculated assuming 14 g of coffee powder per 0.2 L cup (according to ISO 6668 [46]) and assuming 100% extraction yield, except for furan for which data from brewed beverage analyses were available. Average bodyweight 73.9 kg [47]. The data for furan were probabilistically calculated and taken from [33]. All other values were own calculations using point estimates. [2] NOEL: no-observed effect level; BMDL10: benchmark dose lower confidence limit for 10% response. [3] MOE = Toxicological threshold/exposure. Values pessimistically rounded to significance. The higher the MOE, the lower the risk. A MOE > 10,000 is typically interpreted as low risk for genotoxic carcinogens, while >100 is used for non-genotoxic compounds with thresholded effects.

4. Discussion

Roasting properties of coffee are basically dependent on the amount of heat transferred into the coffee beans during roasting and on the roasting time [17]. A good indicator for the achieved heat transfer rate is the area under the curve of the roasting profile. These values show a negative correlation with acrylamide during our roasting experiment, confirming the inverse relationship of roasting energy and acrylamide [10,14,20–24]. In contrast, the other contaminants under study (furfuryl alcohol, furan and HMF) appear to be positively related to the roasting energy, meaning the highest contents are typically found in the strongest roasts (espresso).

Interestingly, despite early findings that acrylamide in coffee decreases with the roasting degree, there is still considerable misinformation about this topic. Some small artisanal coffee roasters even advertise on their webpages that their "mild" roasting process with temperatures rising only up to 200 °C would result in lower acrylamide contents. The contrary being clearly the case, however.

Compared to results from our institutes published in 2002 (average acrylamide content in coffee: 303 μg/kg, median 313 μg/kg; 90% percentile 461 μg/kg) [11], the contents found during this study were lower. In Germany, the minimization of acrylamide has been most advanced of all EU member states [25]. Manufacturers should therefore not be challenged, even if the current benchmark level should become the new legal maximum limit [25]. Our results confirm this assumption, since none of our official samples exceeded the current benchmark level.

Some authors have questioned the influence of species, e.g., Mojksja and Gielecinska [22], who found no significant difference in acrylamide contents between Arabica and Robusta coffee. Our restricted results of two pure *C. canephora* coffees (260–270 μg/kg) lie actually above the average acrylamide contents of all coffee samples (196 μg/kg), which is consistent with the majority of literature [14,24,48]. However, in our case a comparison is confined due to the fact that the species is unknown in most of the analyzed commercial samples. It may be speculated that the difference is caused by the lower quality of commercial *C. canephora* coffee with a higher degree of defective beans. However, the comparison of our high-quality single estate terrace coffees (Table 1) also points to higher levels of acrylamide in *C. canephora*.

There are only few studies available on the correlation of other contaminants with acrylamide. Kocadagli et al. [49] studied the kinetics of both acrylamide formation and HMF formation and found similar tendencies, meaning both acrylamide and HMF are reduced by more intense roasts. This is in contrast to our results, which detected this behavior only for acrylamide but not for HMF. An explanation may the different methodology in Kocadagli et al. [49], which did not apply a commercial coffee roaster but only an oven at 220 °C for 5–60 min. We therefore believe that our results may have a higher relevance for commercial coffee roasting. Nevertheless, there remains some uncertainty in HMF exposure from coffee. For example, the survey reported by Arribas-Lorenzo [39] from Spain found higher HMF levels than our study with less samples. According to the German Federal Institute for Risk Assessment (BfR) evaluation, the levels of HMF in foods were suggested to exhibit no identifiable health risk for the consumer [50]. However, the BfR did not include coffee in its evaluation of HMF due to a lack of food monitoring data necessary for exposure assessment.

For other heat-induced contaminants besides acrylamide, no action has been typically taken to reduce levels and there are also no EU benchmark or maximum levels for heat-induced contaminants besides acrylamide. Therefore, focus and research activity have been mainly aimed at acrylamide in the past. The Codex Code of Practice to reduce acrylamide in foods currently does not provide guidance for coffee because to date "no commercial measures for reducing acrylamide in coffee are currently available" [15,51]. While this opinion is probably outdated, as various measures have shown to be effective (see introduction), our findings suggest that indeed no measures should be implemented that solely focus on acrylamide. Using a holistic risk assessment approach, all major heat-induced contaminants in coffee need to be modelled prior to pointing out any measure. Otherwise it could well mean that the benefit gained by reduction of acrylamide might be outweighed by the elevated risk of other contaminants such as furfuryl alcohol that are concomitantly increased by the applied measure. As other authors have shown [7,20], holistic risk-benefit analysis would be most preferable as the mitigation of acrylamide might not only lead to increased formation of other contaminants such as furfuryl alcohol [36], but may also lead to reduced contents in beneficial compounds in coffee such as antioxidants.

Compared to other lifestyle factors such as tobacco smoking or alcohol drinking, the cancer risk from coffee (if any exists) appears to be rather low. According to IARC, epidemiological studies even suggest a lack of carcinogenicity of drinking coffee for cancer of the liver [52,53], which is the major target organ of heat-induced contaminants. Bladder cancer was the only cancer site for which an increased risk had been observed in some earlier epidemiological studies, leading to an IARC grouping as 2B in 1991 [54]. However, more recent well-conducted epidemiologic studies were unable to replicate the association with bladder cancer, and coffee consumption has been removed from the classification as a possible/probable human carcinogen [52,53].

Author Contributions: Conceptualization, D.W.L.; Methodology, D.W.L., S.S., M.H. and C.M.B.-U.; Validation, D.W.L., T.K., M.H. and C.M.B.-U.; Formal Analysis, D.W.L.; Investigation, D.W.L., M.H. and C.M.B.-U.; Resources, S.S. and S.G.W.; Data Curation, D.W.L.; Writing-Original draft preparation, D.W.L.; Writing-Review and Editing, C.M.B.-U., S.S., M.H., S.G.W., T.K., J.T.; Visualization, D.W.L., S.S.; Supervision, S.G.W.

Acknowledgments: The laboratory teams of the NMR and the GC-MS department at CVUA Karlsruhe and beverages department at CVUA Stuttgart are thanked for excellent technical assistance. The team at Coffee Consulate is thanked for conducting the roasting experiment.

Appendix A

Table A1. Full analytical results of samples measured between 2012–2018 for acrylamide, furfuryl alcohol and HMF.

Sample ID	Sample Description	Category according to EU Regulation 2017/2158	Year	AA (μg/kg)	FA (mg/kg)	Furan (mg/kg)	HMF (mg/kg)
12119400	Coffee substitute, soluble	Coffee substitutes from a mixture of cereals and chicory	2012	803	-	-	-
12119400-1	Coffee substitute, soluble	Coffee substitutes from a mixture of cereals and chicory	2012	792	-	-	-
12119400-2	Coffee substitute, soluble	Coffee substitutes from a mixture of cereals and chicory	2012	806	-	-	-
12119400-3	Coffee substitute, soluble	Coffee substitutes from a mixture of cereals and chicory	2012	759	-	-	-
130122855	Coffee substitute, soluble	Coffee substitutes from a mixture of cereals and chicory	2013	664	-	-	-
130123291	100% soluble coffee	Instant (soluble) coffee	2013	866	-	-	-
130123334	Coffee, soluble	Instant (soluble) coffee	2013	495	-	-	-
130124497	Coffee substitute, soluble	Coffee substitutes exclusively from cereals	2013	436	-	-	-
130124499	Coffee, soluble	Instant (soluble) coffee	2013	744	-	-	-
130127835	100% soluble coffee	Instant (soluble) coffee	2013	796	-	-	-
130128813	Coffee, soluble	Instant (soluble) coffee	2013	325	-	-	-
130128818	100% soluble coffee	Instant (soluble) coffee	2013	628	-	-	-
130130022	Coffee substitute, soluble	Coffee substitutes from a mixture of cereals and chicory	2013	591	-	-	-
130132127	Coffee substitute, soluble	Coffee substitutes from a mixture of cereals and chicory	2013	214	-	-	-
130132150	Coffee substitute, soluble	Coffee substitutes exclusively from cereals	2013	619	-	-	-
130237309	Coffee substitute, soluble	Coffee substitutes from a mixture of cereals and chicory	2013	387	-	-	-
150231135	Coffee, soluble	Instant (soluble) coffee	2015	1135	-	-	-
150231200	Coffee, soluble	Instant (soluble) coffee	2015	199	-	-	-
150231825	Coffee substitute, soluble	Coffee substitutes exclusively from cereals	2015	508	-	-	-
150231835	Turkish coffee	Roast coffee	2015	127	-	-	-

Table 1. *Cont.*

Sample ID	Sample Description	Category according to EU Regulation 2017/2158	Year	AA (μg/kg)	FA (mg/kg)	Furan (mg/kg)	HMF (mg/kg)
150309974	Coffee	Roast coffee	2015	70	-	-	-
150309977	Coffee, decaffeinated	Instant (soluble) coffee	2015	335	-	-	-
150334870	Espresso Italiano	Roast coffee	2015	132	-	-	-
150334875	Coffee, soluble	Instant (soluble) coffee	2015	356	-	-	-
150334880	Coffee, soluble	Instant (soluble) coffee	2015	320	-	-	-
150337674	Coffee	Roast coffee	2015	141	-	-	-
150337675	Coffee, decaffeinated	Instant (soluble) coffee	2015	585	-	-	-
150337676	Coffee	Instant (soluble) coffee	2015	452	-	-	-
160450717	Malt coffee	Coffee substitutes exclusively from cereals	2016	370	-	-	-
160451307	Coffee, soluble	Instant (soluble) coffee	2016	223	-	-	-
160451426	Coffee, soluble	Instant (soluble) coffee	2016	273	-	-	-
160451967	Coffee substitute, soluble	Coffee substitutes from a mixture of cereals and chicory	2016	361	-	-	-
160452173	100% soluble coffee, 100% Arabica	Instant (soluble) coffee	2016	269	-	-	-
160452684	Coffee substitute, soluble	Coffee substitutes exclusively from cereals	2016	74	-	-	-
160454844	Coffee, soluble	Instant (soluble) coffee	2016	206	-	-	-
160472527	Coffee substitute, soluble	Coffee substitutes from a mixture of cereals and chicory	2016	407	-	-	-
160472529	Coffee substitute, soluble	Coffee substitutes from a mixture of cereals and chicory	2016	431	-	-	-
160509176	Coffee, soluble	Instant (soluble) coffee	2016	925	-	-	-
160509194	Espresso	Roast coffee	2016	447	-	-	-
160574673	Coffee substitute, soluble	Coffee substitutes from a mixture of cereals and chicory	2016	347	-	-	-
180352602	Coffee substitute, soluble	Coffee substitutes from a mixture of cereals and chicory	2018	650	-	-	-
180352683	Coffee	Roast coffee	2018	160	108	1.7	46
180365240	Costa Rica Arabica coffee	Roast coffee	2018	170	108	2.0	46
180379248	India Monsooned	Roast coffee	2018	150	74	5.4	34

Table 1. *Cont.*

Sample ID	Sample Description	Category according to EU Regulation 2017/2158	Year	AA (µg/kg)	FA (mg/kg)	Furan (mg/kg)	HMF (mg/kg)
180398113	Coffee	Roast coffee	2018	210	156	2.3	52
180409591	Arabica-Robusta-mixture, coffee	Roast coffee	2018	130	133	2.6	48
180420281	coffee, organic	Roast coffee	2018	110	122	-	57
180420519	Coffee	Roast coffee	2018	95	104	3.5	46
180433746	Coffee	Roast coffee	2018	310	76	-	53
180439193	coffee, Ethiopia	Roast coffee	2018	150	-	-	-
180444177	Coffee	Roast coffee	2018	170	121	-	42
180447473	Coffee	Roast coffee	2018	110	116	-	57
180453665	Coffee Sumatra	Roast coffee	2018	120	119	-	52
180468077	Coffee	Roast coffee	2018	130	122	-	51
180478363	Coffee substitute, soluble	Instant (soluble) coffee	2018	730	-	-	-
180481476	Coffee substitute, soluble	Instant (soluble) coffee	2018	110	64	-	40
180486743	Coffee substitute, soluble	Instant (soluble) coffee	2018	670	-	-	-
180486745	Coffee substitute, soluble	Instant (soluble) coffee	2018	460	-	-	-
180489109	Coffee substitute, soluble	Instant (soluble) coffee	2018	420	-	-	-
180489247	Coffee substitute, soluble	Instant (soluble) coffee	2018	440	-	-	-
180492032	100% soluble coffee, 100% Arabica	Instant (soluble) coffee	2018	870	-	-	-
180492048	Coffee, soluble	Instant (soluble) coffee	2018	660	-	-	-
180494672	Coffee, soluble	Instant (soluble) coffee	2018	910	-	-	-
180504580	Coffee, soluble	Instant (soluble) coffee	2018	420	-	-	-
180520043	100% soluble coffee, 100% Arabica	Instant (soluble) coffee	2018	690	-	-	-
180533580	Pure Canephora coffee	Roast coffee	2018	260	42	-	40
180533581	Pure Canephora coffee	Roast coffee	2018	270	37	-	49
180533582	Arabica coffee	Roast coffee	2018	270	78	-	46
180533583	Arabica coffee	Roast coffee	2018	460	55	-	43

Table 1. *Cont.*

Sample ID	Sample Description	Category according to EU Regulation 2017/2158	Year	AA (µg/kg)	FA (mg/kg)	Furan (mg/kg)	HMF (mg/kg)
180533584	Arabica coffee	Roast coffee	2018	330	60	-	35
180539910	Coffee, soluble	Instant (soluble) coffee	2018	600	-	-	-
180552699	100% Organic Arabica coffee	Roast coffee	2018	120	-	-	-
180619399	Coffee substitute, soluble	Coffee substitutes exclusively from cereals	2018	400	-	-	-
180619831	Coffee, soluble	Instant (soluble) coffee	2018	240	-	-	-
180628887	Espresso	Roast coffee	2018	210	-	-	-
180631120	Coffee substitute, soluble	Coffee substitutes from a mixture of cereals and chicory	2018	460	-	-	-
180631416	Coffee	Roast coffee	2018	250	-	-	-
180638375	Espresso	Roast coffee	2018	100	99	-	38
180643299	Coffee substitute, soluble	Coffee substitutes from a mixture of cereals and chicory	2018	220	-	-	-
180643300	Coffee substitute, soluble	Coffee substitutes from a mixture of cereals and chicory	2018	1500	-	-	-

AA: Acrylamide; FA: Furfuryl alcohol; HMF: 5-Hydroxymethylfurfural; "-": parameter not analyzed in that sample.

References

1. Lachenmeier, D.W. Carcinogens in food: Opportunities and challenges for regulatory toxicology. *Open Toxicol. J.* **2009**, *3*, 30–34. [CrossRef]
2. Wenzl, T.; Lachenmeier, D.W.; Gökmen, V. Analysis of heat-induced contaminants (acrylamide, chloropropanols and furan) in carbohydrate-rich food. *Anal. Bioanal. Chem.* **2007**, *389*, 119–137. [CrossRef] [PubMed]
3. Lachenmeier, D.W.; Przybylski, M.C.; Rehm, J. Comparative risk assessment of carcinogens in alcoholic beverages using the margin of exposure approach. *Int. J. Cancer* **2012**, *131*, E995–E1003. [CrossRef] [PubMed]
4. Pflaum, T.; Hausler, T.; Baumung, C.; Ackermann, S.; Kuballa, T.; Rehm, J.; Lachenmeier, D.W. Carcinogenic compounds in alcoholic beverages: An update. *Arch. Toxicol.* **2016**, *90*, 2349–2367. [CrossRef] [PubMed]
5. IARC Working Group on the Evaluation of Carcinogenic Risks to Humans. Acrylamide. *IARC Monogr. Eval. Carcinog. Risks Hum.* **1994**, *60*, 389–433. [PubMed]
6. EFSA Panel on Contaminants in the Food Chain. Scientific Opinion on acrylamide in food. *EFSA J.* **2015**, *13*, 4104. [CrossRef]
7. Guenther, H.; Anklam, E.; Wenzl, T.; Stadler, R.H. Acrylamide in coffee: Review of progress in analysis, formation and level reduction. *Food Addit. Contam.* **2007**, *24* (Suppl. 1), 60–70. [CrossRef] [PubMed]
8. Tareke, E.; Rydberg, P.; Karlsson, P.; Eriksson, S.; Tornqvist, M. Analysis of acrylamide, a carcinogen formed in heated foodstuffs. *J. Agric. Food Chem.* **2002**, *50*, 4998–5006. [CrossRef]
9. Tareke, E.; Rydberg, P.; Karlsson, P.; Eriksson, S.; Tornqvist, M. Acrylamide: A cooking carcinogen? *Chem. Res. Toxicol.* **2000**, *13*, 517–522. [CrossRef]
10. Deutscher Kaffeeverband. *Jahresbericht 2004*; Deutscher Kaffeeverband e.V.: Hamburg, Germany, 2005.
11. Gutsche, B.; Weisshaar, R.; Buhlert, J. Acrylamide in food—Screening results from food control in Baden-Württemberg. *Dtsch. Lebensm. Rundsch.* **2002**, *98*, 437–443.
12. Soares, C.M.D.; Alves, R.C.; Oliveira, M.B. Acrylamide in coffee: Influence of processing. In *Processing and Impact on Active Components in Food*; Preedy, V., Ed.; Academic Press: San Diego, CA, USA, 2015; pp. 575–582. [CrossRef]
13. Weisshaar, R.; Gutsche, B. Formation of acrylamide in heated potato products—Model experiments pointing to asparagine as precursor. *Dtsch. Lebensm. Rundsch.* **2002**, *98*, 397–400.
14. Bagdonaite, K.; Derler, K.; Murkovic, M. Determination of acrylamide during roasting of coffee. *J. Agric. Food Chem.* **2008**, *56*, 6081–6086. [CrossRef] [PubMed]
15. Stadler, R.H.; Theurillat, V. Acrylamide in coffee. In *Coffee: Emerging Health Effects and Disease Prevention*; Chu, Y., Ed.; John Wiley & Sons: Hoboken, NJ, USA, 2012. [CrossRef]
16. Soares, C.M.D.; Alves, R.C.; Oliveira, M.B. Factors affecting acrylamide levels in coffee beverages. In *Coffee in Health and Disease Prevention*; Preedy, V.R., Ed.; Academic Press: San Diego, CA, USA, 2015; pp. 217–224. [CrossRef]
17. Lantz, I.; Ternite, R.; Wilkens, J.; Hoenicke, K.; Guenther, H.; van der Stegen, G.H. Studies on acrylamide levels in roasting, storage and brewing of coffee. *Mol. Nutr. Food Res.* **2006**, *50*, 1039–1046. [CrossRef] [PubMed]
18. Banchero, M.; Pellegrino, G.; Manna, L. Supercritical fluid extraction as a potential mitigation strategy for the reduction of acrylamide level in coffee. *J. Food Eng.* **2013**, *115*, 292–297. [CrossRef]
19. Anese, M.; Nicoli, M.C.; Verardo, G.; Munari, M.; Mirolo, G.; Bortolomeazzi, R. Effect of vacuum roasting on acrylamide formation and reduction in coffee beans. *Food Chem.* **2014**, *145*, 168–172. [CrossRef] [PubMed]
20. Summa, C.A.; de la Calle, B.; Brohee, M.; Stadler, R.H.; Anklam, E. Impact of the roasting degree of coffee on the in vitro radical scavenging capacity and content of acrylamide. *LWT Food Sci. Technol.* **2007**, *40*, 1849–1854. [CrossRef]
21. Senyuva, H.Z.; Gokmen, V. Study of acrylamide in coffee using an improved liquid chromatography mass spectrometry method: Investigation of colour changes and acrylamide formation in coffee during roasting. *Food Addit. Contam.* **2005**, *22*, 214–220. [CrossRef]
22. Mojska, H.; Gielecinska, I. Studies of acrylamide level in coffee and coffee substitutes: Influence of raw material and manufacturing conditions. *Roczniki Państwowego Zakładu Higieny* **2013**, *64*, 173–181.
23. Granby, K.; Fagt, S. Analysis of acrylamide in coffee and dietary exposure to acrylamide from coffee. *Anal. Chim. Acta* **2004**, *520*, 177–182. [CrossRef]

24. Alves, R.C.; Soares, C.; Casal, S.; Fernandes, J.O.; Oliveira, M.B.P. Acrylamide in espresso coffee: Influence of species, roast degree and brew length. *Food Chem.* **2010**, *119*, 929–934. [CrossRef]
25. Rexroth, A. Acrylamide. The new EU regulation 2017/2158. *Dtsch. Lebensm. Rundsch.* **2018**, *114*, 142–151.
26. European Commission. Commission regulation (EU) 2017/2158 of 20 November 2017 establishing mitigation measures and benchmark levels for the reduction of the presence of acrylamide in food. *Off. J. EU* **2017**, *L304*, 24–44.
27. IARC Working Group on the Evaluation of Carcinogenic Risks to Humans. Furan. *IARC Monogr. Eval. Carcinog. Risks Hum.* **1995**, *63*, 393–407. [PubMed]
28. Grosse, Y.; Loomis, D.; Guyton, K.Z.; El, G.F.; Bouvard, V.; Benbrahim-Tallaa, L.; Mattock, H.; Straif, K. Some chemicals that cause tumours of the urinary tract in rodents. *Lancet Oncol.* **2017**, *18*, 1003–1004. [CrossRef]
29. IARC Working Group on the Evaluation of Carcinogenic Risks to Humans. Furfuryl alcohol. *IARC Monogr. Eval. Carcinog. Risks Hum.* **2019**, *119*, 83–113.
30. Monakhova, Y.B.; Lachenmeier, D.W. The Margin of exposure of 5-hydroxymethylfurfural (HMF) in alcoholic beverages. *Environ. Health Toxicol.* **2012**, *27*, e2012016. [CrossRef] [PubMed]
31. NTP. NTP toxicology and carcinogenesis studies of 5-(hydroxymethyl)-2-furfural (CAS No. 67-47-0) in F344/N rats and B6C3F1 mice (gavage studies). *Natl. Toxicol. Program Tech. Rep. Ser.* **2010**, *554*, 1–180. [CrossRef]
32. EFSA. Update on furan levels in food from monitoring years 2004–2010 and exposure assessment. *EFSA J.* **2011**, *9*, 2347. [CrossRef]
33. Lachenmeier, D.W. Furan in Coffee Products: A Probabilistic Exposure Estimation. In *Coffee in Health and Disease Prevention*; Preedy, V.R., Ed.; Academic Press: San Diego, CA, USA, 2015; pp. 887–893. [CrossRef]
34. Morehouse, K.M.; Nyman, P.J.; Mcneal, T.P.; Dinovi, M.J.; Perfetti, G.A. Survey of furan in heat processed foods by headspace gas chromatography/mass spectrometry and estimated adult exposure. *Food Addit. Contam.* **2008**, *25*, 259–264. [CrossRef]
35. Okaru, A.O.; Lachenmeier, D.W. The food and beverage occurrence of furfuryl alcohol and myrcene—Two emerging potential human carcinogens? *Toxics* **2017**, *5*, 9. [CrossRef]
36. Okaru, A.O.; Lachenmeier, D.W. Processing Contaminants: Furfuryl Alcohol. In *Encyclopedia of Food Chemistry*; Melton, L., Shahidi, F., Varelis, P., Eds.; Elsevier: Amsterdam, The Netherlands, 2019; pp. 543–549. [CrossRef]
37. Albouchi, A.; Russ, J.; Murkovic, M. Parameters affecting the exposure to furfuryl alcohol from coffee. *Food Chem. Toxicol.* **2018**, *118*, 473–479. [CrossRef] [PubMed]
38. Albouchi, A.; Murkovic, M. Formation kinetics of furfuryl alcohol in a coffee model system. *Food Chem.* **2018**, *243*, 91–95. [CrossRef] [PubMed]
39. Arribas-Lorenzo, G.; Morales, F.J. Estimation of dietary intake of 5-hydroxymethylfurfural and related substances from coffee to Spanish population. *Food Chem. Toxicol.* **2010**, *48*, 644–649. [CrossRef] [PubMed]
40. CEN/TC 275. *EN 16618:2015. Food Analysis—Determination of Acrylamide in Food by Liquid Chromatography Tandem Mass Spectrometry (LC-ESI-MS/MS)*; European Committee for Standardization (CEN): Brussels, Belgium, 2015.
41. Waizenegger, J.; Winkler, G.; Kuballa, T.; Ruge, W.; Kersting, M.; Alexy, U.; Lachenmeier, D.W. Analysis and risk assessment of furan in coffee products targeted to adolescents. *Food Addit. Contam. Part A* **2012**, *29*, 19–28. [CrossRef] [PubMed]
42. Monakhova, Y.B.; Ruge, W.; Kuballa, T.; Ilse, M.; Winkelmann, O.; Diehl, B.; Thomas, F.; Lachenmeier, D.W. Rapid approach to identify the presence of Arabica and Robusta species in coffee using ^{1}H NMR spectroscopy. *Food Chem.* **2015**, *182*, 178–184. [CrossRef] [PubMed]
43. Mueller, U.; Agudo, A.; Carrington, C.; Doerge, D.; Hellenäs, K.E.; Leblanc, J.C.; Rao, M.; Renwick, A.; Slob, W.; Wu, Y. Acrylamide (Addendum). In *WHO Food Additives Series 63. Safety Evaluation of Certain Contaminants in Food. Prepared by the Seventy-Second Meeting of the Joint FAO/WHO Expert Committee on Food Additives (JECFA)*; WHO and FAO: Geneva, Switzerland, 2011; pp. 1–151.
44. JECFA. Furfuryl alcohol and related substances. In *Evaluation of Certain Food Additives/Seventy-Sixth Report of the Joint FAO/WHO Expert Committee on Food Additives (JECFA) (WHO Technical Report Series 974)*; World Health Organization: Geneva, Switzerland, 2012; pp. 67–69.
45. Carthew, P.; DiNovi, M.; Setzer, R.W. Application of the margin of exposure (MoE) approach to substances in

food that are genotoxic and carcinogenic: Example: Furan (CAS No. 110-00-9). *Food Chem. Toxicol.* **2010**, *48*, S69–S74. [CrossRef] [PubMed]
46. ISO. *ISO 6668:2008. Green Coffee—Preparation of Samples for Use in Sensory Analysis*; International Organization for Standardization: Geneva, Switzerland, 2008.
47. EFSA Scientific Committee. Guidance on selected default values to be used by the EFSA Scientific Committee, Scientific Panels and Units in the absence of actual measured data. *EFSA J.* **2012**, *10*, 2579. [CrossRef]
48. Bagdonaite, K.; Murkovic, M. Factors affecting the formation of acrylamide in coffee. *Czech J. Food Sci.* **2004**, *22*, 22–24. [CrossRef]
49. Kocadagli, T.; Göncuöglu, N.; Hamzalioglu, A.; Gökmen, V. In depth study of acrylamide formation in coffee during roasting: Role of sucrose decomposition and lipid oxidation. *Food Funct.* **2012**, *3*, 970–975. [CrossRef]
50. BfR. *5-HMF-Gehalte in Lebensmitteln Sind Nach Derzeitigem Wissenschaftlichen Kenntnisstand Gesundheitlich Unproblematisch. Stellungnahme No 030/2011 des BfR vom 15. Mai 2011 [According to the Current State of Scientific Knowledge 5-HMF Concentrations Occurring in Foods Do Not Give Rise to Safety Concerns. BfR Opinion Nr. 030/2011, 15 May 2011]*; Bundesinstitut für Risikobewertung: Berlin, Germany, 2011.
51. Codex Alimentarius. *Code of Practice for the Reduction of Acrylamide in Foods (CAC/RCP 67-2009)*; Joint FAO/WHO Food Standards Programme: Rome, Italy, 2009.
52. IARC Working Group on the Evaluation of Carcinogenic Risks to Humans. Coffee, mate, and very hot beverages. *IARC Monogr. Eval. Carcinog. Risks Hum.* **2018**, *116*, 1–501.
53. Loomis, D.; Guyton, K.Z.; Grosse, Y.; Lauby-Secretan, B.; El, G.F.; Bouvard, V.; Benbrahim-Tallaa, L.; Guha, N.; Mattock, H.; Straif, K.; et al. Carcinogenicity of drinking coffee, mate, and very hot beverages. *Lancet Oncol.* **2016**, *17*, 877–878. [CrossRef]
54. IARC Working Group on the Evaluation of Carcinogenic Risks to Humans. Coffee. *IARC Monogr. Eval. Carcinog. Risks Hum.* **1991**, *51*, 41–206. [PubMed]

7

Analysis of Volatile Constituents in *Platostoma palustre* (Blume) using Headspace Solid-Phase Microextraction and Simultaneous Distillation-Extraction

Tsai-Li Kung [1], Yi-Ju Chen [2], Louis Kuoping Chao [2], Chin-Sheng Wu [2], Li-Yun Lin [3,*] and Hsin-Chun Chen [2,*]

1. Taoyuan District Agricultural Research and Extension Station, Council of Agriculture, Executive Yuan, Taoyuan 327, Taiwan; tlkung@tydais.gov.tw
2. Department of Cosmeceutics, China Medical University, Taichung 404, Taiwan; yc9429@hotmail.com (Y.-J.C.); kuoping@mail.cmu.edu.tw (L.K.C.); cswu@mail.cmu.edu.tw (C.-S.W.)
3. Department of Food Science and Technology, Hungkuang University, Taichung 433, Taiwan

* Correspondence: lylin@sunrise.hk.edu.tw (L.-Y.L.); d91628004@ntu.edu.tw (H.-C.C.)

Abstract: Hsian-tsao (*Platostoma palustre* Blume) is a traditional Taiwanese food. It is admired by many consumers, especially in summer, because of its aroma and taste. This study reports the analysis of the volatile components present in eight varieties of Hsian-tsao using headspace solid-phase microextraction (HS-SPME) and simultaneous distillation-extraction (SDE) coupled with gas chromatography (GC) and gas chromatography-mass spectrometry (GC/MS). HS-SPME is a non-heating method, and the results show relatively true values of the samples during flavor isolation. However, it is a kind of headspace analysis that has the disadvantage of a lower detection ability to relatively higher molecular weight compounds; also, the data are not quantitative, but instead are used for comparison. The SDE method uses distillation 2 h for flavor isolation; therefore, it quantitatively identifies more volatile compounds in the samples while the samples withstand heating. Both methods were used in this study to investigate information about the samples. The results showed that Nongshi No. 1 had the highest total quantity of volatile components using HS-SPME, whereas SDE indicated that Taoyuan Mesona 1301 (TYM1301) had the highest volatile concentration. Using the two extraction methods, 120 volatile components were identified. Fifty-six volatile components were identified using HS-SPME, and the main volatile compounds were α-pinene, β-pinene, and limonene. A total of 108 volatile components were identified using SDE, and the main volatile compounds were α-bisabolol, β-caryophyllene, and caryophyllene oxide. Compared with SDE, HS-SPME sampling extracted a significantly higher amount of monoterpenes and had a poorer detection of less volatile compounds, such as sesquiterpenes, terpene alcohols, and terpene oxide.

Keywords: Hsian-tsao; *Platostoma palustre* (Blume); headspace solid-phase microextraction (SPME); volatile components; simultaneous distillation-extraction (SDE)

---

1. Introduction

Hsian-tsao (*Platostoma palustre* Blume, also known as *Mesona procumbens* Hemsl. [1]), also called Liangfen Cao or black cincau, belongs to the family Lamiaceae. It is an annual plant that is mainly distributed in tropical and subtropical regions, including Taiwan, southern China, Indonesia, Vietnam, and Burma [2]. Hsian-tsao tea, herbal jelly, and sweet soup with herbal jelly are popular during the summer, and heated herbal jelly is admired by many Taiwanese, especially in winter, because of its aroma and taste. In Indonesia, janggelan (*Mesona palustris* BL) has also been made into a herbal drink and a jelly-type dessert [3]. Hsian-tsao is also used as a remedy herb in folk medicine and is supposed to be effective in treating heat-shock, hypertension, diabetes, liver diseases, and muscle and joint pains [4,5].

Hsian-tsao contains polysaccharides (gum) with a unique aroma and texture. Most research has investigated the gum of Hsian-tsao [2,6–8]; however, there are only a few studies of Hsian-tsao aroma. Wei et al. [9] identified 59 volatile compounds in *Mesona* Benth extracted using solvent extraction. They also reported that the important constituents were caryophyllene oxide, α-caryophyllene, eugenol, benzene acetaldehyde, and 2,3-butanedione. Deng et al. [10] reported the chemical constituents of essential oil from *Mesona chinensis* Benth (also known as *P. palustris* Blume [1]) using GC/MS. The major constituents were *n*-hexadecanoic acid, linoleic acid, and linolenic acid. Lu et al. [11] analyzed the volatile oil from *Mesona chinensis* Benth using GC/MS. The results indicated the main components were chavibetol, *n*-hexadecanoic acid, and α-cadinol.

Simultaneous distillation-extraction (SDE) is a traditional extraction method that was introduced by Likens and Nickerson in 1946. SDE combines the advantages of liquid–liquid extraction and steam distillation methods. It is widely used for the extraction of essential oils and volatile compounds [12,13]. In the flavor field, this technique is recognized as a superior extraction method compared to other methods, such as solvent extraction or distillation. Moreover, Gu et al. [14] indicated that SDE has excellent reproducibility and high efficiency compared with traditional extraction methods.

Headspace solid-phase microextraction (HS-SPME) is a non-destructive and non-invasive method that avoids artifact formation and solvent impurity contamination [15]. HS-SPME is a fast, simple, and solventless technique [16–18]. HS-SPME can integrate sampling, extraction, concentration and sample introduction into a single uninterrupted process, resulting in high sample throughput [19].

This study aimed to identify the volatile constituents in different varieties of Hsian-tsao and the differences in extraction methods (HS-SPME and SDE). The differences in volatile compounds caused by heating are discussed. The results from this study provide a reference for the food, horticultural, and flavor industries.

2. Materials and Methods

2.1. Plant Materials

A total of eight varieties of Hsian-tsao from throughout Taiwan were used in this study (Table 1): Nongshi No. 1 from Tongluo Township in Miaoli County; Taoyuan No. 2 from Shoufeng Township in Hualien County; Chiayi strain from Shuishang Township in Chiayi County; Taoyuan No. 1, TYM1301, and TYM1302 from Guanxi Township in Hsinchu County; and TYM1303 and TYM1304 from Shuangxi District in New Taipei City. Eight varieties of Hsian-tsao were grown at the Sinpu Branch Station (Sinpu Township in Hsinchu County) of Taoyuan District Agricultural Research and Extension Station. The identities of the plants were confirmed by Tsai-Li Kung (Chief of the Sinpu Branch Station). After shade drying, dried samples were stored at room temperature for one year before the experiment was conducted.

Table 1. The study of collections of taxa currently assigned to Hsian-tsao.

Varieties	Origin	Growing Locality
Nongshi No.1	Tongluo Township, Miaoli County	Sinpu Township, Hsinchu County
Taoyuan No.1	Guanxi Township, Hsinchu County	Sinpu Township, Hsinchu County
Taoyuan No.2	Shoufeng Township, Hualien County	Sinpu Township, Hsinchu County
Chiayi strain	Shuishang Township, Chiayi County	Sinpu Township, Hsinchu County
TYM1301	Guanxi Township, Hsinchu County	Sinpu Township, Hsinchu County
TYM1302	Guanxi Township, Hsinchu County	Sinpu Township, Hsinchu County
TYM1303	Shuangxi District, New Taipei City	Sinpu Township, Hsinchu County
TYM1304	Shuangxi District, New Taipei City	Sinpu Township, Hsinchu County

2.2. Methods

2.2.1. Optimization of the HS-SPME Procedure

The method used was modified from those of Yeh et al. [20]:

1. Comparisons of SPME fiber coatings: five different coated SPME fibers, 85 μm polyacrylate (PA), 100 μm polydimethylsiloxane (PDMS), 65 μm polydimethylsiloxane/divinylbenzene (PDMS/DVB), 75 μm carboxen/polydimethylsiloxane (CAR/PDMS), and 50/30 μm divinylbenzene/carboxen/polydimethylsiloxane (DVB/CAR/PDMS), (Supelco, Inc., Bellefonte, PA, USA) were used for the aroma extraction. Samples (Nongshi No. 1) were placed into a homogenizer (WAR7012S 7-Speed Blender 1 Qt. 120 V) (Waring commercial, Torrington, CT, USA). After being homogenized for 30 s, 1 g of homogenized samples was put into a 7 mL vial (Hole Cap Polytetrafluoroethene/Silicone Septa) (Supelco, Inc., Bellefonte, PA, USA) and sealed. The SPME method was used to extract the aroma components. The extraction temperature was 25 ± 2 °C and the extraction time was 40 min. This experiment and all other experiments in this study were performed in triplicates.
2. Comparisons of the extraction times: The above-mentioned optimal extraction fiber was used in the comparison of the extraction times. The tested extraction times were 10 min, 20 min, 30 min, 40 min, or 50 min, and the extraction temperature was maintained at 25 ± 2 °C. Sample preparation steps were the same as above.

2.2.2. Analysis of the Volatile Compounds

1. Analysis of the volatile compounds using HS-SPME extraction: a 50/30 μm divinylbenzene/carboxen/polydimethylsiloxane (DVB/CAR/PDMS) fiber (Supelco, Inc., Bellefonte, PA, USA) was used for aroma extraction. The eight different Hsian-tsao varieties were used as samples. Each sample was homogenized as described above in Section 2.2.1 (1 g was placed in a 7 mL vial (hole cap PTFE/silicone septa)). The SPME fiber was exposed to each sample for 40 min at 25 ± 2 °C; then, each sample was injected into a gas chromatograph injection unit. The injector temperature was maintained at 250 °C and the fiber was held for 10 min. The peak area of a volatile compound or total volatile compounds from the integrator was used to calculate the relative contents.
2. Analysis of volatile compounds by SDE extraction: 100 g samples of Hsian-tsao were cut with scissors into pieces approximately 1–3 cm in size and were then homogenized for 2 min with 2 L of deionized water and were placed into a 5-L round-bottom flask. The flask was attached to a simultaneous distillation-extraction apparatus and 100 °C steam was used as the heat source and passed through the sample. A 50 mL volume of solvent was prepared by mixing *n*-pentane/diethyl ether (1:1, *v*/*v*) into a pear-shaped flask, then placing it in a 40–45 °C water bath. This distillation circulation continued for 2 h, and the collected solvent extract was added to 200 μL of an internal standard solution of cyclohexyl acetate, and an internal standard was used to obtain the weight concentration of volatile compound in the sample; also, anhydrous sodium sulfate was used to

remove the water. Lastly, the distillation column (40 °C, 1 h, 100 cm glass column) was used to volatilize the solvent and the condensed volatile component extracts were collected.

3. GC analysis of the volatile compounds was conducted using a 7890A GC (Agilent Technologies, Palo Alto, CA, USA) equipped with a DB-1 (60 m × 0 .25 mm i.d. × 0.25 μm film thickness, Agilent Technologies) capillary column and a flame ionization detector. The injector and detector temperatures were maintained at 250 °C and 300 °C, respectively. The oven temperature was held at 40 °C for 1 min and then raised to 150 °C at 5 °C/min and held for 1 min, and then increased from 150 to 200 °C at 10 °C/min and held for 11 min. The carrier gas (nitrogen) flow rate was 1 mL/min. The Kovats indices were calculated for the separated components relative to a C_5–C_{25} *n*-alkanes mixture [21]. The purpose gas chromatography-flame ionization detector (GC-FID) was used both for retention indices (RI) comparison and quantitation of peak areas.
4. GC-MS analysis of volatile compounds were identified using an Agilent 7890B GC equipped with DB-1 (60 m × 0.25 mm i.d. × 0.25 μm film thickness) fused silica capillary column coupled to an Agilent model 5977 N MSD mass spectrometer (MS). The GC conditions in the GC-MS analysis were the same as in the GC analysis. The carrier gas (helium) flow rate was 1 mL/min. The electron energy was 70 eV at 230 °C. The constituents were identified by matching their spectra with those recorded in an MS library (Wiley 7N, John Wiley & Sons, Inc. New Jersey, NJ, USA). In addition, the constituents were confirmed by comparing the Kovats indices or GC retention time data with those of authentic standards or data published in the literature. The GC and GC-MS methods used were modified from those of Yeh et al. [20].
5. Statistical Analysis: Each sample was extracted in triplicate and the concentration of volatile compounds was determined as the mean value of three repetitions. The data were subjected to a monofactorial variance analysis with Duncan's multiple range method with a level of significance of $p < 0.05$ (SPSS Base 12.0, SPSS Inc., Chicago, IL, USA).

3. Results

3.1. Comparisons of the Optimization Conditions of HS-SPME

3.1.1. SPME Fiber Selection

The performance of five commercially available SPME fibres: 50/30-μm DVB/CAR/PDMS, 65-μm PDMS/DVB, 75-μm CAR/PDMS, 100-μm PDMS, and 85-μm PA were used to extract the volatile components of Nongshi No. 1. The 50/30-μm DVB/CAR/PDMS fiber extracted more total volatile components than the other fibers (Figure 1).

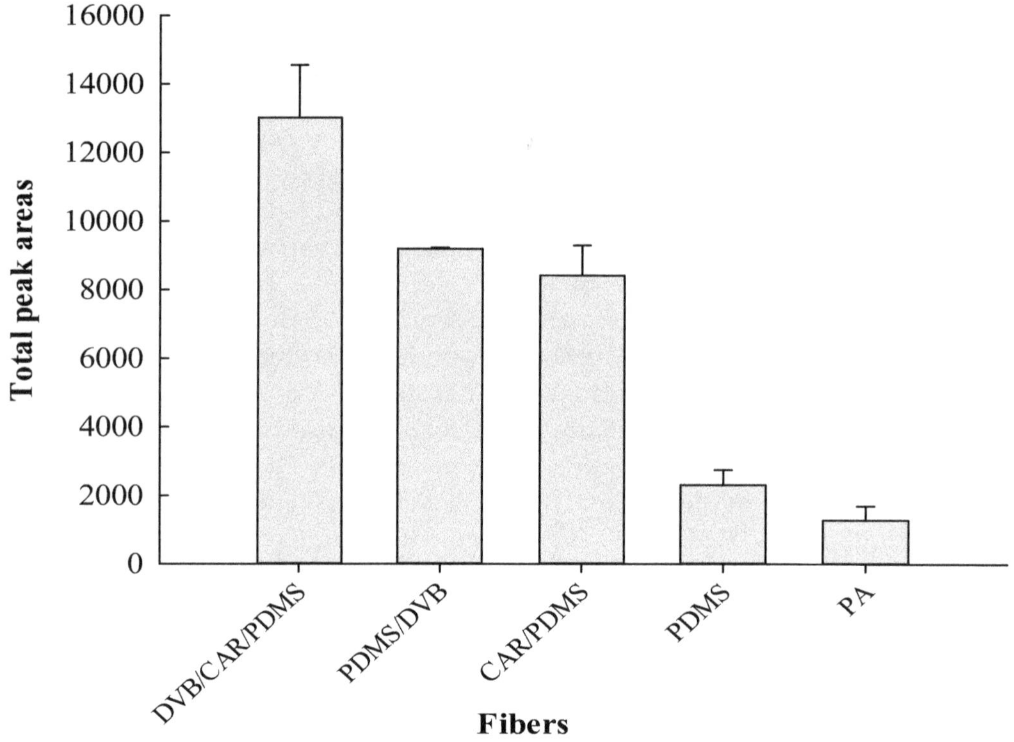

Figure 1. Comparisons of the total peak areas of total volatile compounds detected in the headspace of Nongshi No. 1 using different headspace solid-phase microextraction (HS-SPME) fibers.

Ducki et al. [22] evaluated four different types of SPME fibers (100-μm PDMS, 65-μm PDMS/DVB, 75-μm CAR/PDMS, and 50/30-μm DVB/CAR/PDMS) for the headspace analysis of volatile compounds in cocoa products. The SPME fiber coated with 50/30-μm DVB/CAR/PDMS afforded the highest extraction efficiency. Silva et al. [23] compared the performance of six fibers (PDMS, PDMS/DVB, CW/DVB, PA, CAR/PDMS, and DVB/CAR/PDMS) and found that DVB/CAR/PDMS was the most effective SPME fiber for isolating the volatile metabolites from *Mentha* × *piperita* L. fresh leaves based on the total peak areas, reproducibility, and number of extracted metabolites. Yeh et al. [20] reported the volatile components in *Phalaenopsis* Nobby's Pacific Sunset, and the optimal extraction conditions were obtained using a DVB/CAR/PDMS fiber.

The 50/30-μm DVB/CAR/PDMS was revealed to be the most suitable and was subsequently used in all further experiments.

3.1.2. HS-SPME Extraction Time

The optimal SPME fiber (50/30-μm DVB/CAR/PDMS) was used to extract Nongshi No. 1 at 25 ± 2 °C, and the extraction times from 10 to 50 min were investigated. The total peak area increased from 10–40 min and reached the peak at 40 min (Figure 2). Silva and Câmara [23] promoted the higher extraction efficiency, corresponding to the higher GC peak areas and the number of identified metabolites. This higher extraction efficiency was achieved using: DVB/CAR/PDMS coating fiber, and 40 °C and 60 min as the extraction temperature and extraction time, respectively. Zhang et al. [24] also obtained optimum extraction conditions, which were using 50/30-μm DVB/CAR/PDMS fiber for 40 min at 90 °C. According to the obtained results, 40 min was selected as the optimal extraction time.

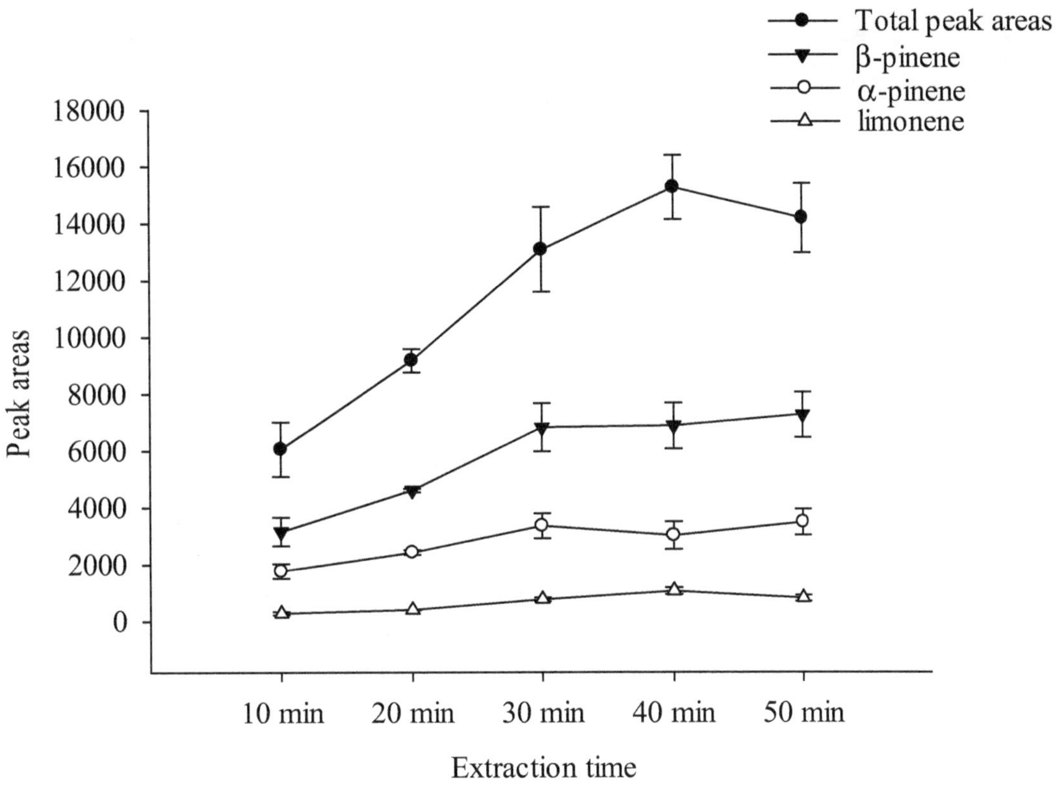

Figure 2. Comparisons of the peak areas of total volatile compounds and main components detected in the headspace of Nongshi No. 1 for different SPME extraction times at 25 °C using a DVB/CAR/PDMS fiber.

3.2. Analyses of the Volatiles of Eight Varieties of Hsian-Tsao Using HS-SPME

As shown in Figure 3, the total peak areas of volatile components was the highest in Nongshi No. 1 and lowest in TYM1304. Volatile compounds in eight varieties of Hsian-tsao were analyzed using headspace solid-phase microextraction (HS-SPME), which was coupled with GC and GC/MS. Table 2 shows a total of 56 compounds that were identified. Monoterpene compounds were the most abundant compounds in the Hsian-tsao analyzed using HS-SPME/GC. The main volatile components of Nongshi No. 1, Chiayi strain, TYM1302, and TYM1303 were β-pinene (43–50%), α-pinene (10–24%), and limonene (4–9%). β-Pinene (36–42%), α-pinene (15–17%), and β-caryophyllene (11%) were the main components from Taoyuan No. 2 and TYM1301. The main components of Taoyuan No. 1 were β-pinene (23%), limonene (21%), and α-pinene (11%). Limonene (32%), β-caryophyllene (13%), and sabinene (7%) were the main components from TYM1304. TYM1304 contained the highest content of limonene (32%), followed by Taoyuan No. 1 (21%). Limonene is a citrus note, having a pungent green and lemon-like aroma [25,26]. The peak areas of α-pinene and β-pinene were the highest in TYM1302 (24% and 50%), whereas TYM1304 was lower than the other varieties. α-Pinene was described as having a fruity, piney, and turpentine-like aroma [27,28], and β-pinene was described as having a dry-woody, pine-like, and citrus aroma [29,30]. β-Caryophyllene was described as having a dry-woody, pine-like, and spicy aroma, and TYM1304 contained the highest content (13%), followed by TYM1301, and Taoyuan No. 2 (11%).

Table 2. Comparisons of volatile compounds from eight varieties of Hsian-tsao extracted using HS-SPME.

Compound [a]	RI [b]	Relative Content (%) [c]							
		Nongshi No. 1	Taoyuan No. 1	Taoyuan No. 2	Chiayi Strain	TYM1301	TYM1302	TYM1303	TYM1304
Aliphatic alcohol									
1-Octen-3-ol [f]	962	0.29 ± 0.11	0.69 ± 0.24	1.16 ± 0.19	0.17 ± 0.03	2.25 ± 0.40	1.39 ± 0.28	1.61 ± 0.22	3.01 ± 0.82
Aliphatic aldehydes									
Hexanal [f]	777	3.63 ± 0.52	1.85 ± 0.68	0.29 ± 0.10	0.39 ± 0.15	1.19 ± 0.26	1.15 ± 0.22	1.81 ± 0.24	1.74 ± 0.19
(*E*)-2-heptenal [f]	931	tr [d]	1.06 ± 0.20	tr	- [e]	tr	tr	tr	tr
Decanal [f]	1181	-	-	-	tr	-	-	-	-
Aliphatic ketones									
1-Octen-3-one [f]	956	tr	0.14 ± 0.02	0.12 ± 0.02	tr	0.29 ± 0.03	tr	0.24 ± 0.02	0.47 ± 0.07
3-Octanone [f]	965	-	-	-	tr	-	-	tr	tr
Aliphatic ester									
Linalyl isobutyrate	1387	0.32 ± 0.05	-	-	-	-	-	-	-
Aromatic alcohol									
Eugenol [g]	1328	-	-	-	0.08 ± 0.02	-	-	-	-
Aromatic aldehyde									
Benzaldehyde [f]	933	tr	-	-	tr	-	-	-	-
Aromatic hydrocarbon									
p-Cymene	1004	-	0.68 ± 0.05	-	tr	tr	tr	tr	tr
Terpene alcohol									
α-Bisabolol	1680	0.11 ± 0.05	0.22 ± 0.09	0.09 ± 0.03	0.13 ± 0.03	0.36 ± 0.04	0.08 ± 0.01	0.08 ± 0.01	-
Monoterpenes									
α-Thujene	927	1.02 ± 0.11	0.12 ± 0.02	1.28 ± 0.17	0.86 ± 0.21	1.16 ± 0.11	1.10 ± 0.05	1.10 ± 0.04	3.19 ± 0.24
α-Pinene [f]	937	19.73 ± 1.72	11.14 ± 1.10	16.65 ± 1.26	22.67 ± 0.45	14.76 ± 0.68	24.20 ± 0.40	20.20 ± 0.69	4.72 ± 0.19
Camphene	949	0.79 ± 0.04	0.46 ± 0.06	0.63 ± 0.02	1.22 ± 0.09	0.70 ± 0.02	0.93 ± 0.02	0.82 ± 0.05	-
sabinene	967	2.75 ± 0.25	3.60 ± 0.15	3.56 ± 0.46	1.46 ± 0.41	2.87 ± 0.40	2.86 ± 0.10	3.07 ± 0.21	7.37 ± 0.62
β-Pinene [f]	972	44.99 ± 2.09	23.19 ± 2.17	42.28 ± 1.31	46.60 ± 1.89	36.18 ± 0.83	49.51 ± 1.45	43.24 ± 0.19	2.15 ± 0.36
β-Myrcene	983	0.97 ± 0.02	2.26 ± 0.16	0.41 ± 0.02	0.28 ± 0.10	0.64 ± 0.09	0.37 ± 0.01	0.82 ± 0.07	2.40 ± 0.09
δ-3-Carene	998	-	0.85 ± 0.04	-	0.43 ± 0.04	-	-	0.46 ± 0.06	1.92 ± 0.09
α-Terpinene	1007	0.52 ± 0.05	0.05 ± 0.00	0.61 ± 0.07	0.65 ± 0.16	0.63 ± 0.06	0.56 ± 0.03	0.60 ± 0.02	1.64 ± 0.09
Limonene [f]	1015	7.06 ± 1.17	20.98 ± 1.01	6.45 ± 0.31	4.66 ± 1.96	7.47 ± 1.68	4.29 ± 0.08	9.05 ± 1.05	32.08 ± 2.81
γ-Terpinene	1045	-	0.55 ± 0.09	0.60 ± 0.07	0.38 ± 0.12	-	0.52 ± 0.04	-	1.63 ± 0.09
α-Terpinolene	1081	-	0.39 ± 0.05	-	0.26 ± 0.04	0.46 ± 0.01	0.32 ± 0.01	-	1.14 ± 0.07
Sesquiterpenes									
α-Cubebene	1353	-	0.16 ± 0.01	0.16 ± 0.05	0.07 ± 0.02	-	-	-	-
α-Ylangene	1378	tr	tr	tr	0.41 ± 0.23	tr	tr	tr	tr
α-Copaene	1382	0.20 ± 0.03	1.45 ± 0.17	0.24 ± 0.05	0.25 ± 0.14	0.62 ± 0.09	0.32 ± 0.02	0.09 ± 0.01	0.24 ± 0.01
β-Elemene	1392	1.13 ± 0.19	3.26 ± 0.12	3.05 ± 0.35	1.27 ± 0.17	2.60 ± 0.30	0.58 ± 0.06	1.04 ± 0.14	4.38 ± 0.28
β-Bourbonene	1393	-	-	tr	0.08 ± 0.00	-	tr	-	-
α-Cedrene	1401	-	0.16 ± 0.01	-	-	-	-	-	-
cis-α-Bergamotene	1414	-	0.49 ± 0.07	-	-	-	-	-	-
α-Gurgujene	1418	-	-	-	0.07 ± 0.00	-	-	-	-
β-Caryophyllene [f,g,h]	1427	2.68 ± 0.40	8.49 ± 1.16	10.71 ± 1.56	2.74 ± 0.74	10.81 ± 1.03	2.76 ± 0.52	2.61 ± 0.25	12.53 ± 1.51
Aromadendrene	1433	-	-	-	tr	-	-	-	-
α-Bergamotene [h]	1436	0.77 ± 0.10	1.31 ± 0.25	-	0.55 ± 0.08	0.90 ± 0.01	-	-	0.71 ± 0.04
cis-Thujopsene	1437	-	0.15 ± 0.02	-	0.09 ± 0.00	-	-	-	-
β-Gurjunene	1442	-	-	0.37 ± 0.06	-	-	-	-	-
cis-β-Farnesene [h]	1445	-	0.69 ± 0.05	-	0.39 ± 0.11	0.59 ± 0.08	-	-	-
trans-β-Farnesene	1452	-	0.20 ± 0.02	-	-	0.18 ± 0.01	-	0.05 ± 0.01	-
α-Caryophyllene [f,h]	1461	0.87 ± 0.10	1.33 ± 0.18	1.51 ± 0.20	0.49 ± 0.10	2.16 ± 0.31	0.50 ± 0.08	0.59 ± 0.25	1.84 ± 0.33

Table 2. *Cont.*

Compound [a]	RI [b]	Relative Content (%) [c]							
		Nongshi No. 1	Taoyuan No. 1	Taoyuan No. 2	Chiayi Strain	TYM1301	TYM1302	TYM1303	TYM1304
α-Muurolene	1479	-	0.36 ± 0.13	-	0.16 ± 0.02	-	-	0.10 ± 0.02	-
γ-Muurolene	1479	-	-	-	-	-	-	-	0.17 ± 0.03
Valencene	1486	0.12 ± 0.01	-	-	-	-	-	-	-
Germacrene D	1487	-	-	0.18 ± 0.00	0.11 ± 0.02	-	-	-	-
β-Selinene	1494	0.35 ± 0.06	0.65 ± 0.17	0.56 ± 0.23	2.72 ± 0.33	1.01 ± 0.11	-	0.38 ± 0.17	1.43 ± 0.12
α-Selinene	1501	-	-	-	1.50 ± 0.17	-	-	-	1.17 ± 0.09
β-Bisabolene	1505	0.42 ± 0.10	0.89 ± 0.09	-	tr	1.20 ± 0.16	-	0.34 ± 0.02	-
α-Chamigrene	1514	-	-	-	0.07 ± 0.01	-	-	-	-
α-Amorphene	1518	-	-	-	0.07 ± 0.02	-	-	-	-
δ-Cadinene	1523	-	-	0.15 ± 0.04	0.08 ± 0.02	-	-	-	-
trans-γ-Bisabolene	1526	-	0.16 ± 0.00	0.06 ± 0.00	-	-	-	-	-
trans-α-Bisabolene	1537	0.18 ± 0.04	0.31 ± 0.05	0.10 ± 0.04	0.43 ± 0.07	-	-	-	-
Terpene oxide									
Caryophyllene oxide [f,g]	1585	-	0.04 ± 0.01	-	-	-	-	-	-
Hydrocarbons									
2-Methyl-octane	869	-	-	0.53 ± 0.05	-	-	-	-	-
Undecane	1098	0.51 ± 0.07	0.44 ± 0.11	0.27 ± 0.02	0.12 ± 0.01	0.24 ± 0.02	-	0.25 ± 0.00	-
Dodecane	1197	0.35 ± 0.07	0.19 ± 0.04	-	tr	0.23 ± 0.00	-	-	0.06 ± 0.01
Tridecane	1294	-	0.12 ± 0.06	0.08 ± 0.01	tr	0.09 ± 0.01	-	0.08 ± 0.01	-
Furan									
2-Pentylfuran	978	tr	tr	tr	tr	tr	tr	tr	tr

[a] Identification of components based on the GC/MS library (Wiley 7N). [b] Retention indices, using paraffin (C_5–C_{25}) as references. [c] Relative percentages from GC-FID, values are means ± standard deviation (SD) of triplicates. [d] Trace. [e] Undetectable. [f] Published in the literature (Wei et al. [9]). [g] Published in the literature (Lu et al. [11]). [h] Published in the literature (Deng et al. [10]).

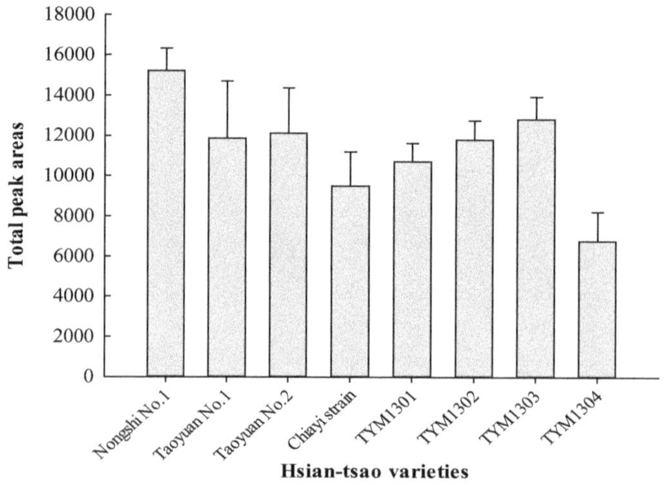

Figure 3. Comparisons of the total peak areas of total volatile compounds of eight varieties of Hsian-tsao (1 g) extracted at 25 °C for 40 min using a DVB/CAR/PDMS fiber. The peak area of a volatile compound or total volatile compounds from the integrator was used to calculate the relative contents using gas chromatography-flame ionization detector (GC-FID). The data corresponds to the mean ± standard deviation (SD) of triplicates.

The eight varieties of Hsian-tsao shared 15 volatile components; the differences in percentage were: 1-octen-3-ol (trace–3%), hexanal (trace–4%), 1-octen-3-one (trace–1%), α-thujene (trace–3%), α-pinene (5–24%), sabinene (2–7%), β-pinene (2–50%), β-myrcene (trace–2%), α-terpinene (trace–2%), limonene (4–32%), α-ylangene (trace), α-copaene (trace–2%), β-elemene (1–4%), β-caryophyllene (3–13%), and α-caryophyllene (1–2%). Among them, hexanal was described as having a green and cut-grass aroma [31], and was responsible for green, apple, and green fruit perceptions [32]. Nongshi No. 1 contained the highest content of hexanal (4%).

3.3. *Analysis of the Volatiles of Eight Varieties (Clones) of Hsian-Tsao Using SDE*

As shown in Table 3, the volatile components content peaked in TYM1301 and was the lowest in TYM1303. Table 4 shows the results of the SDE analysis: 108 components were identified, including 11 aliphatic alcohols, 14 aliphatic aldehydes, 9 aliphatic ketones, 1 aliphatic ester, 3 aromatic alcohols, 2 aromatic aldehydes, 1 aromatic ketones, 2 aromatic esters, 4 aromatic hydrocarbons, 8 terpene alcohols, 2 terpene aldehydes, 2 terpene ketones, 10 monoterpenes, 22 sesquiterpenes, 1 terpene oxide, 7 hydrocarbons, 3 straight-chain acids, 3 furans, 2 methoxy-phenolic compounds, and 1 nitrogen-containing compound. Sesquiterpene compounds, terpene alcohols, and terpene oxide were the main compounds in the Hsian-tsao analyzed using SDE. The major components of Hsain-tsao (Nongshi No. 1, Taoyuan No. 1, and TYM1301) were α-bisabolol (59–144 mg/kg), caryophyllene oxide (9–28 mg/kg), and β-caryophyllene (21–54 mg/kg). β-Caryophyllene (21–56 mg/kg) and caryophyllene oxide (32–50 mg/kg) were the main compounds of Taoyuan No. 2 and TYM1302. α-Bisabolol (116 mg/kg), β-bisabolene (24 mg/kg), β-selinene (23 mg/kg), and β-caryophyllene (19 mg/kg) were the main components of the Chiayi strain. α-Bisabolol (144 mg/kg), α-bisabolol (43 mg/kg), and caryophyllene oxide (15 mg/kg) were the main components of TYM1303. β-Caryophyllene (53 mg/kg), β-elemene (35 mg/kg), α-selinene (15 mg/kg), and α-caryophyllene (13 mg/kg) were the main compounds in TYM1304. Nongshi No. 1, Taoyuan No. 1, Chiayi strain, TYM1301, and TYM1303 contained a higher content of α-bisabolol; Taoyuan No. 2 and TYM1304 contained a higher content of β-caryophyllene; and TYM1302 contained a higher content of caryophyllene oxide. Wei et al. [9] analyzed the volatile components of *Mesona* Benth, where they reported that the main components were caryophyllene oxide and caryophyllene, similar with these experimental results. β-Caryophyllene had a woody aroma, and Taoyuan No. 2 had the highest concentration, followed by TYM1301 and TYM1304 (53–56 mg/kg).

Table 3. Comparisons of the total volatile compounds in eight varieties of Hsian-tsao extracted using SDE.

Varieties	Concentration (mg/kg) [a]
Nongshi No. 1	230.71 ± 89.56 [b]
Taoyuan No. 1	206.90 ± 50.00 [bc]
Taoyuan No. 2	194.20 ± 50.37 [bc]
Chiayi Strain	264.61 ± 22.87 [ab]
TYM1301	329.82 ± 82.32 [a]
TYM1302	205.43 ± 69.26 [bc]
TYM1303	124.67 ± 79.29 [c]
TYM1304	207.50 ± 53.42 [bc]

[a] The 100 g samples of Hsian-tsao were extracted using SDE for 2 h, quantification using cyclohexyl acetate as an internal standard. The data correspond to the mean ± SD of triplicates. Values having different superscripts were significantly ($p < 0.05$) different.

Table 4. Comparisons of volatile compounds from eight varieties of Hsian-tsao extracted using SDE.

Compound [a]	RI [b]	Concentration (mg/kg) [c]							
		Nongshi No. 1	Taoyuan No. 1	Taoyuan No. 2	Chiayi Strain	TYM1301	TYM1302	TYM1303	TYM1304
Aliphatic alcohols									
Isobutanol	643	tr [d]	- [e]	-	-	-	-	-	-
1-Penten-3-ol	693	tr	tr	tr	tr	tr	tr	tr	tr
1-Pentanol [f]	739	-	tr	-	-	-	-	-	-
Isoamyl alcohol [f]	767	0.77 ± 0.89	0.46 ± 0.29	0.16 ± 0.09	0.47 ± 0.29	1.28 ± 0.12	3.10 ± 0.69	1.31 ± 0.82	0.87 ± 0.20
3-Hexanol	791	-	-	0.04 ± 0.01	-	-	-	-	-
(*Z*)-hex-3-en-1-ol	845	1.39 ± 0.53	0.59 ± 0.09	0.77 ± 0.26	0.61 ± 0.03	1.12 ± 0.01	1.29 ± 0.13	0.53 ± 0.22	0.77 ± 0.09
(*E*)-2-hexen-1-ol	855	0.02 ± 0.00	tr	0.02 ± 0.03	0.01 ± 0.02	-	0.20 ± 0.11	0.05 ± 0.04	-
Hexanol [f]	856	0.34 ± 0.26	0.22 ± 0.05	0.36 ± 0.08	0.41 ± 0.54	0.19 ± 0.03	0.45 ± 0.52	0.23 ± 0.03	0.17 ± 0.01
1-Octen-3-ol [f]	962	0.44 ± 0.25	0.32 ± 0.00	0.31 ± 0.07	0.50 ± 0.12	1.14 ± 0.15	0.69 ± 0.02	0.23 ± 0.15	1.05 ± 0.06
3-Octanol [f]	979	0.07 ± 0.03	tr	0.06 ± 0.03	0.07 ± 0.01	-	0.12 ± 0.10	0.14 ± 0.06	0.16 ± 0.02
Aliphatic aldehydes									
Pentanal [f]	700	tr	tr	tr	tr	tr	tr	tr	tr
Hexanal [f]	777	0.69 ± 0.59	tr	0.03 ± 0.03	0.14 ± 0.04	0.22 ± 0.14	5.07 ± 4.73	0.37 ± 0.40	0.07 ± 0.03
(*E*)-2-hexenal [f]	832	2.63 ± 1.36	0.48 ± 0.08	0.19 ± 0.06	0.22 ± 0.02	0.86 ± 0.01	0.69 ± 0.11	0.54 ± 0.27	0.96 ± 0.08
(*Z*)-4-heptenal	877	0.02 ± 0.01	tr	0.04 ± 0.01	0.04 ± 0.00	0.01 ± 0.01	0.16 ± 0.06	tr	0.11 ± 0.08
(*E,E*)-2,4-hexadienal	879	0.05 ± 0.02	tr	tr	tr	0.05 ± 0.02	tr	tr	0.04 ± 0.01
(*E*)-2-heptenal	931	tr	tr	tr	0.03 ± 0.00	0.01 ± 0.01	0.05 ± 0.00	tr	tr
(*E,E*)-2,4-heptadienal [f]	979	tr	-	-	tr	0.14 ± 0.02	tr	-	-
(*E*)-2-octenal [f]	1024	0.39 ± 0.09	0.45 ± 0.08	0.43 ± 0.08	0.56 ± 0.07	0.37 ± 0.04	0.72 ± 0.04	0.42 ± 0.21	0.50 ± 0.09
Nonanal [f]	1078	3.99 ± 1.18	0.11 ± 0.00	0.18 ± 0.00	0.16 ± 0.02	1.07 ± 0.17	0.93 ± 0.03	tr	2.20 ± 0.27
(*E*)-2-nonenal [f]	1130	0.34 ± 0.22	0.23 ± 0.01	0.06 ± 0.11	0.65 ± 0.08	0.51 ± 0.03	0.82 ± 0.05	0.45 ± 0.17	0.69 ± 0.12
Safranal	1175	0.80 ± 0.22	tr	0.73 ± 0.20	-	0.54 ± 0.03	-	-	0.76 ± 0.17
Decanal [f]	1181	-	0.34 ± 0.30	0.32 ± 0.04	0.21 ± 0.03	0.12 ± 0.03	0.34 ± 0.01	0.23 ± 0.07	0.16 ± 0.02
(*E*)-2-decenal	1241	0.40 ± 0.21	0.19 ± 0.03	0.32 ± 0.07	-	0.12 ± 0.05	0.64 ± 0.01	0.33 ± 0.05	0.17 ± 0.08
(*E,E*)-2,4-decadienal [f]	1285	0.20 ± 0.16	0.56 ± 0.58	1.17 ± 0.49	1.07 ± 0.14	0.42 ± 0.08	1.48 ± 0.04	0.85 ± 0.52	0.89 ± 0.25

Table 4. *Cont.*

Compound [a]	RI [b]	Concentration (mg/kg) [c]							
		Nongshi No. 1	Taoyuan No. 1	Taoyuan No. 2	Chiayi Strain	TYM1301	TYM1302	TYM1303	TYM1304
Aliphatic ketones									
3-Hexen-2-one	820	-	-	0.07 ± 0.04	-	-	-	-	-
2-Heptanone [f]	872	0.18 ± 0.04	tr	0.10 ± 0.03	0.26 ± 0.02	0.17 ± 0.08	0.36 ± 0.05	0.22 ± 0.13	0.17 ± 0.09
1-Octen-3-one [f]	956	4.79 ± 1.63	1.97 ± 0.03	1.89 ± 0.76	2.19 ± 0.31	6.47 ± 0.43	3.49 ± 0.14	2.66 ± 0.93	8.29 ± 0.94
3-Octanone	965	2.43 ± 2.56	tr	0.25 ± 0.05	2.97 ± 1.25	tr	5.51 ± 1.15	0.45 ± 0.31	tr
3,5-Octadien-2-one	1066	0.16 ± 0.02	0.07 ± 0.00	0.10 ± 0.00	0.09 ± 0.01	0.08 ± 0.03	0.30 ± 0.12	0.22 ± 0.13	0.11 ± 0.04
2-Nonanone	1070	0.28 ± 0.08	-	-	0.37 ± 0.05	-	0.51 ± 0.05	0.29 ± 0.13	-
6-Methyl-3,5-heptadien-2-one [f]	1074	1.26 ± 0.17	1.08 ± 0.51	1.04 ± 0.39	2.07 ± 0.30	0.73 ± 0.02	1.30 ± 0.07	1.63 ± 0.43	0.93 ± 0.14
3-Nonen-2-one	1110	0.18 ± 0.06	0.13 ± 0.01	0.15 ± 0.02	-	-	0.26 ± 0.04	0.17 ± 0.07	-
β-Damascenone [f,g]	1365	1.17 ± 0.45	0.70 ± 0.02	1.50 ± 0.33	0.97 ± 0.09	0.63 ± 0.55	1.87 ± 0.02	1.03 ± 0.37	0.51 ± 0.16
Aliphatic ester									
Ethyl acetate	631	-	-	tr	tr	tr	tr	tr	tr
Aromatic alcohols									
Benzyl alcohol [f]	999	0.18 ± 0.06	tr	0.10 ± 0.02	0.18 ± 0.01	0.12 ± 0.02	0.57 ± 0.05	0.35 ± 0.23	0.11 ± 0.06
1-Octanol	1048	0.18 ± 0.05	tr	0.10 ± 0.01	0.13 ± 0.01	-	0.36 ± 0.12	-	-
Eugenol [g]	1328	2.16 ± 0.96	1.04 ± 0.28	0.62 ± 0.44	2.17 ± 0.25	1.27 ± 0.29	0.73 ± 0.02	0.35 ± 0.18	1.00 ± 0.35
Methyleugenol	1369	0.23 ± 0.19	0.15 ± 0.06	2.43 ± 0.49	0.44 ± 0.07	-	-	-	-
Aromatic aldehydes									
Benzaldehyde [f]	933	tr	tr	tr	tr	tr	tr	tr	tr
Benzeneacetaldehyde	1002	tr	0.11 ± 0.01	tr	tr	0.20 ± 0.01	0.80 ± 0.33	tr	0.28 ± 0.04
Aromatic ketone									
Acetophenone	1033	0.46 ± 0.07	0.29 ± 0.03	0.38 ± 0.03	0.31 ± 0.04	0.03 ± 0.02	0.62 ± 0.05	0.26 ± 0.12	0.05 ± 0.02
Aromatic esters									
Hexyl acetate	991	tr	-	-	-	-	-	-	-
Methyl salicylate	1170	1.29 ± 0.49	0.82 ± 0.11	1.78 ± 0.25	-	0.56 ± 0.08	-	1.70 ± 0.48	0.42 ± 0.09
Aromatic hydrocarbons									
p-Cymene	1004	0.30 ± 0.08	0.29 ± 0.01	0.27 ± 0.03	0.31 ± 0.07	-	-	0.34 ± 0.25	0.24 ± 0.05
Aromatic hydrocarbons									
β-Methylnaphthalene	1275	0.30 ± 0.20	-	-	0.39 ± 0.05	-	-	-	0.25 ± 0.14
α-Ionene [g]	1397	-	0.45 ± 0.11	-	-	-	-	-	-
Cuparene	1514	-	-	-	-	-	-	2.14 ± 2.60	10.00 ± 5.81
Terpene alcohols									
Linalool [f]	1082	0.20 ± 0.07	-	0.21 ± 0.00	0.18 ± 0.06	0.15 ± 0.03	0.34 ± 0.04	0.21 ± 0.11	0.13 ± 0.05
Borneol [g]	1149	0.63 ± 0.24	0.26 ± 0.12	0.35 ± 0.09	-	-	-	0.35 ± 0.29	-
Menthol	1157	-	-	-	-	-	-	0.19 ± 0.07	-
4-Terpineol [f]	1163	0.48 ± 0.23	0.17 ± 0.01	0.26 ± 0.03	0.58 ± 0.09	0.08 ± 0.03	0.83 ± 0.17	0.28 ± 0.17	0.13 ± 0.04
α-Terpineol [f]	1173	0.27 ± 0.17	0.27 ± 0.16	tr	1.58 ± 0.28	0.17 ± 0.04	1.45 ± 0.22	tr	0.17 ± 0.05
Myrtenol	1184	0.18 ± 0.13	-	0.11 ± 0.09	0.19 ± 0.03	tr	1.14 ± 0.04	0.64 ± 0.27	-
Gossonorol	1617	0.99 ± 0.77	2.03 ± 0.11	-	-	1.51 ± 0.46	-	1.81 ± 0.83	1.28 ± 0.51
α-Bisabolol	1680	58.69 ± 18.60	72.25 ± 12.19	4.23 ± 0.16	116.29 ± 3.79	144.20 ± 32.40	2.78 ± 0.10	42.90 ± 15.95	0.67 ± 0.29
Terpene aldehydes									
β-Cyclocitral	1198	0.83 ± 0.42	-	1.41 ± 0.13	0.84 ± 0.15	0.91 ± 0.07	2.60 ± 0.06	0.65 ± 0.32	1.11 ± 0.22
β-Citral [f]	1217	-	-	-	-	-	-	-	0.07 ± 0.06
Terpene ketones									
Camphor	1128	-	0.43 ± 0.07	0.65 ± 0.26	-	-	-	-	-
β-Ionene [g]	1469	4.84 ± 3.53	-	3.10 ± 0.55	3.29 ± 0.28	2.46 ± 0.46	8.28 ± 0.48	1.30 ± 1.06	4.09 ± 1.17

Table 4. *Cont.*

Compound [a]	RI [b]	Concentration (mg/kg) [c]							
		Nongshi No. 1	Taoyuan No. 1	Taoyuan No. 2	Chiayi Strain	TYM1301	TYM1302	TYM1303	TYM1304
Monoterpenes									
α-Thujene	927	0.63 ± 0.83	0.47 ± 0.24	0.34 ± 0.32	0.80 ± 0.65	4.45 ± 0.54	1.28 ± 0.52	1.00 ± 1.46	0.62 ± 0.09
α-Pinene [f]	937	0.02 ± 0.02	tr	tr	0.03 ± 0.00	0.05 ± 0.05	0.04 ± 0.00	tr	0.03 ± 0.01
Camphene	949	0.87 ± 0.32	0.96 ± 0.02	0.67 ± 0.21	1.18 ± 0.04	4.16 ± 0.30	2.48 ± 0.17	1.18 ± 0.74	7.17 ± 0.46
Sabinene	967	-	1.56 ± 0.52	1.83 ± 1.43	-	12.57 ± 1.41	-	3.38 ± 4.04	1.41 ± 0.23
β-Pinene [f]	972	2.39 ± 0.66	0.73 ± 0.26	1.12 ± 0.26	0.97 ± 0.20	0.59 ± 0.05	1.66 ± 0.10	0.95 ± 0.53	-
β-Myrcene	983	-	tr	-	-	-	-	-	0.06 ± 0.02
α-Phellandrene	998	-	0.13 ± 0.00	0.21 ± 0.01	-	-	0.22 ± 0.20	-	-
Limonene [f]	1015	0.70 ± 0.23	1.43 ± 0.13	0.63 ± 0.12	0.90 ± 0.12	1.41 ± 1.14	0.57 ± 0.03	0.59 ± 0.35	2.43 ± 2.05
γ-Terpinene	1045	0.28 ± 0.07	0.23 ± 0.01	0.27 ± 0.11	0.29 ± 0.03	0.23 ± 0.03	0.37 ± 0.00	-	0.16 ± 0.02
α-Terpinolene	1081	-	0.11 ± 0.01	-	-	-	-	-	-
Sesquiterpenes									
α-Cubebene	1353	0.28 ± 0.22	0.26 ± 0.09	0.51 ± 0.13	0.48 ± 0.01	-	1.10 ± 0.27	0.30 ± 0.22	-
α-Ylangene	1378	0.82 ± 0.39	1.80 ± 0.13	0.80 ± 0.07	1.17 ± 0.11	1.04 ± 0.60	2.66 ± 0.14	0.31 ± 0.20	0.99 ± 0.31
α-Copaene	1382	0.47 ± 0.33	0.69 ± 0.20	1.60 ± 0.20	1.04 ± 0.09	-	1.65 ± 0.00	0.75 ± 0.41	1.76 ± 1.77
β-Elemene	1392	8.07 ± 3.08	7.73 ± 0.54	16.89 ± 4.16	-	11.94 ± 2.44	-	4.48 ± 2.02	35.25 ± 8.75
β-Bourbonene	1393	-	-	-	6.78 ± 0.57	-	6.40 ± 0.07	-	-
β-Caryophyllene [f,g,h]	1427	24.11 ± 9.73	20.57 ± 3.12	55.71 ± 16.29	18.99 ± 2.29	53.59 ± 10.44	20.68 ± 18.88	4.13 ± 1.55	52.80 ± 10.04
α-Bergamotene [h]	1436	7.64 ± 1.99	3.51 ± 0.16	-	5.36 ± 0.30	5.90 ± 1.07	-	-	5.91 ± 1.86
cis-Thujopsene	1437	-	-	-	-	-	-	1.32 ± 1.01	-
β-Gurjunene	1442	tr	-	-	-	-	-	-	5.90 ± 1.83
cis-β-Farneseneh	1445	2.57 ± 0.89	1.86 ± 0.51	-	-	0.79 ± 0.04	-	-	-
α-Caryophyllene [f,h]	1461	6.73 ± 3.56	5.50 ± 0.77	3.85 ± 3.46	5.68 ± 0.12	11.83 ± 2.62	8.57 ± 2.49	2.59 ± 1.53	12.51 ± 2.53
γ-Muurolene	1479	-	2.55 ± 0.22	-	-	-	-	-	-
Germacrene D	1487	4.77 ± 3.40	5.12 ± 0.77	7.63 ± 1.31	6.03 ± 0.26	5.73 ± 1.74	-	-	6.40 ± 1.65
β-Selinene	1494	11.48 ± 2.20	5.92 ± 2.25	5.20 ± 1.26	22.45 ± 2.48	8.32 ± 6.00	7.31 ± 0.89	5.42 ± 2.70	tr
α-Selinene	1501	5.68 ± 0.88	3.96 ± 2.85	10.22 ± 1.00	tr	12.66 ± 8.28	8.14 ± 1.48	5.91 ± 2.72	14.45 ± 3.13
β-Bisabolene	1505	10.74 ± 2.11	8.08 ± 3.25	-	23.72 ± 0.53	8.76 ± 3.09	6.39 ± 2.50	-	-
δ-Guaiene	1510	-	-	-	-	-	-	-	3.93 ± 0.92
α-Chamigrene	1514	2.14 ± 1.82	-	-	4.60 ± 0.02	-	-	2.37 ± 3.11	-
γ-Cadinene	1517	11.17 ± 3.59	3.98 ± 3.80	7.72 ± 4.82	6.38 ± 1.42	-	13.13 ± 1.01	-	-
δ-Cadinene	1523	-	8.54 ± 9.04	4.23 ± 1.65	6.14 ± 0.39	-	8.52 ± 3.31	4.40 ± 6.68	2.15 ± 1.45
trans-γ-Bisabolene	1526	5.81 ± 7.90	3.43 ± 3.46	-	4.47 ± 4.18	0.56 ± 0.17	-	-	2.68 ± 1.83
trans-α-Bisabolene	1537	3.38 ± 2.03	3.49 ± 0.37	-	3.97 ± 0.22	6.90 ± 1.60	-	2.99 ± 3.22	-
Terpene oxide									
Caryophyllene oxide [f,g]	1585	24.16 ± 5.73	27.53 ± 1.85	32.15 ± 3.37	2.13 ± 0.21	8.67 ± 4.24	49.59± 20.74	15.07 ± 9.85	9.51 ± 1.81
Hydrocarbons									
Heptane	721	tr	-	-	-	-	-	-	-
Octane	805	-	-	-	0.03 ± 0.00	-	0.05 ± 0.05	-	-
Nonane	898	-	tr	0.06 ± 0.01	0.07 ± 0.01	0.04 ± 0.00	0.11 ± 0.02	-	0.02 ± 0.01
Undecane	1098	0.21 ± 0.06	0.21 ± 0.07	0.13 ± 0.03	-	0.06 ± 0.03	-	-	-
Dodecane	1197	-	-	-	-	0.24 ± 0.02	-	-	-
Tridecane	1294	-	-	-	-	0.13 ± 0.07	-	-	-
Pentadecane	1498	-	-	-	-	-	12.41 ± 6.36	-	-
Straight-chain acids									
Nonanoic acid	1253	tr	-	0.22 ± 0.04	-	-	-	0.22 ± 0.17	-
Decanoic acid	1344	-	-	5.32 ± 4.16	-	-	1.25 ± 0.06	tr	-

Table 4. *Cont.*

Compound [a]	RI [b]	Concentration (mg/kg) [c]							
		Nongshi No. 1	Taoyuan No. 1	Taoyuan No. 2	Chiayi Strain	TYM1301	TYM1302	TYM1303	TYM1304
Dodecanoic acid	1542	tr	-	tr	-	-	-	1.22 ± 1.14	-
Furans									
2-Ethylfuran	712	tr	tr	-	-	tr	tr	tr	tr
Furfural	795	0.02 ± 0.00	tr	tr	-	tr	-	0.04 ± 0.04	tr
2-Pentylfuran	978	0.15 ± 0.04	-	0.02 ± 0.02	-	0.90 ± 0.17	-	-	1.28 ± 0.46
Methoxy-phenolic compounds									
2-Methoxy-phenol	1058	-	0.09 ± 0.00	0.14 ± 0.02	-	0.04 ± 0.02	-	-	-
2-Methoxy-4-vinylphenol	1279	tr	tr	0.19 ± 0.02	-	0.18 ± 0.05	-	-	0.35 ± 0.16
Nitrogen-containing compound									
Indole	1260	0.22 ± 0.16	0.23 ± 0.01	0.25 ± 0.03	-	0.17 ± 0.11	-	0.23 ± 0.16	0.15 ± 0.10

[a] Identification of components based on the GC/MS library (Wiley 7N); [b] Retention indices, using paraffin (C_5–C_{25}) as references. [c] Relative percentages from GC-FID, values are means ± SD of triplicates; [d] Trace; [e] Undetectable. [f] Published in the literature (Wei et al. [9]); [g] Published in the literature (Lu et al. [11]); [h] Published in the literature (Deng et al. [10]).

3.4. *Comparisons of the Differences between HS-SPME and SDE*

Similar to Table 2, Table 4 shows the eight different Hsian-tsao varieties, along with the 120 components identified using HS-SPME and SDE, of which, 44 were found using both extraction methods, 12 (mainly α-terpinene, δ-3-carene, and *cis*-α-bergamotene) were identified using HS-SPME but not detected using SDE, and 64 (mainly nonanal, 6-methyl-3,5-heptadien-2-one, and gossonorol) were identified using SDE but not detected using HS-SPME.

Table 5 and Figure 4 show that the monoterpene relative content was higher than that of sesquiterpene. Table 6 and Figure 5 show that the SDE samples had a high content of sesquiterpenes, terpene oxide, and terpene alcohols, but a lower content of monoterpenes than the SPME samples. Tersanisni and Berry [33] reported that certain hydrocarbon compounds, such as linalool and α-terpineol, as well as their hydrocarbon interactions, can be interrupted by heat stress, resulting in the induction of volatilization. We detected α-terpineol using SDE but by using HS-SPME. However, both methods identified terpene hydrocarbons as the major components. HS-SPME extracted more terpene hydrocarbons, and the majority was highly volatile monoterpenes with a low molecular weight. SDE extracted mainly sesquiterpenes with higher molecular weights. SDE also identified components that HS-SPME was unable to identify, such as straight-chain acids, aromatic ketones, aromatic esters, terpene aldehydes, terpene ketones, methoxy phenols, and nitrogen-containing compounds. Montserrat et al. [34] analyzed the volatile composition of white salsify (*Tragopogon porrifolius* L.) and found that SDE used high temperature and a long extraction time, and large quantities of volatile components were lost during the extraction process. Therefore, the SDE method may increase the low volatile compounds with a high molecular weight, such as sesquiterpenes and straight-chain acids. HS-SPME used shorter extraction times, so it was able to extract highly volatile monoterpenes with lower molecular weights. As such, HS-SPME is more appropriate for quality control. This study found that although HS-SPME was more rapid and SDE had a higher temperature and longer extraction time, SDE was able to extract more Hsian-tsao compounds; therefore, both methods can be used to complement each other. Yang et al. [35] compared HS-SPME with traditional methods in the analysis of *Melia azedarach* and reported that the HS-SPME method is a powerful analytic tool and is complementary to traditional methods for the determination of the volatile compounds in herbs. Comparing both techniques, HS-SPME samples were smaller (1 g) and did not require heating, the data was accurate, and involved less chemical reactions and changes, but the yield of larger molecules were lower, and the identified components were fewer, while SDE needed the use of 100 g of plant material and heating (2 h). The popularity of this method comes from the fact that volatiles with medium to high boiling points are recovered well. The aroma profile can be greatly altered via the formation of artifacts due to heating the sample during isolation. However, Hsian-tsao food needs to be processed using heat; therefore, by combining the HS-SPME and SDE methods of volatile compounds isolation, each isolation technique provides a part of the overall Hsian-tsao profile.

Table 5. Percentages of extracted chemical groups of Hsian-tsao analyzed using HS-SPME.

	Nongshi No. 1	Taoyuan No.1	Taoyuan No. 2	Chiayi Strain	TYM1301	TYM1302	TYM1303	TYM1304
Aliphatic alcohol	0.29	0.69	1.16	0.17	2.25	1.39	1.61	3.01
Aliphatic aldehydes	3.63	2.91	0.29	0.39	1.19	1.15	1.81	1.74
Aliphatic ketones	tr	0.14	0.12	tr	0.29	tr	0.24	0.47
Aliphatic ester	0.32	-	-	-	-	-	-	-
Aromatic aldehyde	tr	-	-	tr	-	-	-	-
Aromatic alcohol	-	-	-	0.08	-	-	-	-
Aromatic hydrocarbon	-	0.68	-	tr	tr	tr	tr	tr
Terpene alcohol	0.11	0.22	0.09	0.13	0.36	0.08	0.08	-
Monoterpenes	77.83	63.59	72.47	79.47	64.87	84.63	79.36	58.24
Sesquiterpenes	6.72	20.06	16.93	11.28	20.5	4.16	5.2	22.47
Terpene oxide	-	0.04	-	-	-	-	-	-
Hydrocarbons	0.86	0.75	0.88	0.12	0.56	-	0.33	0.06
Furan	tr	tr	tr	tr	tr	tr	tr	tr

All the definitions of the symbols used in Table 2 mean values were also used in Table 5.

Table 6. Concentrations of chemical groups of overall extracted Hsian-tsao analyzed using SDE.

	Nongshi No. 1	Taoyuan No. 1	Taoyuan No. 2	Chiayi Strain	TYM1301	TYM1302	TYM1303	TYM1304
Aliphatic alcohols	3.03	1.59	1.72	2.07	3.73	5.85	2.49	3.02
Aliphatic aldehydes	9.51	2.36	3.47	3.08	4.44	10.90	3.19	6.55
Aliphatic ketones	10.45	3.95	5.10	8.92	8.08	13.60	6.67	10.01
Aliphatic ester	-	-	tr	tr	tr	tr	tr	tr
Aromatic alcohols	2.75	1.19	3.25	2.92	1.39	1.66	0.7	1.11
Aromatic aldehydes	tr	0.11	tr	tr	0.2	0.8	tr	0.28
Aromatic ketone	0.46	0.29	0.38	0.31	0.03	0.62	0.26	0.05
Aromatic esters	1.29	0.82	1.78	-	0.56	-	1.7	0.42
Aromatic hydrocarbons	0.60	0.74	0.27	0.7	-	-	2.48	10.49
Terpene alcohols	61.44	74.98	5.16	118.82	146.11	6.54	46.38	2.38
Terpene aldehydes	0.83	-	1.41	0.84	0.91	2.6	0.65	1.18
Terpene ketones	4.84	0.43	3.75	3.29	2.46	8.28	1.3	4.09
Monoterpenes	4.89	5.62	5.07	4.17	23.46	6.62	7.1	11.88
Sesquiterpenes	105.86	86.99	124.36	117.26	128.02	84.55	34.97	144.73
Terpene oxide	24.16	27.53	32.15	2.13	8.67	49.59	15.07	9.51
Hydrocarbons	0.21	0.21	0.19	0.10	0.47	12.57	0	0.02
Straight-chain acids	tr	-	5.54	-	-	1.25	1.44	-
Furans	0.17	tr	0.02	-	0.9	-	0.04	1.28
Methoxy-phenolic compounds	tr	0.09	0.33	-	0.22	-	-	0.35
Nitrogen-containing compound	0.22	-	0.25	-	0.17	-	0.23	0.15

Notes: All the definitions of the symbols used in Table 3 mean values were also used in Table 4.

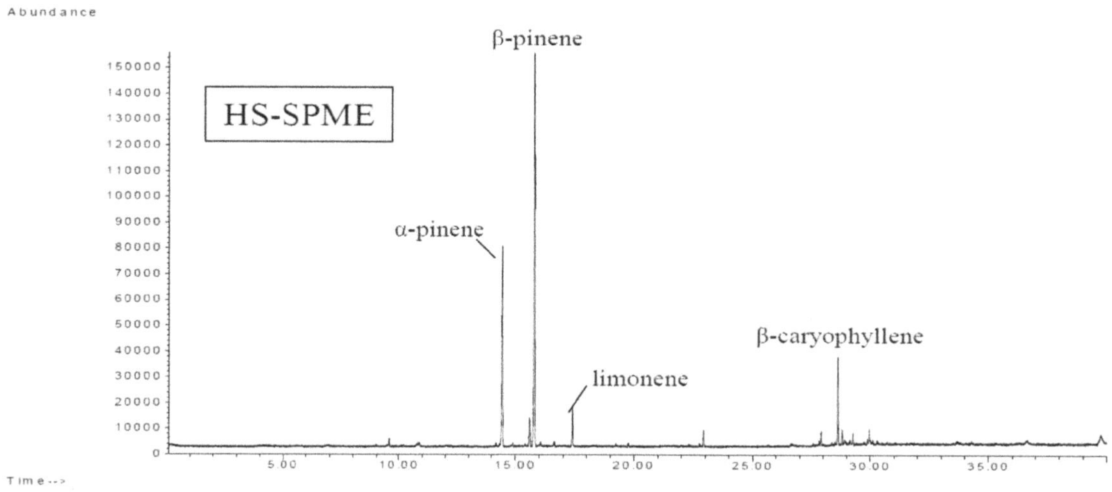

Figure 4. Total ion chromatogram of volatile components of Nongshi No. 1 determined using HS-SPME.

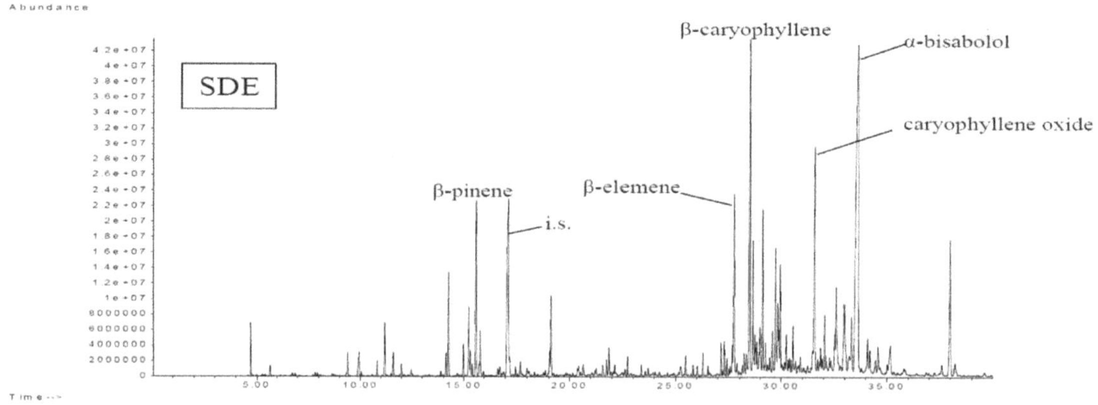

Figure 5. Total ion chromatogram of volatile components of Nongshi No. 1 determined using SDE.

4. Conclusions

This study determined the volatile components present in eight varieties of Hsian-tsao using HS-SPME and SDE methods. A total of 120 volatile components were identified, of which, 56 were verified using HS-SPME and 108 using SDE. HS-SPME extracted more monoterpenes; however, SDE extracted more sesquiterpenes and terpene alcohols, and a terpene oxide, such as β-caryophyllene, α-bisabolol, and caryophyllene oxide. SDE was able to detect more components, but HS-SPME analysis was more convenient. In the future, the two extraction methods can be used in a complementary manner for Hsian-tsao analysis and research.

Author Contributions: Conceptualization, T.-L.K., L.-Y.L., and H.-C.C.; methodology, Y.-J.C. and T.-L.K.; validation, L.K.C., C.-S.W., and H.-C.C.; formal analysis, Y.-J.C. and L.K.C.; investigation, T.-L.K. and L.-Y.L.; writing—original draft preparation, Y.-J.C. and H.-C.C.; writing—review and editing, L.-Y.L. and H.-C.C.

Acknowledgments: Financial support from from the Council of Agriculture, Executive Yuan (Taiwan) (108AS-7.2.5-FD-Z1 (2), Ministry of Science and Technology (Taiwan) (107-2221-E-039-008-) and Ministry of Education (Taiwan) (1038142*) are gratefully acknowledged.

References

1. Paton, A.J. Classification and species of *Platostoma* and its relationship with *Haumaniastrum* (Labiatae). *Kew Bull.* **1997**, *52*, 257–292. [CrossRef]
2. Feng, T.; Ye, R.; Zhuang, H.; Fang, Z.; Chen, H. Thermal behavior and gelling interactions of *Mesona Blumes* gum and rice starch mixture. *Carbohydr. Polym.* **2012**, *90*, 667–674. [CrossRef]
3. Widyaningsih, T.D. Cytotoxic effect of water, ethanol and ethyl acetate extract of black cincau (*Mesona Palustris* BL) against HeLa cell culture. *APCBEE Procedia* **2012**, *2*, 110–114. [CrossRef]
4. Yen, G.-C.; Hung, C.-Y. Effects of alkaline and heat treatment on antioxidative activity and total phenolics of extracts from Hsian-tsao (*Mesona procumbens* Hemsl.). *Food Res. Int.* **2000**, *33*, 487–492. [CrossRef]
5. Hung, C.-Y.; Yen, G.-C. Antioxidant activity of phenolic compounds isolated from *Mesona procumbens* Hemsl. *J. Agric. Food Chem.* **2002**, *50*, 2993–2997. [CrossRef]
6. Lai, L.-S.; Liao, C.-L. Dynamic rheology of structural development in starch/decolourised hsian-tsao leaf gum composite systems. *J. Sci. Food Agric.* **2002**, *82*, 1200–1207. [CrossRef]
7. Lai, L.-S.; Liu, Y.-L.; Lin, P.-H. Rheological/textural properties of starch and crude hsian-tsao leaf gum mixed systems. *J. Sci. Food Agric.* **2003**, *83*, 1051–1058. [CrossRef]
8. Zhuang, H.; Feng, T.; Xie, Z.; Toure, A.; Xu, X.; Jin, Z.; Su, Q. Effect of *Mesona* Blumes gum on physicochemical and sensory characteristics of rice extrudates. *Int. J. Food Sci. Technol.* **2010**, *45*, 2415–2424. [CrossRef]
9. Wei, J.; Zheng, E.-L.; Cai, X.-K.; Ji, X.-D.; Xu, C.-H. Preparation of water-soluble extracts from *Mesona Benth* and analysis of the volatile aroma components by GC-MS. *Food Sci. Technol.* **2014**, *39*, 190–192.
10. Deng, C.; Li, R. Analysis of the chemical constituents of the essential oils from *Mesona chinensis* benth by gas chromatography-mass spectrometry. *China Modern Med.* **2012**, *19*, 68–69.
11. Lu, S.-P.; Tian, Y.-K. Analysis of the chemical constituents of the volatile oil from *Mesona chinensis* Benth in Guangdong Yangjiang. *Strait. Pharm. J.* **2014**, *26*, 33–36.
12. Orav, A.; Stulova, I.; Kailas, T.; Müürisepp, M. Effect of storage on the essential oil composition of *Piper nigrum* L. fruits of different ripening states. *J. Agric. Food Chem.* **2004**, *52*, 2582–2586. [CrossRef] [PubMed]
13. Blanch, G.P.; Reglero, G.; Herraiz, M. Rapid extraction of wine aroma compounds using a new simultaneous distillation-solvent extraction device. *Food Chem.* **1996**, *56*, 439–444. [CrossRef]
14. Gu, X.; Zhang, Z.; Wan, X.; Ning, J.; Yao, C.; Shao, W. Simultaneous distillation extraction of some volatile flavor components from Pu-erh tea samples comparison with steam distillation-liquid/liquid extraction and soxhlet extraction. *Int. J. Anal. Chem.* **2009**, 276713. [CrossRef] [PubMed]
15. Heath, H.B.; Reineccius, G. *Flavour Chemistry and Technology*; Van Nostrand Reinhold Co.: New York, NY, USA, 1986.
16. Holt, R.U. Mechanisms effecting analysis of volatile flavour components by solid-phase microextraction and gas chromatography. *J. Chromatogr. A.* **2001**, *937*, 107–114. [CrossRef]
17. Kataoka, H.; Lord, H.L.; Pawliszyn, J. Applications of solid-phase microextraction in food analysis. *J. Chromatogr. A.* **2000**, *880*, 35–62. [CrossRef]
18. Zhang, Z.; Yang, M.J.; Pawliszyn, J. Solid-phase microextraction. A solvent-Free Alternative for Sample Preparation. *Anal. Chem.* **1994**, *66*, 844A–853A. [CrossRef]
19. Adam, M.; Juklová, M.; Bajer, T.; Eisner, A.; Ventura, K. Comparison of three different solid-phase microextraction fibres for analysis of essential oils in yacon (*Smallanthus sonchifolius*) leaves. *J. Chromatogr. A* **2005** *1084*, 2–6. [CrossRef] [PubMed]

20. Yeh, C.-H.; Tsai, W.-Y.; Chiang, H.-M.; Wu, C.-S.; Lee, Y.-I.; Lin, L.-Y.; Chen, H.-C. Headspace solid-phase microextraction analysis of volatile components in *Phalaenopsis* Nobby's Pacific Sunset. *Molecules* **2014**, *19*, 14080–14093. [CrossRef] [PubMed]
21. Schomburg, G.; Dielmann, G. Identification by means of retention parameters. *J. Chromatogr. Sci.* **1973**, *11*, 151–159. [CrossRef]
22. Ducki, S.; Miralles-Garcia, J.; Zumbe, A.; Tornero, A.; Storey, D.M. Evaluation of solid-phase micro-extraction coupled to gas chromatography-mass spectrometry for the headspace analysis of volatile compounds in cocoa products. *Talanta* **2008**, *74*, 1166–1174. [CrossRef] [PubMed]
23. Silva, C.L.; Câmara, J.S. Profiling of volatiles in the leaves of Lamiaceae species based on headspace solid phase microextraction and mass spectrometry. *Food Res. Int.* **2013**, *51*, 378–387. [CrossRef]
24. Zhang, C.; Qi, M.; Shao, Q.; Zhou, S.; Fu, R. Analysis of the volatile compounds in Ligusticum chuanxiong Hort. using HS-SPME-GC-MS. *J. Pharm. Biomed Anal.* **2007**, *44*, 464–470. [CrossRef] [PubMed]
25. Minh Tu, N.T.; Onishi, Y.; Choi, H.S.; Kondo, Y. Characteristic Odor Components of Citrus sphaerocarpa Tanaka (Kabosu) Cold-Pressed Peel Oil. *J. Agric. Food Chem.* **2002**, *50*, 2908–2913. [CrossRef] [PubMed]
26. Guillot, S.; Peytavi, L.; Bureau, S.; Boulanger, R.; Lepoutre, J.; Crouzet, J.; Schorrgalindo, S. Aroma characterization of various apricot varieties using headspace-solid phase microextraction combined with gas chromatography—mass spectrometry and gas chromatography-olfactometry. *Food Chem.* **2006**, *96*, 147–155. [CrossRef]
27. Högnadóttir, Á.; Rouseff, R.L. Identification of aroma active compounds in orange essence oil using gas chromatography-olfactometry and gas chromatography-mass spectrometry. *J. Chromatogr. A.* **2003**, *998*, 201–211. [CrossRef]
28. Ravi, R.; Prakash, M.; Bhat, K.K. Aroma characterization of coriander (*Coriandrum sativum* L.) oil samples. *Eur. Food Res. Technol.* **2007**, *225*, 367–374. [CrossRef]
29. Rega, B.; Fournier, N.; Guichard, E. Solid phase microextraction (SPME) of orange juice flavor: Odor representativeness by direct gas chromatography olfactometry (D-GC-O). *J. Agric. Food Chem.* **2003**, *51*, 7092–7099. [CrossRef]
30. Jirovetz, L.; Buchbauer, G.; Ngassoum, M.B.; Geissler, M. Aroma compound analysis of *Piper nigrum* and *Piper guineense* essential oils from cameroon using solid-phase microextraction-gas chromatography, solid-phase microextraction-gas chromatography-mass spectrometry and olfactometry. *J. Chromatogr. A.* **2002**, *976*, 265–275. [CrossRef]
31. Friedrich, J.E.; Acree, T.E. Gas chromatography olfactometry (GC/O) of dairy products. *Int. Dairy J.* **1998**, *8*, 235–241. [CrossRef]
32. Morales, M.T.; Aparicio, R. Effect of extraction conditions on sensory quality of virgin olive oil. *J. Am. Oil Chem. Soc.* **1999**, *76*, 295–300. [CrossRef]
33. Tersanisni, P.; Berry, R.G. Chemical changes in flavor components. In *Chemical Changes in Food During Processing*; Richardson, T., Finely, J.W., Eds.; Springer: Berlin, Germany, 1985; pp. 327–346.
34. Montserrat, R.-A.; Vargas, L.; Vichi, S.; Guadayol, J.M.; López-Tamames, E.; Buxaderas, S. Characterisation of volatile composition of white salsify (*Tragopogon porrifolius* L.) by headspace solid-phase microextraction (HS-SPME) and simultaneous distillation–extraction (SDE) coupled to GC–MS. *Food Chem.* **2011**, *129*, 557–564.
35. Yang, Y.; Xiao, Y.; Liu, B.; Fang, X.; Yang, W.; Xu, J. Comparison of headspace solid-phase microextraction with conventional extraction for the analysis of the volatile components in *Melia azedarach*. *Talanta* **2011**, *86*, 356–361. [CrossRef] [PubMed]

8

Release of Indospicine from Contaminated Camel Meat following Cooking and Simulated Gastrointestinal Digestion

Saira Sultan [1], Cindy Giles [2], Gabriele Netzel [1], Simone A. Osborne [3], Michael E. Netzel [1] and Mary T. Fletcher [1,*]

1 Queensland Alliance for Agriculture and Food Innovation (QAAFI), The University of Queensland, Health and Food Sciences Precinct, Coopers Plains, QLD 4108, Australia; saira.sultan@uq.net.au (S.S.); g.netzel@uq.edu.au (G.N.); m.netzel@uq.edu.au (M.E.N.)

2 Department of Agriculture and Fisheries, Queensland Government, Health and Food Sciences Precinct, Coopers Plains, QLD 4108, Australia; cindy.giles@daf.qld.gov.au

3 Commonwealth Scientific and Industrial Research Organisation, Agriculture and Food, St. Lucia, QLD 4067, Australia; simone.osborne@csiro.au

* Correspondence: mary.fletcher@uq.edu.au

Abstract: Indospicine, a hepatotoxic arginine analog, occurs in leguminous plants of the *Indigofera* genus and accumulates in the tissues of grazing animals that consume these plants. Furthermore, indospicine has caused toxicity in dogs following consumption of indospicine-contaminated meat; however, the potential impact on human health is unknown. The present study was designed to determine the effect of simulated human gastrointestinal digestion on the release and degradation of indospicine from contaminated camel meat following microwave cooking. Results showed no significant ($p > 0.05$) indospicine degradation during cooking or in vitro digestion. However, approximately 70% indospicine was released from the meat matrix into the liquid digesta during the gastric phase (in the presence of pepsin) and increased to >90% in the intestinal phase (with pancreatic enzymes). Following human consumption of contaminated meat, this soluble and more bioaccessible fraction of intact indospicine could be readily available for absorption by the small intestine, potentially circulating indospicine throughout the human body to tissues where it could accumulate and cause detrimental toxic effects.

Keywords: indospicine; hepatotoxicity; meat; in vitro digestion; human

Key Contribution: Indospicine is released from naturally contaminated camel meat following cooking and in vitro human digestion but does not undergo significant breakdown. Therefore, hepatotoxic indospicine could be readily available for absorption across the small intestine following human consumption and digestion of naturally contaminated meat.

---

1. Introduction

The non-proteinogenic amino acid indospicine (L-6-amidino-2-amino-hexanoic acid) is a hepatotoxic arginine analog (Figure 1) found widely in plants of the *Indigofera* genus [1]. This genus contains over 700 species distributed across tropical Africa, Asia, Australia, and North and South America, and includes species such as *I. spicata*, *I. hendecaphylla*, *I. linnaei*, *I. lespedezioides*, *I. vicioides* and *I. volkensii* that have all been reported to contain in excess of 500 mg indospicine/kg dry matter of foliage. Indospicine has been found to be directly toxic to livestock [2], and to also act as a secondary toxin due to its unusual ability to accumulate in tissues of livestock grazing on these plants [3–5].

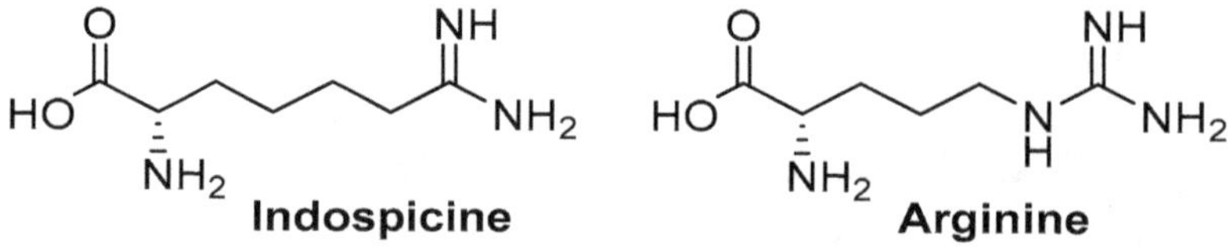

Figure 1. Chemical structures of the amino acids indospicine and arginine.

Indospicine hepatotoxicity has been demonstrated in rats following a single dose of indospicine that inhibited protein synthesis and induced fatty changes in, and enlargement of, the liver [6]. Livestock consuming *Indigofera* develop similar indospicine-induced liver hepatotoxicity; however, symptoms of toxicity frequently become apparent only after extended periods of *Indigofera* consumption [7]. Before the toxicity of *Indigofera* to livestock was recognized in the 1950s, several *Indigofera* species (notably *I. spicata* and *I. hendacphylla*) were introduced as pasture legumes in the USA and Australia. As a result, these species are now widely found in tropical regions far beyond their native range, across Africa, Asia, Australia, the Americas, and islands of both the Pacific and Indian Oceans [2]. *I. linnaei* and *I. lespedezioides* are also regionally abundant with extensive native ranges in Australia and South America respectively [2]. *Indigofera* plants are palatable legumes that are readily consumed by livestock leading to the reported indospicine accumulation in the meat of cattle [3], camels [4], and horses [5,8].

Indospicine is non-proteinogenic and instead occurs in both plant and animal tissues as a free amino acid [3–5]. It is a competitive inhibitor of arginase [9] and DNA synthesis [10], and has been shown to cause liver degeneration [11] and abortion [12]. In fact, indospicine hepatotoxicity has been demonstrated in all animal species investigated to date with reports of acute and sub-chronic hepatotoxic evidence for rats, [13] mice, [14] rabbits, [15] guinea pigs, [7] sheep, [7] dogs [5], and cattle [7]. However, the severity of the toxicity appears to vary considerably between species with dogs being particularly vulnerable [16]. Indeed, secondary poisoning of dogs has been reported in dogs consuming meat naturally contaminated with indospicine arising from livestock that had grazed on *Indigofera* plants [5,17].

With respect to digestion of indospicine following consumption of *Indigofera* plants, previous in vitro rumen studies have shown that indospicine can be metabolized by the rumenal microbial system [18]. However, little is known about the stability, release, and potential degradation of indospicine within monogastric digestive systems, such as the human gastro-intestinal tract, in which a single-chambered stomach secretes enzymes and acid to facilitate digestion prior to passage into the small and large intestines. When considering the digestive stability of one component, it is important to consider not only the chemical structure of a compound but also the nature of its bond to the food matrix. Therefore, static and/or dynamic in vitro digestion models that mimic the human gastrointestinal digestion process are a common approach to determine the matrix release (bioaccessibility) and stability of food components (like nutrients and toxins) as an initial measure to predict their potential bioavailability [19–22].

In the present study, an in vitro model was utilized to investigate stability, release, and degradation of indospicine to better understand the potential human exposure following the consumption of indospicine-contaminated meat. Specifically, the aims of this research were to investigate the effect of cooking (microwave) on indospicine stability in camel meat naturally containing indospicine, and most importantly, predict the bioaccessibility and digestive fate of indospicine using a static in vitro digestion model mimicking the human gastric and small intestinal digestion process. To date and to the authors' knowledge, this is the first study investigating the bioaccessibility of indospicine within a monogastric model system.

2. Results and Discussion

The hepatotoxic amino acid indospicine is found only in plant species of the *Indigofera* genus [1] and has been linked with poisoning of grazing livestock [2], and the deaths of dogs consuming contaminated meat from livestock that had grazed on these plants [5,17]. These canine deaths in Australia following the consumption of indospicine-contaminated camel [17] and horse meat [5] have raised both industry and consumer concern with regard to the potential contamination of meat for human consumption. The possible impact on human health is particularly concerning as all livestock grazing pastures that contain *Indigofera* plant species have the potential to accumulate indospicine as a free amino acid in their meat [3,4,23,24]. These indospicine residues are not readily excreted and can persist in tissues for up to 6 months after the cessation of *Indigofera* consumption, suggesting a strong affinity of indospicine with the meat matrix. Surveys of camel meat collected in Australia from animals killed in the field (in situ) and in abattoirs have demonstrated significant levels of indospicine residues above detectable limits (0.05 mg/kg), with meat from individual camels having levels up to 3.73 mg/kg [25]. Canine poisonings have occurred in the past following repeated consumption of indospicine-contaminated meat from the same or similar source repeatedly over several months [5,17]. The Australian supply of camel meat is sourced from more than 350,000 rangeland animals grazing arid inland regions of Australia [26], where *Indigofera* is also seasonally prevalent. These animals are slaughtered in abattoirs located within these regions before being supplied to growing niche markets in urban Australian Middle East and North African communities, as well as Central Australian local populations [26]. It is plausible that this lack of supply chain diversity could contribute to the same repeated exposure in these human consumers, and the potential health risk needs to be considered.

There are currently no identified studies providing epidemiological evidence or observational data indicating indospicine-induced adverse effects in humans. In the absence of such data, exposure risk can only be derived from available toxicity studies in dogs (often considered a model for human studies). The most substantial data available relates to thirteen different indospicine doses derived from two dog feeding experiments conducted for between 4 and 70 days [5,16]. A "lowest observed adverse effect level" (LOAEL) of 0.13 mg indospicine/kg bw/day is suggested based on observational data from the 70-day sub-chronic animal feeding experiment [5]. In this study, only minor histological liver lesions were observed in four dogs fed diets containing between 0.13 and 0.25 mg indospicine/kg bw/day [5]. A guidance value for human consumption of 1.3 μg indospicine/kg bw/day is thus proposed by dividing the selected LOAEL for dogs by an uncertainty factor of 10 to take into account mild degenerative changes to the liver in low dose dogs in the 70-day feeding trial, and by an additional factor of 10 to take into account intra species variation [27]. However, it can be postulated that a person with an average bodyweight of 70 kg [27] consuming the average Australian total daily meat intake of 143 g [28] could potentially consume 7.6 μg indospicine/kg bw/day if the dietary meat source was camel meat containing the reported 3.73 mg indospicine/kg [25]. This calculated intake of indospicine exceeds the proposed guidance value by a factor of 5. An additional factor of 10 could also be considered in the derived guidance value [27] due to the short duration of the literature study (70 days), and if this was implemented, then the calculated intake of indospicine could exceed the guidance value by a factor much higher than 5. Also, based on in vitro assessments of indospicine absorption using human intestinal cells [29], indospicine exhibits a 2-fold higher apparent permeability across an in vitro intestinal barrier compared to arginine (the amino acid analogue of indospicine). These findings indicate that indospicine is more readily absorbed than dietary arginine, suggesting preferential uptake that could potentiate further risks of toxicity.

Indospicine has been investigated in plant material by amino acid analyzer [5,30], high performance liquid chromatography (HPLC) with derivatization and UV detection [8,24] and liquid chromatography–tandem mass spectrometry (LC−MS/MS) [1,31]. However, the analysis of indospicine in meat is challenging due to low levels of contamination together with the complexity of the meat matrix. The incorporation of D_3-L-indospicine as an internal standard in sample extracts as used in this study can be beneficial in LC−MS/MS analysis as it overcomes the matrix effects observed in previous studies [17].

In the present study, indospicine-contaminated camel meat (microwave cooked) was subjected to in vitro digestion through sequential addition of pepsin in 0.1 M HCl and pancreatin-bile solutions with an appropriate adjustment of pH to mimic human in vitro digestion (Figure 2). Liquid and solid digesta from the in vitro gastric and small intestinal digestion were separated by centrifugation, prior to determination of indospicine concentration by LC-MS/MS (Figure 3), utilizing the previously reported and validated method [32]. Indospicine concentration was also measured in uncooked and cooked camel meat using the same method. All studies were carried out in triplicate with results shown in Figure 4 and expressed as the mean and standard deviation (SD). To enable a comparison of the liquid and solid phases, results are presented as indospicine content (μg) in each phase rather than concentration.

Results from this study indicate that there were no significant changes ($p > 0.05$) in the total indospicine content, suggesting that indospicine was not degraded during microwave cooking or gastrointestinal digestion in vitro (Figure 4). Cooking causes shrinkage of collagen fibres [33] and also increases meat protein surface hydrophobicity [34]. Indospicine is a water-soluble free amino acid, and this meat matrix breakdown during cooking and subsequent in vitro digestion resulted in an almost complete release of this amino acid from the solid phase into the liquid phase (Figure 4). This is evident from the observed release of approximately 70% indospicine from solid to liquid phase after the incubation of cooked meat with pepsin (during the gastric phase). Moreover, digestion with pancreatin and bile in the small intestinal phase resulted in a total release of more than 90% indospicine into the liquid digesta.

In contrast to the observed lack of indospicine degradation in the present in vitro model of gastrointestinal digestion in a monogastric system, indospicine was almost 100% degraded when *Indigofera* plant material was incubated in camel foregut fluid for 48 h [18]. This differing result is indicative of the presence of microbes able to degrade indospicine in the camel gastric system. Indospicine was similarly degraded when incubated with bovine ruminal fluid [18], and further studies are underway to isolate the responsible microbes with the potential to be utilized as a preventive probiotic. However, the observed accumulation of indospicine in camel tissues suggest that even though indospicine can be degraded by foregut fermentation, complete degradation does not occur before passage of the digesta into the intestine and a significant portion of indospicine is then available for absorption [18].

It must be noted that in vitro digestion models have several limitations that should be considered when interpreting the results. For example, no current in vitro model is capable of replicating all aspects of in vivo digestion, absorption, distribution, biodegradation (including the metabolic activity of the gut microbiota), and elimination [35,36]. Nevertheless, our results indicate that indospicine is released from the meat matrix and appears resistant to human gastrointestinal conditions, potentially making it available for absorption in the small intestine from liquid digesta. Postprandial indospicine may circulate throughout the human body to tissues and organs, such as the liver, where it could accumulate over time and cause detrimental, toxic effects.

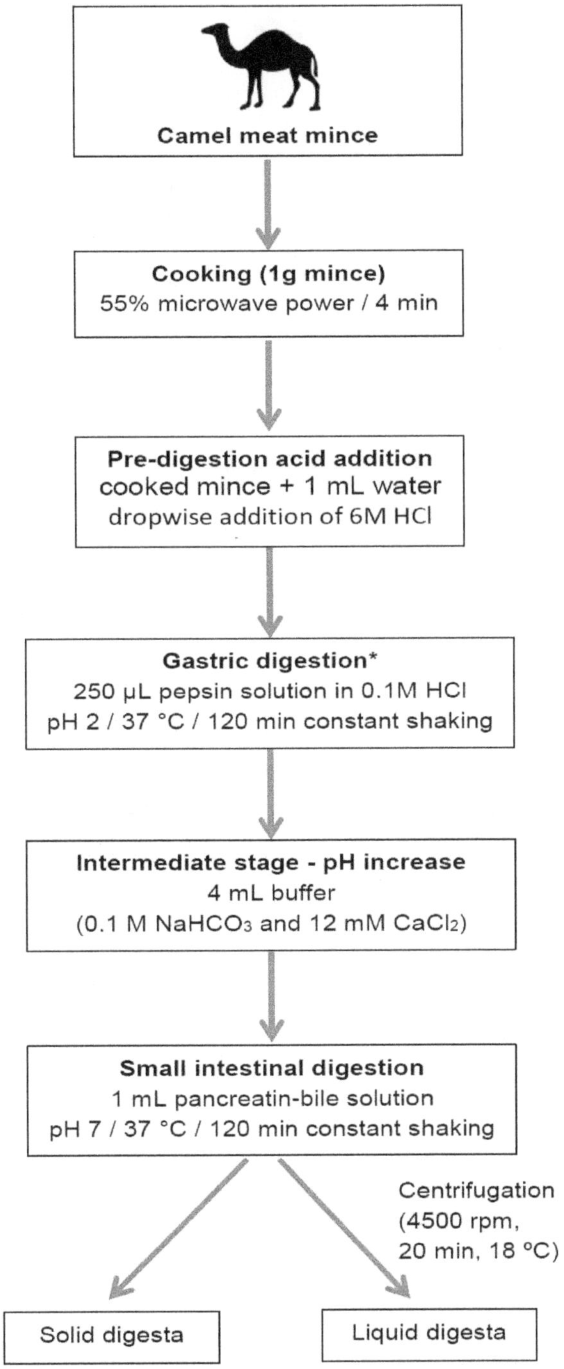

Figure 2. Schematic depiction of the stepwise in vitro digestion of camel mince under simulated human gastro-small intestinal conditions. (* Collection of gastric solid and liquid digesta as depicted for small intestinal digestion).

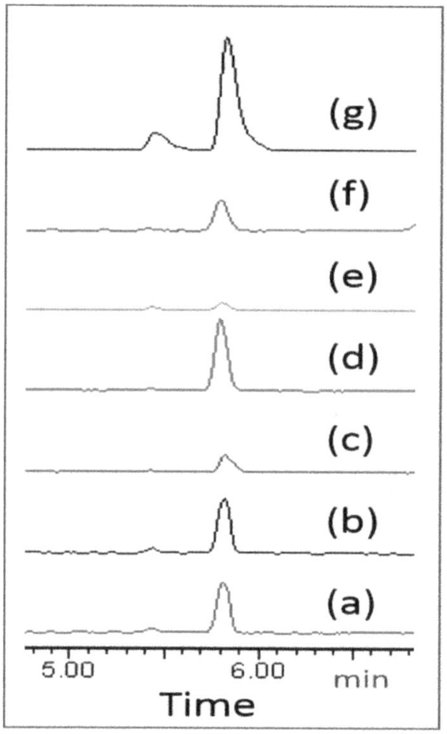

Figure 3. LC-MS/MS analysis of indospicine on Waters Micromass Quattro Premier triple quadrupole mass spectrometer utilizing a selected reaction monitoring (SRM) transitions of m/z 174 → 111 in (**a**) uncooked camel meat, (**b**) cooked camel meat, (**c**) solid phase of gastric digesta, (**d**) liquid phase of gastric digesta, (**e**) solid phase of small intestinal digesta, (**f**) liquid phase of small intestinal digesta, and (**g**) 0.05 mg/L standard indospicine solution.

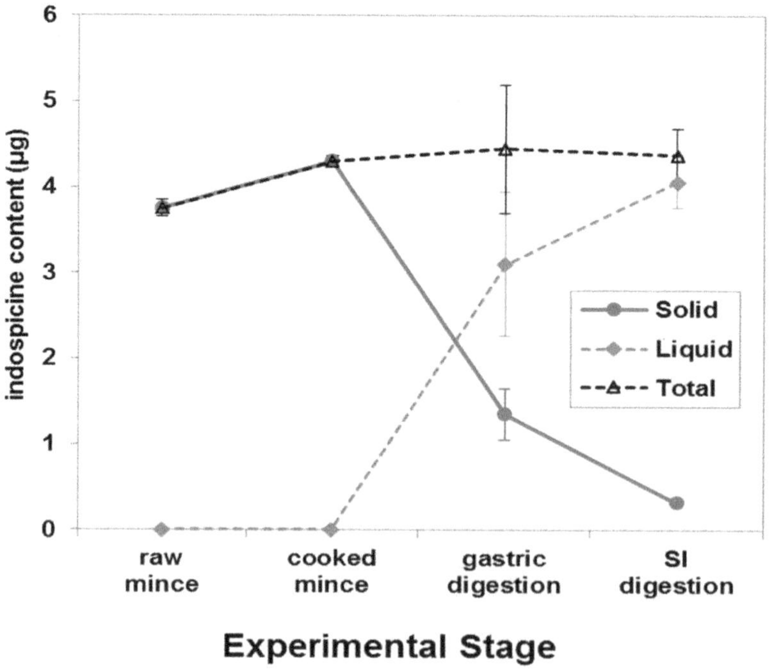

Figure 4. Indospicine content in raw, microwave cooked, and in vitro gastric and small intestinal (SI) digested indospicine-contaminated camel meat as determined by LC-MS/MS analysis. (All points are the mean of analysis of three replicates. Error bars show the SD).

There is no known mammalian enzyme that can degrade the amidino group of indospicine [37], and the preferred route to avoid indospicine toxicity is thus to prevent digestive uptake through degradation of the toxin during food processing. Our results suggest that the human digestive system does not have the capacity to degrade indospicine. Tan et al. [38] have recently reported that microwaving indospicine-contaminated camel meat under mild alkaline conditions (0.05% sodium bicarbonate, pH 8.8, 15 min) achieved 100% degradation of indospicine, with products identified as 2-aminopimelamic acid (major) and 2-aminopimelic acid (minor) (Figure 5). Such processing treatments may have ready applicability in the pet food industry given the recorded sensitivity of dogs to indospicine-contaminated meat, but are perhaps not appropriate in the processing of food for human consumption. Additionally, the metabolic fate and toxicity of indospicine hydrolysis products remains unknown and requires further investigation [38].

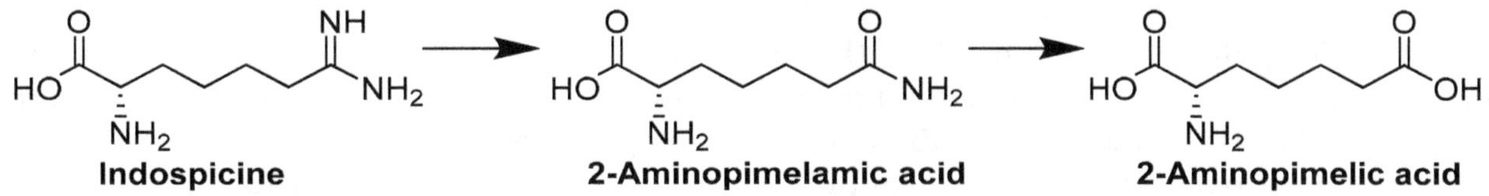

Figure 5. Hydrolysis of the amidino group of indospicine to corresponding amide (2-aminopimelamic acid) and acid (2-aminopimelic acid) under mild alkaline conditions as reported by Tan et al. [38].

3. Conclusions

Simple cooking of contaminated camel meat in a microwave, as carried out in the present study, does not degrade indospicine. Moreover, in vitro human gastrointestinal digestion conditions also had no effect on indospicine degradation and only helped to release the toxin from the solid meat matrix into the liquid digesta. These observations imply that following human consumption and digestion of contaminated meat, indospicine could be readily bioaccessible for absorption across the small intestine. The toxicity of indospicine for humans is uncertain [39], but the known toxicity in dogs (often considered a model for human studies) is particularly concerning. Camel meat is not commonly consumed by the broader Australian population, but is eaten by local indigenous populations and some immigrant ethnic groups within Australia. Further risk assessments, particularly for these high exposure groups, need to be undertaken with additional consideration given to possible indospicine contamination of other red meat supply chains.

4. Materials and Methods

4.1. Reagents

Unless otherwise stated, all chemicals were purchased from Sigma-Aldrich (Castle Hill, Sydney, NSW, Australia), and were of analytical or HPLC grade. De-ionized water was used throughout all experiments.

4.2. Study Design

In this study, cooked indospicine-contaminated camel meat (both meat and juices) was subjected to in vitro digestion through the sequential addition of pepsin and pancreatin-bile solutions with an appropriate adjustment of pH as outlined in Figure 2 to simulate the human gastro-small intestinal digestion process. All samples were digested in triplicate for both gastric digestion alone and for gastric plus small intestinal digestion. Liquid and solid digesta after gastric and small intestinal digestion were separated by centrifugation, and the concentration of indospicine in the digesta (liquid and solid) and uncooked and cooked camel meat was measured by LC-MS/MS.

4.3. Camel Meat Samples

Camel meat samples were obtained from a previously described experimental feeding trial in which camels were fed a diet containing the pasture legume *Indigofera spicata* with consequential accumulation of indospicine in meat tissues [3,4]. This feeding trial was conducted under approval of the Animal Ethics Committee of the University of Queensland, QLD, Australia (AEC Approval No. SAFS/047/14/SLAI; Date of approval: 19 March 2014). Indospicine-contaminated camel meat samples collected at autopsy [4] were utilized in this present study. Camel meat samples were minced using a commercial meat mincer (PRO 1400 meat grinder, Kenwood, Prestons, NSW, Australia) to provide a homogenous sample and stored frozen at −20 °C until used for further analysis.

4.4. In Vitro Digestion of Camel Meat

The in vitro digestion of camel meat samples was performed following the method described by Netzel et al. [36] with some modifications (Figure 2). The gastric phase was 120 min to account for reported variations in gastric emptying following consumption of meals that produce larger particle sizes like meat [40,41]. A 120 min intestinal phase was also employed.

Prior to in vitro digestion, camel meat (minced, 1 g each) in 15 mL screw–cap Falcon tubes was cooked in a microwave oven (Panasonic Genius NN5752–750 Watts, Sydney, NSW, Australia) for 4 min at medium heat (55% power, approximate temperature 70 °C).

4.4.1. Gastric Digestion

After cooking, samples were allowed to cool to room temperature and water (1 mL) was added to form a slurry. To prepare the samples for gastric digestion, the pH was lowered to 2.0 by the dropwise addition of HCl (6 M). To perform gastric digestion, 250 μL of pepsin solution (40 mg/mL pepsin from porcine gastric mucosa (1:2500 U/mg, Chem-Supply, Gillman, SA, Australia) dissolved in 0.1 M HCl) was added into the meat slurry and shaken manually to mix well. The mixture was incubated with continuous shaking at 37 °C for 120 min using an orbital mixer (RATEK Instruments, Boronia, VIC, Australia) placed in an incubator (Clayson IM550, Sydney, NSW, Australia). After 120 min constant shaking, tubes for gastric digestion were immediately centrifuged (4500 rpm, 20 min, 18 °C) to separate solid and liquid digesta.

4.4.2. Small Intestine Digestion

The sample tubes identified for small intestinal digestion were processed further. To these tubes, 4 mL of buffer containing 0.1 M $NaHCO_3$ and 12 mM $CaCl_2$ was added dropwise to slowly raise the pH to 5.7. The digesta samples were mixed well and incubated for a further 30 min at 37 °C under constant shaking. This intermediate step was integrated to mimic the transition from the gastric to the small intestine environment. To start the small intestinal digestion, the pH of the mixture was further raised to 7.0 by the dropwise addition of 1 M NaOH followed by the addition of 1 mL of pancreatin-bile solution (8 mg/mL pancreatin from porcine pancreas (102557, USP Grade, MP Biomedicals, LLC, Illkirch, France) and 12 mg/mL porcine bile extract (B8631, Sigma-Aldrich, St. Louis, MO, USA) in 0.1 M $NaHCO_3$). The digesta was again incubated at 37 °C for 120 min with constant shaking.

4.4.3. Separation of Liquid and Solid Digesta

After completion of each digestion step (gastric and small intestinal) sample tubes were centrifuged with a Sigma 6K15 centrifuge (Sigma Centrifuges, Osterode, Germany; 4500 rpm, 20 min, 18 °C) to separate solid and liquid digesta. Liquid and solid digesta were stored separately at −40 °C until extracted and analyzed for indospicine.

4.5. Preparation of External and Internal Standards for LC-MS/MS Analysis

Indospicine analysis of all samples was conducted by LC-MS/MS utilizing synthetic indospicine as an external standard for preparation of a calibration curve and deuterium-labelled D_3-L-indospicine as a stable isotopically labeled internal standard added to all samples and standard solutions to overcome matrix effects.

Synthesized indospicine as the external standard (>99% pure) and deuterium-labelled D_3-L-indospicine (>99% pure) as the internal standard were kindly provided by Dr. Robert Lang and Prof. James De Voss, School of Chemistry and Molecular Biosciences, The University of Queensland, St. Lucia, QLD, Australia [42]. Stock solutions for both internal and external standards were prepared in de-ionized H_2O with 0.1% heptafluorobutyric acid (HFBA) and were frozen at −20 °C until used. Internal (1 mg/L) and external (0.002–2 mg/L) standard solutions for indospicine LC-MS/MS quantification were prepared from the stock solutions and were stored frozen at −20 °C for no longer than a month before used.

4.6. Extraction of Indospicine from Camel Meat Samples and Digesta

Camel meat samples (uncooked, cooked, and both solid and liquid digesta) were extracted and analyzed by a previously validated and published liquid chromatography-tandem mass spectrometry method [32]. Prior to analysis, centrifugal filter units (Amicon® Ultra 0.5 mL 3K, Merck, Bayswater, VIC, Australia) were pre-rinsed and centrifuged (Microcentrifuge 5424, Eppendorf, North Ryde, NSW, Australia) at 10,000 rpm for 20 min with de-ionized water (2 × 300 μL) to remove glycerine, then inverted and spun for 1 min at 1000 rpm.

Minced un-cooked camel meat was thawed, weighed (0.5 g), and mixed with 0.1% HFBA (25 mL), followed by homogenization (Polytron T25, Labtek, Brendale, QLD, Brendale, Australia) for 15 s. The homogenized samples were chilled (4 °C) for 20 min and then centrifuged at 4500 rpm for 20 min at 18 °C. Aliquots of 1.0 mL of the resulting supernatants were spiked with 100 μL internal standard (D_3-L-indospicine, 1 mg/L in 0.1% HFBA), vortexed for 10 s, and a 450 μL portion was transferred into pre-rinsed centrifugal filters. The filtered sample mixture was then centrifuged (10,000 rpm, 20 min) and transferred to a limited volume insert (≈350 μL) for LC-MS/MS analysis.

Cooked meat and solid digesta were extracted and processed in a similar fashion to the raw meat. Liquid digesta (500 μL) was mixed with 0.1% HFBA (5 mL) and processed in a similar manner. All quantitations were calculated back to total indospicine content (μg) in either solid or liquid phase.

4.7. LC-MS/MS Analysis of Samples

Separation of the indospicine was achieved using a Waters ACQUITY UPLC® system (Waters, Lane Cove, NSW, Australia) equipped with a Waters BEH C18 column (1.7 μm, 100 mm length, 2.1 mm i.d.) at 30 °C and a flow rate of 0.2 mL/ min. The mobile phase was a mixture of (A) H_2O with 0.1% HFBA (*v*/*v*; pH 2.15) and (B) acetonitrile with 0.1% HFBA with the following gradient: 99% A to 70% A in 4 min, 70% A isocratic for 3 min, 70% A to 99% A in 1 min and 99% A for 2 min.

MS/MS detection was carried out using a Waters Micromass Quattro Premier triple quadrupole mass spectrometer with an electrospray ionisation (ESI) source operated in positive mode as previously described [32]. Eluted indospicine was quantitated utilizing selected reaction monitoring (SRM) transitions of m/z 174 → 111 (verified by transition of m/z 174 → 157) for indospicine, and m/z 177 → 114 (verified by transition of m/z 177 → 113) for D_3-L-indospicine as internal standard. The capillary voltage was 2.79 kV; cone gas flow was 50 L/h; desolvation gas flow was 600 L/h. The source and desolvation temperatures were set at 150 °C and 350 °C, respectively. Argon gas collision energy of indospicine (15 and 12 eV) and D_3-L-indospicine (15 and 15 eV) were set with cone voltage at 25 V.

4.8. Statistics

The data generated were processed using Microsoft Excel® 2010 (Microsoft, Redmond, WA, USA). Statistical analysis was conducted using ANOVA (GraphPad Prism, Version 6, La Jolla, CA, USA) with a completely randomized design. Differences were considered significant when p-values were below 0.05.

Author Contributions: Study design, M.T.F., S.A.O., G.N. and M.E.N.; Investigation and Formal Analysis, S.S. and C.G.; Writing-Original Draft Preparation, S.S.; Writing-Review & Editing, M.T.F., S.A.O., G.N., and M.E.N.; Project Administration, M.T.F.; Funding Acquisition, M.T.F.

References

1. Tan, E.T.T.; Materne, C.M.; Silcock, R.G.; D'Arcy, B.R.; Al Jassim, R.; Fletcher, M.T. Seasonal and species variation of the hepatotoxin indospicine in Australian *Indigofera* legumes as measured by UPLC-MS/MS. *J. Agric. Food Chem.* **2016**, *64*, 6613–6621. [CrossRef] [PubMed]
2. Fletcher, M.T.; Al Jassim, R.A.M.; Cawdell-Smith, A.J. The occurrence and toxicity of indospicine to grazing animals. *Agriculture* **2015**, *5*, 427–440. [CrossRef]
3. Fletcher, M.T.; Reichmann, K.G.; Ossedryver, S.M.; McKenzie, R.A.; Carter, P.D.; Blaney, B.J. Accumulation and depletion of indospicine in calves (*Bos taurus*) fed creeping indigo (*Indigofera spicata*). *Anim. Prod. Sci.* **2018**, *58*, 568–576. [CrossRef]
4. Tan, E.T.T.; Al Jassim, R.; Cawdell-Smith, A.J.; Ossedryver, S.M.; D'Arcy, B.R.; Fletcher, M.T. Accumulation, persistence, and effects of indospicine residues in camels fed *Indigofera* plant. *J. Agric. Food Chem.* **2016**, *64*, 6622–6629. [CrossRef] [PubMed]
5. Hegarty, M.P.; Kelly, W.R.; McEwan, D.; Williams, O.J.; Cameron, R. Hepatotoxicity to dogs of horse meat contaminated with indospicine. *Aust. Vet. J.* **1988**, *65*, 337–340. [CrossRef] [PubMed]
6. Christie, G.S.; Madsen, N.P.; Hegarty, M.P. Acute biochemical changes in rat liver induced by the naturally-occurring amino acid indospicine. *Biochem. Pharmacol.* **1969**, *18*, 693–700. [CrossRef]
7. Nordfeldt, S.; Henke, L.A.; Morita, K.; Matsumoto, H.; Takahash, M.; Younge, O.R.; Willers, E.H.; Cross, R.F. Feeding tests with *Indigofera endecaphylla* Jacq. (Creeping indigo) and some observations on its poisonous effects on domestic animals. *Univ. Hawaii Agric. Exp. Stat. Technol. Bull.* **1952**, *15*, 5–23.
8. Ossedryver, S.M.; Baldwin, G.I.; Stone, B.M.; McKenzie, R.A.; Eps, A.W.; Murray, S.; Fletcher, M.T. *Indigofera spicata* (creeping indigo) poisoning of three ponies. *Aust. Vet. J.* **2013**, *91*, 143–149. [CrossRef] [PubMed]
9. Madsen, N.P.; Christie, G.S.; Hegarty, M.P. Effect of indospicine on incorporation of L-arginine-14C into protein and transfer ribonucleic acid by cell-free systems from rat liver. *Biochem. Pharmacol.* **1970**, *19*, 853–857. [CrossRef]
10. Christie, G.S.; De Munk, F.G.; Madsen, N.P.; Hegarty, M.P. Effects of an arginine antagonist on stimulated human lymphocytes in culture. *Pathology* **1971**, *3*, 139–144. [CrossRef] [PubMed]
11. Hegarty, M.P.; Pound, A.W. Indospicine, a hepatotoxic amino acid from *Indigofera spicata*: Isolation, structure, and biological studies. *Aust. J. Biol. Sci.* **1970**, *23*, 831–842. [CrossRef]
12. Pearn, J.H.; Hegarty, M.P. Indospicine—The teratogenic factor from *Indigofera spicata* extract causing cleft palate. *Br. J. Exp. Pathol.* **1970**, *51*, 34–36. [PubMed]
13. Christie, G.S.; Wilson, M.; Hegarty, M.P. Effects on the liver in the rat of ingestion of *Indigofera spicata*, a legume containing an inhibitor of arginine metabolism. *J. Pathol.* **1975**, *117*, 195–205. [CrossRef] [PubMed]
14. Hutton, E.M.; Windrum, G.M.; Kratzing, C.C. Studies on the toxicity of *Indigofera endecaphylla*: II. Toxicity for mice. *J. Nutr.* **1958**, *65*, 429–440. [CrossRef] [PubMed]
15. Hutton, E.M.; Windrum, G.M.; Kratzing, C.C. Studies on the toxicity of *Indigofera endecaphylla*: I. Toxicity for rabbits. *J. Nutr.* **1958**, *64*, 321–337. [CrossRef] [PubMed]
16. Kelly, W.R.; Young, M.P.; Hegarty, M.P.; Simpson, G.D. The hepatotoxicity of indospicine in dogs. In *Poisonous Plants*; James, L.F., Keeler, R.F., Bailey, E.M., Cheeke, P.R., Hegarty, M.P., Eds.; Iowa State University Press: Ames, IA, USA, 1992; pp. 126–130.
17. FitzGerald, L.M.; Fletcher, M.T.; Paul, A.E.; Mansfield, C.S.; O'Hara, A.J. Hepatotoxicosis in dogs consuming a diet of camel meat contaminated with indospicine. *Aust. Vet. J.* **2011**, *89*, 95–100. [CrossRef] [PubMed]
18. Tan, E.T.T.; Al Jassim, R.; D'Arcy, B.R.; Fletcher, M.T. In vitro biodegradation of hepatotoxic indospicine

in *Indigofera spicata* and its degradation derivatives by camel foregut and cattle rumen fluids. *J. Agric. Food Chem.* **2017**, *65*, 7528–7534. [CrossRef] [PubMed]
19. Bobrich, A.; Fanning, K.J.; Rychlik, M.; Russell, D.; Topp, B.; Netzel, M. Phytochemicals in Japanese plums: Impact of maturity and bioaccessibility. *Food Res. Int.* **2014**, *65*, 20–26. [CrossRef]
20. Braga, A.C.; Alves, R.N.; Maulvault, A.L.; Barbosa, V.; Marques, A.; Costa, P.R. In vitro bioaccessibility of the marine biotoxin okadaic acid in shellfish. *Food Chem. Toxicol.* **2016**, *89*, 54–59. [CrossRef] [PubMed]
21. Ekbatan, S.S.; Sleno, L.; Sabally, K.; Khairallah, J.; Azadi, B.; Rodes, L.; Prakash, S.; Donnelly, D.J.; Kubow, S. Biotransformation of polyphenols in a dynamic multistage gastrointestinal model. *Food Chem.* **2016**, *204*, 453–462. [CrossRef] [PubMed]
22. Versantvoort, C.H.; Oomen, A.G.; Van de Kamp, E.; Rompelberg, C.J.; Sips, A.J. Applicability of an in vitro digestion model in assessing the bioaccessibility of mycotoxins from food. *Food Chem. Toxicol.* **2005**, *43*, 31–40. [CrossRef] [PubMed]
23. Hegarty, M.P. Non-metallic chemical residues in toxic plants with potential importance to animal and human health. In *Vet Update '92'*; Osborne, H.G., Ed.; University of Queensland. Continuing Professional Education: Brisbane, Australia, 1992; pp. 323–332.
24. Pollitt, S.; Hegarty, M.P.; Pass, M.A. Analysis of the amino acid indospicine in biological samples by high performance liquid chromatography. *Nat. Toxins* **1999**, *7*, 233–240. [CrossRef]
25. Tan, E.T.T.; Al Jassim, R.; D'Arcy, B.R.; Fletcher, M.T. Level of natural hepatotoxin (Indospicine) contamination in Australian camel meat. *Food Addit. Contam. Part A* **2016**, *33*, 1587–1595. [CrossRef] [PubMed]
26. Andrews, L.; Clarke, M.; Lethbridge, M.; Sobels, J. *Central Australian Commercial Camel Meat Viability Study, Report to the Northern Territory and South Australian Governments, 2015–2016. 71p*; Agriknowledge: Mylor, South Australia, 2016; Available online: https://agriknowledge2.weebly.com/agribusiness.html (accessed on 20 August 2018).
27. enHealth Council. Environmental Health Risk Assessment: Guidelines for Assessing Human Health Risks from Environmental Hazards. 2012. Available online: http://www.eh.org.au/documents/item/916 (accessed on 2 August 2018).
28. ABS. Australian Health Survey: Nutrition First Results—Foods and Nutrients, 2011-12. 2014. Available online: http://www.abs.gov.au/ausstats/abs@.nsf/detailspage/4364.0.55.0072011-12 (accessed on 2 August 2018).
29. Sultan, S.; Osborne, S.A.; Addepalli, R.; Netzel, G.; Netzel, M.E.; Fletcher, M.T. Indospicine cytotoxicity and transport in human cell lines. *Food Chem.* **2018**, *267*, 119–123. [CrossRef] [PubMed]
30. Aylward, J.; Haydock, K.; Strickland, R.; Hegarty, M. *Indigofera* species with agronomic potential in the tropics. Rat toxicity studies. *Crop Pasture Sci.* **1987**, *38*, 177–186. [CrossRef]
31. Gardner, D.R.; Riet-Correa, F. Analysis of the toxic amino acid indospicine by liquid chromatography-tandem mass spectrometry. *Int. J. Poisonous Plant Res.* **2011**, *1*, 20–27.
32. Tan, E.T.T.; Fletcher, M.T.; Yong, K.W.L.; D'Arcy, B.R.; Al Jassim, R. Determination of hepatotoxic indospicine in Australian camel meat by ultra-performance liquid chromatography–tandem mass spectrometry. *J. Agric. Food Chem.* **2014**, *62*, 1974–1979. [CrossRef] [PubMed]
33. Tornberg, E. Effects of heat on meat proteins—Implications on structure and quality of meat products. *Meat Sci.* **2005**, *70*, 493–508. [CrossRef] [PubMed]
34. Morita, J.-I.; Yasui, T. Involvement of hydrophobic residues in heat-induced gelation of myosin tail subfragments from rabbit skeletal muscle. *Agric. Biol. Chem.* **1991**, *55*, 597–599.
35. Hur, S.J.; Lim, B.O.; Decker, E.A.; McClements, D.J. In vitro human digestion models for food applications. *Food Chem.* **2011**, *125*, 1–12. [CrossRef]
36. Netzel, M.; Netzel, G.; Zabaras, D.; Lundin, L.; Day, L.; Addepalli, R.; Osborne, S.A.; Seymour, R. Release and absorption of carotenes from processed carrots (*Daucus carota*) using in vitro digestion coupled with a Caco-2 cell trans-well culture model. *Food Res. Int.* **2011**, *44*, 868–874. [CrossRef]
37. Hegarty, M.P. Toxic amino acids in foods of animals and man. *Proc. Nutr. Soc. Aust.* **1986**, *11*, 73–81.
38. Tan, E.T.T.; Yong, K.W.L.; Wong, S.-H.; D'Arcy, B.R.; Al Jassim, R.; De Voss, J.J.; Fletcher, M.T. Thermo-alkaline treatment as a practical degradation strategy to reduce indospicine contamination in camel meat. *J. Agric. Food Chem.* **2016**, *64*, 8447–8453. [CrossRef] [PubMed]
39. Pass, M.A. *Contaminated Horsemeat. Assessment and Prevention of Toxicity from Indospicine. RIRDC Report,*

Project No UQ-46A; Rural Industries Research & Development Corporation: Kingston, ACT, Australia, 2000; p. 9.

40. Kong, F.; Singh, R.P. Disintegration of solid foods in human stomach. *J. Food Sci.* **2008**, *73*, R67–R80. [CrossRef] [PubMed]
41. Collins, P.J.; Horowitz, M.; Maddox, A.; Myers, J.C.; Chatterton, B.E. Effects of increasing solid component size of a mixed solid/liquid meal on solid and liquid gastric emptying. *Am. J. Physiol. Gastrointest. Liver Physiol.* **1996**, *271*, G549–G554. [CrossRef] [PubMed]
42. Lang, C.-S.; Wong, S.-H.; Chow, S.; Challinor, V.L.; Yong, K.W.L.; Fletcher, M.T.; Arthur, D.M.; Ng, J.C.; De Voss, J.J. Synthesis of L-indospicine, [5,5,6-$^2H_3$]-L-indospicine and L-norindospicine. *Org. Biomol. Chem.* **2016**, *14*, 6826–6832. [CrossRef] [PubMed]

9

Consumption of Minerals, Toxic Metals and Hydroxymethylfurfural: Analysis of Infant Foods and Formulae

Christian Vella [1] and Everaldo Attard [2,*]

[1] Department of Pharmacy, Faculty of Medicine and Surgery, University of Malta, Msida MSD 2080, Malta; christian.g.vella@gov.mt
[2] Division of Rural Sciences and Food Systems, Institute of Earth Systems, University of Malta, Msida MSD 2080, Malta
* Correspondence: everaldo.attard@um.edu.mt

Abstract: Infant foods and formulae may contain toxic substances and elements which can be neo-formed contaminants or derived from raw materials or processing. The content of minerals, toxic elements, and hydroxymethylfurfural (HMF) in infant foods and formulae were evaluated. The effect of storage temperature on HMF formation in infant formulae and its potential as a quality parameter was also evaluated. Prune-based foods contained the highest HMF content. HMF significantly increased when the storage temperature was elevated to 30 °C for 21 days. All trace elements were present in adequate amounts, while the concentration of nickel was higher when compared to those of other studies. The study indicates that HMF can be used as a quality indicator for product shelf-life and that the concentrations of minerals and toxic elements vary greatly due to the diverse compositions of foods and formulae. Such contaminants need to be monitored as infants represent a vulnerable group compared to adults.

Keywords: infant formulae; infant foods; minerals; toxic metals; hydroxymethylfurfural; storage conditions; safety

---

1. Introduction

Infants are more sensitive than adults to food contaminants due to a higher rate of uptake by the gastrointestinal tract, an incompletely developed blood–brain barrier, an undeveloped detoxification system, and high food consumption relative to body mass [1]. Heavy metals are contaminants which can accumulate in infant foods through the food chain, during food processing or leakage from packaging materials [2]. Their effect on living organisms depends on the nature and concentration of the element concerned. Some elements are an essential part of the human diet, while others can be xenobiotic and highly toxic [3]. Maximum levels for heavy metals in infant foods and formulae are only defined for cadmium, lead, and tin through Regulation (CE) No. 1881/2006 and subsequent updates [1]. Contaminants can also be formed during the heating or preservation of foods and can pose harm to human health. These are termed neo-formed contaminants. Hydroxymethylfurfural (HMF) is a neo-formed contaminant in food, being an intermediate in the Maillard reaction which consists of a series of reactions, starting with a reaction between the carbonyl group of a reducing sugar with a free amino group, or it can result from the direct dehydration of sugars [4]. It is practically not present in fresh food but it is found in variable amounts in processed foods, such as jams, fruit juices, and syrups, as its synthesis depends on the temperature, pH, concentration of saccharides, presence of organic acids, and presence of divalent ions [5].

The aim of the study was to assess the content of minerals, toxic metals (Cr, Cu, Hg, Ni, Zn, Mn and Fe), and HMF in infant foods and formulae. This would provide an insight into the potential effects of undesirable substances within a vulnerable group.

2. Materials and Methods

2.1. Sample Collection

Thirty-two infant foods from four different manufacturers were randomly selected via convenience sampling from local pharmacies and supermarkets, and categorized as apple, pear, prune, fish, poultry, and ruminant-based foods. Six infant formulae from 3 different manufacturers were randomly collected from local pharmacies and were categorized as beginner infant formulae (0–6 months) and follow-on formulae (6–12 months).

2.2. Determination of pH

The pH of samples was measured with a Thermo scientific Orion Star A215 pH meter (Life Technologies Ltd., Paisley, UK). For infant foods, the pH was measured directly using a probe for viscous samples while for the powdered infant formulae, a reconstitution in de-ionized water at a ratio of 1:10 was carried out.

2.3. Determination of HMF

HMF content was determined according to a spectrophotometric method after White [6]. The determination of HMF content was based on the determination of UV absorbance of HMF at 284 nm (SpectroStar-Nano, BMG, Labtech, Ortenberg, Germany). The difference between the absorbance of a clear sample solution and the sample solution after the addition of 0.2% $NaHSO_3$ was determined to avoid the interference of other compounds at this wavelength. Five grams of each of the baby foods and infant formulae were tested for HMF content at a temperature of 18 °C. Furthermore, the infant formulae were incubated and maintained at 30 °C for 21 days in a water bath. The same HMF test procedure was used to determine the effect of temperature on HMF levels. Limits of detection (LOD) and limits of quantification (LOQ) for HMF were calculated as 3 s/m and 10 s/m, respectively, where s refers to the standard deviation of the intensity of blank samples and m refers to the slope of the calibration curve for HMF (Table 1).

Table 1. Hydroxymethylfurfural (HMF), mineral and toxic metal wavelength of detection, regression value (R^2), limits of detection (LOD) and limits of quantification (LOQ).

Method	Element	Wavelength (nm)	R^2	LOD (mg/kg)	LOQ (mg/kg)
White	HMF	284.000	0.99000	0.1122	0.3400
MP-AES	Cr	425.433	0.99999	0.0005	0.0014
MP-AES	Cu	324.754	1.00000	0.0007	0.0022
MP-AES	Hg	253.652	0.99990	0.0789	0.2391
MP-AES	Ni	352.454	0.99998	0.0056	0.0169
MP-AES	Mn	403.076	1.00000	0.0042	0.0127
MP-AES	Fe	259.940	0.99986	0.0037	0.0113
MP-AES	Zn	213.857	1.00000	0.0301	0.0912

2.4. Determination of Trace Elements

For mineral and toxic metal analysis, the samples were mineralized by digesting 1 g of the sample using 5 mL of 5% HNO_3 at 80 °C, followed by 2 mL of 34.5–36.5% H_2O_2 after the acid evaporated. Further mineralization of the sample was carried out by ashing at 500 °C in a muffle furnace (Wisetherm, Wisd, Laboratory Instruments, Germany) for 6 h. The ash was reconstituted in 5 mL of 5% HNO_3 and filtered. Deionized water was added up to 50 mL and the samples were quantitatively analyzed using a Microwave Plasma-Atomic Emission Spectrometer (MP-AES 4100, Agilent Technologies Inc., Santa Clara, CA, USA). The method was validated according to Berg [7]. The LOD and LOQ for each heavy metal were calculated as 3 s/m and 10 s/m, respectively, with respect to the calibration curve for each element (Table 1).

2.5. *Statistical Analysis*

All measurements were conducted in triplicate and average results were reported. The statistical program Prism 5 (GraphPad Software Inc., San Diego, CA, USA) was used for data analysis. The results for the heavy metal elements and hydroxymethylfurfural contents were analyzed by one-way ANOVA with the Bonferroni post hoc test to compare the statistical difference between means of the data sets and their mean difference. The same statistical test was carried out to compare the mean content of hydroxymethylfurfural between infant formulae stored at room temperature and infant formulae stored at 30 °C for 21 days. Principal component analysis and Pearson correlations were conducted on all samples, using XLSTAT v.2014.4.04 (Microsoft, version 19.4.46756, SAS Institute Inc., Marlow, Buckinghamshire, UK) to determine any clustering of minerals and toxic metals. A P value less than 0.05 was considered as statistically significant.

3. Results

A total of 38 samples were assessed for HMF content and selected heavy metal elements. The infant foods ($n = 32$) exhibited variable amounts of HMF, ranging from 0.89 mg/kg to 144 mg/kg, with the lowest content being present in poultry-based infant foods, while the highest content was present in prune-based products (Table 2). The HMF content in infant formulae ($n = 6$) ranged from 0.29 mg/kg to 7.87 mg/kg when examined at room temperature. The HMF content in all types of infant formulae significantly increased ($p \leq 0.05$) after being stored at 30 °C for 21 days and ranged from 1.80 mg/kg to 9.43 mg/kg (Figure 1). The mean heavy metal content of Cr, Cu, Hg, Ni, Fe, Mn, and Zn is shown in Table 3. The trace elements were detected in all infant food and formulae samples analyzed except for Hg, which was detected only in one sample from the pear-based infant food category ($n = 6$).

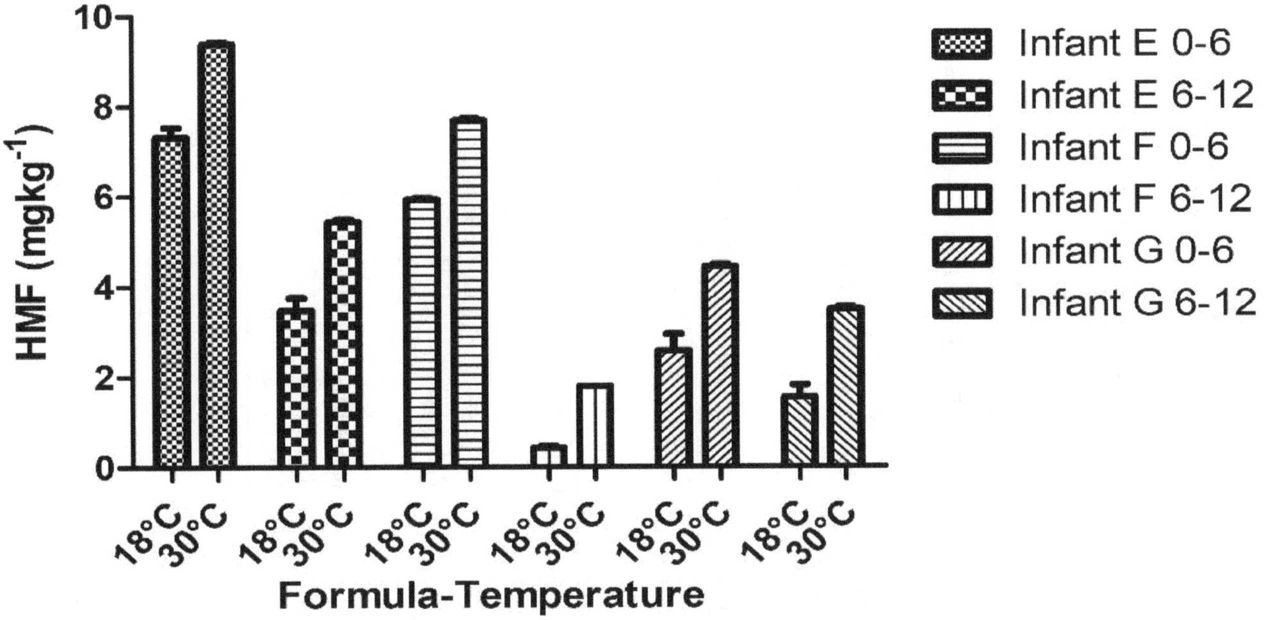

Figure 1. The HMF content in all types of infant formulae.

Table 2. HMF content (mg/kg) and pH of infant foods and formulae.

Mean HMF and pH Values	Prune-Based Food	Pear-Based Food	Apple-Based Food	Fish-Based Food	Poultry-Based Food	Ruminant Meat-Based Food	Formulae 0–6 Months	Formulae 6–12 Months
Mean HMF mg/kg (at 18 °C)	99.10 ± 11.45	6.327 ± 0.4945	9.674 ± 1.004	3.133 ± 0.2191	1.858 ± 0.1807	2.359 ± 0.1171	5.27 ± 1.40	1.81 ± 0.88
Mean HMF mg/kg (at 30 °C)	nd	nd	nd	nd	nd	nd	7.17 ± 1.44	3.57 ± 1.05
Mean pH	3.31 ± 0.05	3.558 ± 0.06	3.31 ± 0.04	5.64 ± 0.82	5.61± 0.16	5.17 ± 0.38	6.76 ± 0.17	6.66 ± 0.14

nd = not determined.

Table 3. Metal content (mg/kg) in infant foods and formulae.

Mean Metal Content (mg/kg)	Apple-Based (*n* = 6)	Pear-Based (*n* = 6)	Prune-Based (*n* = 4)	Fish-Based (*n* = 8)	Poultry-Based (*n* = 4)	Ruminant Meat-Based (*n* = 4)	Formulae 0–6 (*n* = 3)	Formulae 6–12 (*n* = 3)
Cr	0.21 ± 0.06	0.09 ± 0.03	0.18 ± 0.07	0.07 ± 0.02	0.04 ± 0.02	0.02 ± 0.01	0.29 ± 0.05	0.24 ± 0.03
Cu	0.65 ± 0.05	0.93 ± 0.11	0.66 ± 0.07	0.78 ± 0.07	0.68 ± 0.04	0.75 ± 0.07	3.33 ± 0.24	3.37 ± 0.21
Hg	nd	0.12 ± 0.12	nd	nd	nd	nd	nd	nd
Ni	0.63 ± 0.08	0.85 ± 0.03	0.86 ± 0.08	0.81 ± 0.06	1.07 ± 0.22	0.73 ± 0.06	0.76 ± 0.00	0.82 ± 0.06
Fe	0.86 ± 0.08	1.18 ± 0.26	1.67 ± 0.4	1.55 ± 0.14	1.64 ± 0.07	1.67 ± 0.41	18.34 ± 2.51	18.87 ± 3.06
Mn	4.93 ± 0.36	3.54 ± 0.06	3.22 ± 0.12	2.90 ± 0.11	2.37 ± 0.04	3.25 ± 1	2.13 ± 0.41	2.05 ± 0.21
Zn	1.07 ± 0.78	8.54 ± 8.05	1.03 ± 0.34	2.61 ± 1.46	3.19 ± 0.72	5.76 ± 0.69	27.24 ± 2.77	33.00 ± 0.95

nd = not detected.

4. Discussion

Toxic substances may be either present in the raw materials or evolve during the processing of the raw materials into the finished products. Although the assurance of food quality is the responsibility of the producer and manufacturer, authorities worldwide do not control food products for safety. Several reports have shown that baby foods may contain contaminants, some of which include microorganisms [8,9], mycotoxins [10,11], aromatic compounds [12,13], furans [14,15], and metals [16–18].

The HMF content was determined at a temperature of 18°C for the baby foods, and at two temperatures (18 and 30 °C) for the infant formulae. Since baby foods in individual jars are consumed within one meal and the foods have undergone extensive processing in industry, the baby foods were not tested at a temperature of 30 °C for a 21-day period. It is more likely that for infant formulae, repeated quantities are consumed from the same can over a period of time. There is no limit for the HMF content in foods, apart for honey at 40 mg/kg in general environmental conditions, 80 mg/kg for honey produced in tropical climates, and 15 mg/kg for honey with low enzymatic activity [19]. This makes it difficult to ascertain whether acceptable or excessive levels of HMF are found in the studied foods. The results from studies carried out by Kalábová and Večerek [20], and Čížková and coworkers [21], for the determination of the HMF content in infant foods, reported ranges from 2.10 mg/kg to 9.80 mg/kg and 4.10 mg to 28.90 mg/kg, respectively. The current study showed a larger spread of values nearly fifteen times the upper limit, observed by Kalábová and Večerek [20], and seven times the upper limit, observed by Čížková and coworkers [21]. This variability could be related to the type of food tested, since this varied in the different studies. A significant difference in the HMF content of prune-based infant foods compared to other infant foods ($p \leq 0.05$) was observed and these were identified as a potential source of high HMF consumption in children. Products processed from prunes, such as pitted prunes and prune juices, have been reported to have an HMF content as high as 291 mg/kg and 528 mg/L, respectively. The higher HMF value in fruit-based foods is due to greater carbohydrate degradation as a consequence of the Maillard reaction, which is favored by a lower pH (Table 2). On the other hand, a higher furan content is present in vegetable-based foods compared to fruit-based foods. This is related to either a greater furfural content or a greater ascorbic acid degradation [14].

The HMF content in infant formulae observed in the study, ranging from 0.29 mg/kg to 7.87 mg/kg, was comparable with other studies, such as that by Michalak and coworkers [22], reporting an HMF content between 1.22 mg/kg and 8.20 mg/kg. With respect to the changes of HMF content during storage at 30 °C for 21 days, the HMF content in all formulae increased significantly after storage ($p \leq 0.05$). This temperature-dependent effect was shown in various studies, such as that by Chávez-Servín and coworkers [23], where they demonstrated a similar significance and proportional increase after 70 days of storage at 25 °C. However, the relationship between HMF concentration and pH in infant formulae was not significant ($p > 0.05$). Therefore, HMF synthesis was not dependent on the pH of infant formulae. In a study conducted earlier by Chávez-Servín and coworkers [24], it was observed that infant formulae at a neutral pH for a period of 12 months of storage exhibited insignificant formation of HMF.

There was a variation in the absorbance value with respect to the concentration of the heavy metal element, and, therefore, a strong positive linear relationship was present between the two parameters ($r = 0.9986$). The low LOD and LOQ values demonstrate that the MP-AES method for the analysis of heavy metal elements was highly sensitive (Table 1). The heavy metal content varied widely due to many factors, such as differences between food types, the characteristics of the manufacturing practices and processes, and possible contamination during processing. The present study demonstrated wide variations in the concentration of the most essential and toxic elements in infant formulae and foods (Table 3). In the infant formulae, the manufacturer's fortification of essential elements resulted in concentrations many times higher than those found in foods, especially Fe, Zn, and Cu. The concentration of nickel in the samples, ranging from 0.63 mg/kg to 1.07 mg/kg, exceeded the reference value of 5 µg/kg bw/day set by the Food and Agriculture Organization/World Health Organization

(FAO/WHO) Joint Expert Committee on Food Additives (JECFA) [25], as the daily intake of Ni through infant formulae ranged from 7 μg/kg bw day to 19 μg/kg bw day. Mehrnia and Bashti [26] reported daily intake values of nickel through infant formula more than tenfold the reference value set by the JECFA. Nickel toxicity is associated with immediate and delayed hypersensitivity reactions. It has the potential to cause immunological disturbances and act as an immunotoxic agent in humans [27]. Only one sample was contaminated with Hg at a concentration of 0.7 mg/kg. Since Hg was detected in a pear-based food, the presence of methylmercury is excluded, as this bioaccumulates in fish. Therefore, this value cannot be compared to the EFSA [28], which establishes a TWI reference value of 1.3 μg/kg bw for methylmercury. Cruz and coworkers [29] reported infant formulae testing positive for mercury, with levels of 0.63 mg/kg and 0.83 mg/kg.

Factor analysis using principal components was used to identify latent traits within the data. Pearson correlation (Table 4) revealed that there were several correlations between the minerals and toxic metals. There were positive correlations between Cr and Cu, Fe, Zn ($r = 0.718$, 0.725 and 0.631), Cu and Fe, Zn ($r = 0.996$ and 0.984), and Fe with Zn ($r = 0.974$). There were negative correlations between Cu and Mn ($r = -0.636$), Mn with Fe, and Zn ($r = -0.654$ and -0.641). Two latent factors had an eigenvalue greater than 1, which together explained 80.04% of the total variance (Figure 2a). The factor loadings demonstrated the different groups of variables. For the first factor, the factor loadings of Cr, Cu, Fe, and Zn, and the second factor, weighed heavily on Hg, Ni, and Mn. Figure 2b demonstrates the factor scores of the two latent factors. Factor 1, on the horizontal axis, demonstrates the clustering of baby foods on the left hand side of the scatter plot, while the infant formulae scattered more on the right hand side. This demonstrates the distinction of the foods and formulae characteristics with respect to mineral and toxic metal values.

Table 4. Minerals and toxic metals found in the baby foods and infant formulae.

Variables	Cu	Hg	Ni	Fe	Mn	Zn
Cr	**0.718**	−0.217	−0.393	**0.725**	−0.090	**0.631**
Cu		−0.159	−0.108	**0.996**	**−0.636**	**0.984**
Hg			0.106	−0.236	0.211	−0.062
Ni				−0.091	−0.574	−0.073
Fe					**−0.654**	**0.974**
Mn						**−0.641**

Bold values represent significant correlations.

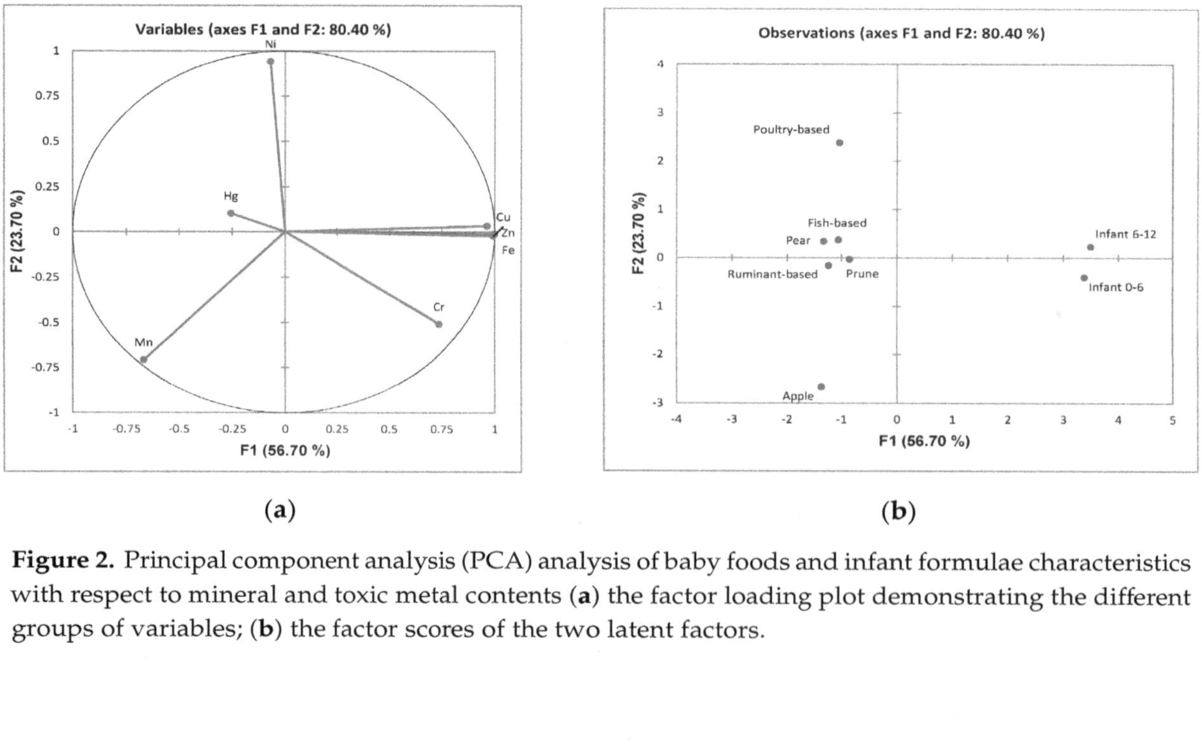

(a) (b)

Figure 2. Principal component analysis (PCA) analysis of baby foods and infant formulae characteristics with respect to mineral and toxic metal contents (**a**) the factor loading plot demonstrating the different groups of variables; (**b**) the factor scores of the two latent factors.

5. Conclusions

Opinions on the cytotoxicity, carcinogenic, and genotoxic potential of hydroxymethylfurfural vary, while certain minerals and toxic metals are known to be deleterious if consumed in large quantities. However, the concentrations of such metals vary depending on the food type used. Infant foods and formulae contained varying amounts of HMF and metals, thus, the total daily intake of these contaminants is affected by individual feeding patterns. Notably, a high HMF content was observed in prune-based infant foods. On the other hand, with regard to the metal contents, it was observed that infant foods contained Mn, Zn, Fe, Cu, and Cr, while infant formulae contained Zn, Fe, Cu, Mn, and Cr in decreasing order. There was a low presence of Ni and negligible quantities of Hg. Infants are within a vulnerable age group and have a restricted diet compared to other age groups, therefore, it is recommended that foods are monitored to ensure safe use. The setting up of limits with respect to this vulnerable group should be considered through further studies, using a greater diversification of samples that are subjected under varying conditions.

Author Contributions: Conceptualization, writing—review, editing and supervision: E.A.; methodology and investigation: C.V.

Acknowledgments: The authors acknowledge Adrian Bugeja-Douglas for technical support.

References

1. Mania, M.; Wojciechowska-Mazurek, M.; Starska, K.; Rebeniak, M.; Szynal, T.; Strzelecka, A.; Postupolski, J. Toxic Elements in Commercial Infant Food, Estimated Dietary Intake, and Risk Assessment in Poland. *Pol. J. Environ. Stud.* **2015**, *24*, 2525–2536. [CrossRef]
2. Ojo, R.J.; Olabode, O.S. Analysis of Heavy Metals and Hydrocyanic Acid in Selected Infant Formula in Abuja, Federal Capital Territory of Nigeria. *Sch. Acad. J. Biosci.* **2013**, *1*, 318–325.
3. Tamás, M.J.; Sharma, S.K.; Ibstedt, S.; Jacobson, T.; Christen, P. Heavy metals and metalloids as a cause for protein misfolding and aggregation. *Biomolecules* **2014**, *4*, 252–267. [CrossRef] [PubMed]
4. Capuano, E.; Fogliano, V. Acrylamide aestand 5-hydroxymethylfurfural (HMF): A review on metabolism, toxicity, occurrence in food and mitigation strategies. *LWT-Food Sci. Tech.* **2011**, *44*, 793–810. [CrossRef]
5. Santonicola, S.; Mercogliano, R. Occurrence and Production of Furan in Commercial Foods. *Ital. J. Food Sci.* **2016**, *28*, 2016–2155.
6. White, J.W., Jr. Spectrophotometric method for hydroxymethylfurfural in honey. *AOAC* **1979**, *62*, 509–514.
7. Berg, I. Validation of MP-AES at the Quantification of Trace Metals in Heavy Matrices with Comparison of Performance to ICP-MS. Available online: http://www.diva-portal.se/smash/get/diva2:853851/FULLTEXT01.pdf (accessed on 26 May 2019).
8. Simmons, B.P.; Gelfand, M.S.; Haas, M.; Metts, L.; Ferguson, J. *Enterobacter sakazakii* infections in neonates associated with intrinsic contamination of a powdered infant formula. *Infect. Control. Hosp. Epidemiol.* **1989**, *10*, 398–401. [CrossRef] [PubMed]
9. Mehall, J.R.; Kite, C.A.; Saltzman, D.A.; Wallett, T.; Jackson, R.J.; Smith, S.D. Prospective study of the incidence and complications of bacterial contamination of enteral feeding in neonates. *J. Pediatr. Surg.* **2002**, *37*, 1177–1182. [CrossRef] [PubMed]
10. Aidoo, K.E.; Mohamed, S.M.; Candlish, A.A.; Tester, R.F.; Elgerbi, A.M. Occurrence of fungi and mycotoxins in some commercial baby foods in North Africa. *Food Nutr. Sci.* **2011**, *2*, 751. [CrossRef]
11. Kabak, B. Aflatoxin M1 and ochratoxin A in baby formulae in Turkey: Occurrence and safety evaluation. *Food Control* **2012**, *26*, 182–187. [CrossRef]
12. Lorán, S.; Bayarri, S.; Conchello, P.; Herrera, A. Risk assessment of PCDD/PCDFs and indicator PCBs contamination in Spanish commercial baby food. *Food Chemical Toxicol.* **2010**, *48*, 145–151. [CrossRef] [PubMed]
13. Pandelova, M.; Piccinelli, R.; Kasham, S.; Henkelmann, B.; Leclercq, C.; Schramm, K.W. Assessment of dietary exposure to PCDD/F and dioxin-like PCB in infant formulae available on the EU market. *Chemosphere* **2010**, *81*, 1018–1021. [CrossRef] [PubMed]
14. Mesías, M.; Guerra-Hernández, E.; García-Villanova, B. Furan content in Spanish baby foods and its relation with potential precursors. *CyTA-J. Food* **2013**, *11*, 1–6. [CrossRef]

15. Madani-Tonekaboni, M.; Kamankesh, M.; Farsani, A.M.M.; Ferdowsi, R.; Mohammad, A. Determination of furfural (F) and hydroxylmethyl furfural (HMF) in baby formulas obtained from Tehran market using dispersive liquid-liquid microextraction (DLLME) followed by high-performance liquid chromatography. *Iran J. Nutr. Sci. Food Technol.* **2015**, *9*, 97–107.
16. Kazi, T.G.; Jalbani, N.; Baig, J.A.; Arain, M.B.; Afridi, H.I.; Jamali, M.K.; Shah, A.Q.; Memon, A.N. Evaluation of toxic elements in baby foods commercially available in Pakistan. *Food. Chem.* **2010**, *119*, 1313–1317. [CrossRef]
17. Al Khalifa, A.S.; Ahmad, D. Determination of key elements by ICP-OES in commercially available infant formulae and baby foods in Saudi Arabia. *Afr. J. Food Sci.* **2010**, *4*, 464–468.
18. Salah, F.A.; Esmat, I.A.; Mohamed, A.B. Heavy metals residues and trace elements in milk powder marketed in Dakahlia Governorate. *Int. Food Res. J.* **2013**, *20*, 1807–1812.
19. Boussaid, A.; Chouaibi, M.; Rezig, L.; Hellal, R.; Donsì, F.; Ferrari, G.; Hamdi, S. Physicochemical and bioactive properties of six honey samples from various floral origins from Tunisia. *Arab. J. Chem.* **2014**, *11*, 265–274. [CrossRef]
20. Kalábová, L.V.I.B.K.; Večerek, V. Hydroxymethylfurfural contents in foodstuffs determined by HPLC method. *J. Food Nutr. Res.* **2006**, *45*, 34–38.
21. Čížková, H.; ŠevČík, R.; Rajchl, A.; Voldřich, M. Nutritional quality of commercial fruit baby food. *Czech J. Food Sci.* **2009**, *27*, 134–137. [CrossRef]
22. Michalak, J.; Kuncewicz, A.; Gujska, E. Monitoring selected quality indicators of powdered infant milk formulae. *Pol. J. Food Nutr. Sci.* **2006**, *15*, 131.
23. Chávez-Servín, J.L.; de la Torre Carbot, K.; García-Gasca, T.; Castellote, A.I.; López-Sabater, M.C. Content and evolution of potential furfural compounds in commercial milk-based infant formula powder after opening the packet. *Food Chem.* **2015**, *166*, 486–491. [CrossRef]
24. Chávez-Servín, J.L.; Romeu-Nadal, M.; Castellote, A.I.; López-Sabater, M.C. Evolution of free mono-and di-saccharide content of milk-based formula powder during storage. *Food Chem.* **2006**, *97*, 103–108. [CrossRef]
25. WHO. Joint FAO/WHO. Expert Committee on Food Additives. In: Summary and Conclusions, 61st Meeting: Methyl Mercury. 2003. Available online: http://www.who.int/pcs/jecfa/Summary61.pdf (accessed on 2 May 2019).
26. Mehrnia, M.A.; Bashti, A. Evaluation of Toxic Element Contents in Infant Foods Commercially. *Iran Bull. Env. Pharmacol. Life Sci.* **2014**, *3*, 249–253.
27. Aguzue, O.C.; Kakulu, S.E.; Thomas, S.A. Flame atomic absorption spectrophotometric determination of heavy metals in selected infant formula in the Nigerian Market. *Arch. Appl. Sci. Res.* **2014**, *6*, 128–132.
28. EFSA. Scientific Opinion of the Panel on Contaminants in the Food Chain (CONTAM) on the risk for public health related to the presence of mercury and methylmercury in food. *EFSA J.* **2012**, *10*, 2985.
29. Cruz, G.C.; Din, Z.; Feri, C.D.; Balaoing, A.M.; Gonzales, E.M.; Navidad, H.M.; Schlaaff, M.M.F.; Winter, J. Analysis of toxic heavy metals (arsenic, lead and mercury) in selected infant formula milk commercially available in the Philippines by AAS. *E-Int. Sci. Res. J.* **2009**, *1*, 40–51.

10

Rapid Solid-Liquid Dynamic Extraction (RSLDE): A Powerful and Greener Alternative to the Latest Solid-Liquid Extraction Techniques

Daniele Naviglio [1], Pierpaolo Scarano [2], Martina Ciaravolo [1] and Monica Gallo [3,*]

1 Department of Chemical Sciences, University of Naples Federico II, via Cintia; Monte S. Angelo Complex, 80126 Naples, Italy
2 Department of Science and Technology, University of Sannio, Via Port'Arsa 11, 82100 Benevento, Italy
3 Department of Molecular Medicine and Medical Biotechnology, University of Naples Federico II, via Pansini 5, 80131 Naples, Italy
* Correspondence: mongallo@unina.it

Abstract: Traditionally, solid-liquid extractions are performed using organic and/or inorganic liquids and their mixtures as extractant solvents in contact with an insoluble solid matrix (e.g., the Soxhlet method) or using sequential atmospheric pressure systems that require long procedures, such as maceration or percolation. The objective of this procedure is the extraction of any compounds that can be carried out from the inner solid material to the outlet, resulting in a solution containing colorants, bioactive compounds, odorous substances, etc. Over the years, in the extraction techniques sector, there have been many important changes from the points of view of production, quality, and human and environmental safety due to improvements in technology. In more recent times, the interest of the scientific community has been aimed at the study of sustainable processes for the valorization of extracts from vegetables and food by-products, through the use of non-conventional (innovative) technologies that represent a valid alternative to conventional methods, generally through saving time and energy and the formation of fewer by-products. Therefore, with the development of principles based on the prevention of pollution, on a lower risk for human health, and on a low environmental impact, new systems have been implemented to reduce extraction times and solvent consumption, to improve efficiency, and to increase the productivity of the extracts. From this point of view, rapid solid-liquid dynamic extraction (RSLDE), performed using the Naviglio extractor, compared to traditional applications, is a technique that is able to reduce extraction times, generally leads to higher yields, does not require heating of the system, allows one to extract the active ingredients, and avoids their degradation. This technique is based on a new solid-liquid extraction principle named Naviglio's principle. In this review, after reviewing the latest extraction techniques, an overview of RSLDE applications in various research and production sectors over the past two decades is provided.

Keywords: solid-liquid extraction; green extraction; RSLDE; bioactive compounds; Naviglio extractor; Naviglio's principle

---

1. Introduction

Solid-liquid extraction processes, both traditional ones (maceration and percolation) and those introduced more recently (e.g., supercritical fluid extraction (SFE) and accelerated solvent extraction (ASE), are based on two fundamental principles: diffusion and/or osmosis. On the basis of these principles, it is possible to make some general forecasts in relation to the extractive system, and it is possible to roughly hypothesize the extraction times and yields with respect to a generic solid matrix (generally vegetable) [1]. Three variables are to be optimized to achieve the best extractive conditions:

by decreasing the "granulometry" of the solid, the extractive yield increases because of an increased surface area of contact between solid and liquid; the raising of the "temperature" of the system reduces the time of extraction due to the increase in diffusion phenomena (Fick's law); the increase of the "affinity" of the extraction liquid towards the compounds to be extracted increases the effectiveness of the extraction process (*similis similia solvuntur*). However, the extractive principles which these techniques are based on have no active effect on the characteristics of the process, such as extraction times, yield, and efficiency. In fact, once the conditions have been set, the system reaches an equilibrium condition that can change only by modifying some parameters, such as the temperature or the addition of other extraction liquid [2]. For this reason, it is suggested that the extractive batch must be mixed during extraction to avoid a partial extraction due to the slow diffusion of compounds extracted.

Though solid-liquid extraction is a technique that has been known for a long time and is still widely used, there are still many unknown aspects that require further investigation to fully understand the mechanism. In the field of solid-liquid extraction techniques, it is possible to distinguish conventional extraction techniques, including maceration, percolation, squeezing, counter-current extraction, extraction through Soxhlet, and distillation, from unconventional (or innovative) ones. Conventional extractions have been used for many years, although they have many drawbacks: they require the use of high quantities of expensive and pure solvents, since during the process they consume a high amount; they have a low selectivity of extraction; they have a high solvent evaporation rate during the process; and they are generally characterized by long extraction times and by the thermal decomposition of thermolabile compounds [3,4]. To overcome all these limitations, new and promising solid-liquid extraction techniques, which are defined as non-conventional, have been introduced, mainly in the industrial field, such as ultrasound-assisted extraction (UAE) [5], supercritical fluid extraction (SFE) [6], microwave-assisted extraction (MAE) [7], extraction with accelerated solvent [8], solid phase microextraction [9], enzyme-assisted extraction [10], and rapid solid-liquid extraction dynamic (RSLDE) via the Naviglio extractor [11]. On the other hand, the interest of the scientific community has recently been aimed at the study of sustainable processes, so all these extraction techniques have common objectives, including the extraction of active ingredients (bioactive compounds) from the vegetable matrices, as well as their by-products for the valorization of waste, to improve the selectivity of the processes, to isolate the bioactive compounds in more suitable forms for detection and separation, and to provide an effective and reproducible method that is independent of the variability of the sample matrices; furthermore, high extraction yields are preferable to promote the economy of the process [12–14].

However, the extraction procedure generally takes place in a single solution (a single-step process), and it is difficult to set two or more extraction stages, because of the rise in extractant volume and time. Only the Soxhlet extractor limits the solvent volume, because it uses the distillation of the solvent, and the process works with fresh solvent. This can be considered a multi-step extractive process. Vice versa, RSLDE is based on a different principle. In fact, "the generation, with a suitable solvent, of a negative pressure gradient between the outside and the inside of a solid matrix containing extractable material, followed by a sudden restoration of the initial equilibrium conditions, induces forced extraction of the compounds not chemically linked to the main structure of which the solid is made" [11]. RSLDE changes the philosophy of solid-liquid extraction; the extraction happens thanks to a negative gradient of pressure between the inner material and the outside of the solid matrix (high pressure inside and low pressure outside; Naviglio's principle). When the gradient of pressure is removed, the liquid flows out of the solid in a very fast manner and carries out all substances not chemically bonded to the main structure of the solid. This means that in this case the extraction is an "active" process because the gradient of pressure forces out the molecules, while techniques based on diffusion and osmosis are "passive" processes because the molecules are not forced out of the matrix.

According to this principle, the solid-liquid extraction process is first of all independent of the affinity that the compounds to be extracted from the solid matrix have towards the extracting solvent: they are, in fact, extracted by a difference of pressure between the liquid inside the matrix and the liquid on the outside of it. They are extracted out of the solid with a suction effect and can therefore also be extracted in solvents with opposite or different polarity. Furthermore, the pressure effect on the solid matrix and following the de-pressure leads to an active action with respect to the extraction process, as a small quantity of material is extracted at each pressure and depression cycle (the "active" solid-liquid extractive process), the extent of which is closely correlated with the pressure difference generated between the inside and the outside of the solid matrix and to the features of the solid matrix. Based on this new and innovative extractive principle, it has been made possible, in many cases, to use water as an extraction solvent, a condition that cannot be achieved with traditional techniques, such as maceration and percolation; in this case, the fermentative process is slower due to the movement of liquid around the solid, and this prevents the microorganisms from growing [15].

This review aims to give a brief overview of the various extraction techniques, focusing mainly on RSLDE and its various fields of application thanks to the introduction of an innovative solid-liquid principle of extraction.

2. State of the Art of Solid-Liquid Extraction Techniques

Solid-liquid extraction techniques are the basis of many analytical procedures for the preparation of samples and are reported in the official methods of analysis [16]. On the other hand, they are applied to the production of small quantities of homemade extracts such as alcoholic beverages and herbal teas [17,18]. These extraction procedures are also applied to industrial production. In fact, in many industrial processes, the initial phase of the preparation of a product requires the application of a solid-liquid extraction technique to isolate the extractable material contained in the most varied solid matrices, mainly vegetables. An important example is represented by medicinal plants, from which active ingredients with pharmacological properties are obtained; related fields are those of herbal medicine, cosmetics, and perfumery, which are the most ancient applications. In other industrial sectors such as the beverage industry, a solid-liquid extraction is used to obtain alcoholic extracts of fruit peels, flowers, leaves, etc., which are then mixed with water and sugar to obtain the finished product. The list could continue by referring to multiple industrial applications that are very similar.

The solid-liquid extraction is based on a simple phenomenon: if a solid matrix containing extractable compounds is immersed in a liquid, the latter begins to enrich itself with certain chemically related substances that move from the inside to the surface of the solid and then from the surface into the liquid. This principle is based on diffusion and osmosis and is performed by maceration, which is the simplest and most economical extraction technique and is therefore widely used. The maceration process requires only a closable glass or stainless steel container in which extractable solid is covered with liquid. To overcome the rapid saturation of liquid strictly around the solid, desultory agitation is required. Unfortunately, it is not always applicable, because it requires long contact times between the solid and the liquid; for example, plants cannot be macerated in water at room temperature for a very long time due to rotting phenomena. The production needs of the industry, which require the obtaining of large volumes of extracts in a short amount of time, have found an application in percolation extraction; in this case it is possible to process large quantities of solid material with large volumes of liquid and obtain the extract quite quickly, albeit sacrificing the efficiency of the extraction, which remains low due to the limited contact between the solid and the extracting liquid [19]. In this case, the solid matrix is not completely exhausted and could be re-extracted with another technique.

For special applications, such as the production of essential oils and, in general, compounds with low vapor pressure, it is possible to resort to steam distillation [20]. This solid-liquid extraction technique is particular in that it requires the transport of volatile compounds through a steam flow; since the isolated product is an essential oil, it can be considered a solid-liquid extraction technique. In any case, the extraction system is subjected to strong heating; therefore, the thermolabile compounds undergo transformations and consequently are not kept intact. As a result of this, steam distillation is not often applicable.

These examples serve to indicate that each of the solid-liquid extraction techniques that are currently used are not universally applicable since they are limited. Moreover, the extractive principle on which they are based is essentially linked to the phenomena of diffusion and osmosis of the substances contained in the solid, which tend to occupy the entire volume of the extracting liquid, after extraction. Therefore, desultory agitation of the extraction batch is necessary. To increase the efficiency of these extraction systems and to reduce the time of extraction, a temperature increase is used, which affects the increase in diffusion (Fick's law), in order to reduce extraction times and increase yields. Generally, this expedient is not often applicable (over 40 °C) to vegetable matrices, because they contain substances that degrade due to heat, especially active principles [21,22].

The use of ultrasounds for the extraction of active ingredients from medicinal plants leads to the same results as extraction by pressing (squeezing). Furthermore, the system heats up due to the prolonged treatment, the solid matrix is completely crushed, and a mixture that is very difficult to separate into its constituents is obtained. Among other things, the use of ultrasound energy of more than 20 kHz may have an effect on the active phytochemicals through the formation of free radicals [23]. However, due to its speed, its economic advantage, and the relatively low-cost technology involved, UAE is one of the techniques used in the industry for bioactive compound extractions. As a result, in many cases, ultrasounds can be a good alternative to pressing because it simplifies the extractive system [24].

An alternative extraction technique is based on the use of supercritical fluids, mainly based on the use of carbon dioxide. In the supercritical phase, carbon dioxide assumes the characteristics of a non-polar solvent and is comparable to liquid n-hexane; with this method, it is therefore possible to extract non-polar compounds from solid matrices. The advantage of this technique is that, at the end of the extraction, the solvent, the carbon dioxide, is removed in the form of gas, enabling the possibility of recovering the concentrated extracted compounds with a very low environmental impact (green extraction). This technique finds applications at an industrial level, such as the extraction of oil from seeds, caffeine from coffee, nicotine from tobacco, etc. [25], but it is still very expensive and not universally applicable due to the difficulty of changing the polarity of carbon dioxide and for the interference of water contained in solids [26].

Another extraction technique is Soxhlet extraction, which is reported as an official extraction method [27] for numerous analytical methods in which an initial preparation of a solid sample extract is expected. The Soxhlet method also uses system heating, since it is based on the principles of diffusion and osmosis, so it cannot be used for substances that degrade due to heat [28]. Soxhlet extraction is a good method for the extraction of high boiling substances such as polycyclic aromatic hydrocarbons (PAH), polychlorobiphenyls (PCBs), dioxine, triglycerides, and so on. Nowadays, an improved method to perform Soxhlet extraction is named Soxtec [29]; this process is based on the same principles; however, thanks to pressure control, it is possible to accelerate the recirculation of the extractant solvent. In this way, the process is about 10 times faster [30].

To increase extraction yields and reduce time, accelerated solvent extraction (ASE) can be used [31]. This technique is based on an increase in diffusion because it is possible to extract solids by using liquids operating above their boiling temperature while being maintained in a liquid state by the increase in pressure. The material to be extracted is placed in a cylindrical steel container, and the extracting solvent is introduced; the temperature of the system is raised above the boiling point of the solvent, which is maintained in the liquid state thanks to a simultaneous increase in pressure (the vial is sealed to resist high pressure values: 100–200 bar). After a short contact period, the solid matrix is completely extracted. With this technique, it is not possible to extract thermally unstable compounds [32].

In this paper, a review of innovative solid-liquid extraction technology is presented, which can be used as a valid alternative to the existing ones, RSLDE, which can be considered a green means of extraction. The application of green technology aims to preserve the natural environment and its resources, and to limit the negative influence of human involvement [33]. The philosophy of green chemistry is to develop and encourage the utilization of procedures that reduce and/or eliminate the use or production of hazardous substances. The extraction takes place for the generation of a negative pressure gradient from the inside towards the outside of the solid matrix, so it can be carried out at room temperature, or even sub-environmental temperatures [11]. The functioning of this innovative system is based on a new solid-liquid extractive principle, as it is not equivalent to others reported in literature. The patent of the instrument named the Naviglio extractor was released in 2000 [34] and registered in 1998. An extractive cycle consists of both static and dynamic phases. During the static phase, the liquid is maintained under pressure at about 10 bar on the solid to be extracted and is left long enough to let the liquid penetrate inside the solid and to balance the pressure between the inside and the outside of the solid (about 1–3 min). After this, at the beginning of the dynamic phase, the pressure immediately drops to atmospheric pressure, causing a rapid flowing of liquid from the inside to the outlet of the solid matrix. At this moment, there is a suction effect of the liquid from the inside towards the outside of the solid. This rapid displacement of the extracting solvent transports the extractable material (compounds not chemically linked) outwards. The cycles can be repeated until the solid runs out. Experimental tests carried out to date on more than 200 vegetables have shown that, working at a pressure of about 10 bar, most solid matrices, regardless of the degree of crumbling, can be extracted using about 30 extractive cycles (two-minute static phase; two-minute dynamic phase) that are completed in two hours. Furthermore, the reproducibility of the extraction on the same matrix in terms of yield was proven, and experiments were carried out to compare this method with other extraction techniques, which showed that RSLDE had a higher recovery and a higher quality of extract, and in no case was the alteration of thermolabile substances induced [11] (Figure 1).

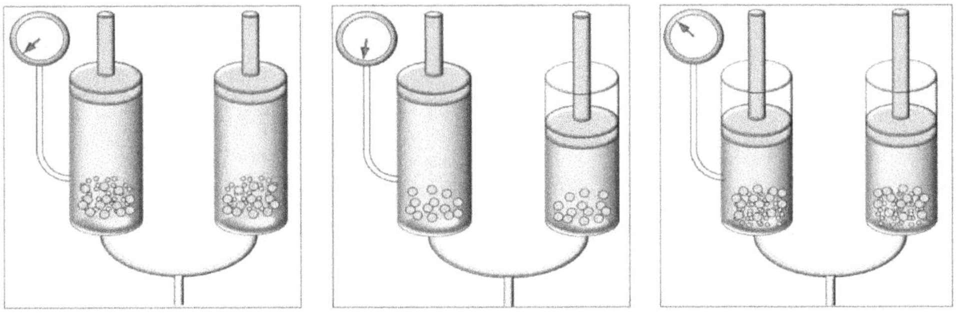

Figure 1. Schematic representation of the Naviglio extractor consisting of two extraction chambers connected via a conduit: the first two images show the dynamic phase, while the third image the static phase.

3. Comparison between the Various Solid-Liquid Extraction Techniques: Pros and Cons

The choice of methods and technologies related to an extraction process based on solid-liquid contact is not simple and this depends largely on the structural complexity and composition of the solid matrix; therefore it is not easy to find universal methods suitable for every type of Solid-Liquid extraction. In choosing the most appropriate techniques, operating conditions, solvents, etc., knowledge of the chemical properties of the compounds to be extracted and their behavior in the presence of different solvents is of fundamental importance. Due to the large extent of vegetables, operating conditions (granulometry of solid, different extractant liquids and their mix, temperature etc.) to the date, numerical and/or mathematical models that could anticipate the time and yield of extraction starting from precise conditions (solid type, solvent, temperature and so on) are not available. Alongside the aforementioned classical techniques, over the years others have been added; more complex and efficient and based on innovative extraction principles, such as extraction with supercritical fluids (SFE), ultrasound extraction (UAE), microwave extraction (MAE), accelerated solvent extraction (ASE) and finally the rapid solid liquid dynamic extraction (RSLDE) that uses the Naviglio extractor, which due to its characteristics of efficiency and improvement compared to other extraction techniques, was the subject of this review. RSLDE is an interesting new and innovative Solid-Liquid technology because it changes the philosophy of extraction; diffusion and osmosis are negligible in respect to the extraction based on difference in pressure between the inner material and the outlet of the solid matrix; this makes the extractive process "active" because it forces molecules out of solid. Below are the positive and negative aspects, pros and cons, of the main Solid-Liquid extraction techniques, nowadays existing, and briefly summarized in Table 1.

Table 1. Comparison and main characteristics of Solid-Liquid extraction techniques are herein presented.

Extraction Technique	Solvent	Granulometry	Time	Yield	Quality Extracted	Extract Stability	References
Squeezing	Indifferent	Not important	Minimum	Exhaustive	Poor	Poor	[35–38]
Maceration	Fundamental	Important	Long	Exhaustive	Great	Great	[39,40]
Decotion	Fundamental	Important	Long	Exhaustive	Great	Great	[41,42]
Percolation	Fundamental	Important	Middle	Partial	Good	Good	[43]
Soxhlet	Fundamental	Important	Long	Exhaustive	Poor	Poor	[28,44]
SCD	Indifferent	Not important	Middle	Partial	Poor	Poor	[45]
MAE	Fundamental	Not important	Middle	Partial	Poor	Poor	[46,47]
UAE	Fundamental	Not important	Middle	Partial	Great	Great	[48,49]
SFE	Indifferent	Not Important	Middle	Exhaustive	Poor	Poor	[50,51]
ASE	Fundamental	Not important	Minimum	Exhaustive	Poor	Poor	[8,52]
RSLDE	Indifferent	Not important	Minimum	Exhaustive	Great	Great	[53]

Abbreviations: SCD: steam current distillation; MAE: microwave-assisted extraction; UAE: ultrasound-assisted extraction; SFE: supercritical fluid extraction; ASE: accelerated Solid-Liquid extraction; RSLDE: rapid Solid-Liquid dynamic extraction.

3.1. *Squeezing*

Squeezing is an ancient extraction technique based on extraction from vegetables of substances such as dyes, perfumes, poisons, and even substances with marked healing properties [35]. The technique is very simple, as it consists of a pressing system based on the application of pressure, by pestles, mortars, mullers, presses, etc., on the mass of a plant material; this mechanical action serves to obtain the exudates of vegetables in which important substances are contained [36]. This technique is unusual because liquid is not used to extract molecules from the inner material of a solid, but in spite of this it is counted among Solid-Liquid extraction techniques for the final effect of obtaining extracts. Squeezing finds its greatest use in the food industry, particularly in the extraction of oil from seeds and oleaginous fruits (olive oil, sunflower oil, etc.) and for the extraction of essential oils from the fruits of the genus *Citrus*. The advantage of squeezing is that it does not use any thermal gradient, which can induce the peroxidation of the extracted oils. Crushing solids releases elements contained in a vegetable contaminated by a series of undesired compounds. The product resulting from this process is rarely used as it is; in most cases, it is necessary to resort to a sophisticated separation processes in order to isolate the desired compounds. For this reason, despite being an ancient technique, applications are limited in number. Because the application of squeezing to other vegetables does not produce useful extracts, the need to discover new technology became a priority. In a paper by Vongsak et al. (2013), varying extraction methods, such as squeezing, decoction, maceration, percolation, and Soxhlet extraction have been used to extract phenolics and flavonoids from fresh and dried leaves of *M. oleifera*. The results show that maceration was more advantageous than other methods for the extraction of phenolics and flavonoids with the highest antioxidant activity [37].

Another particular and specific Solid-Liquid extraction technique is enfleurage, which uses extractant liquid that is not in contact with any solid. It is based on the dissolution of aromatic compounds in a liquid located above a vegetable. In this way, volatile compounds that fill a closed environment solubilize in the liquid (generally n-hexane). This extraction process has been widely used in the cosmetic industry in ancient times. This process was based on the observation that some fats had the peculiarity of absorbing odors from delicate parts of vegetables, generally flowers, which were then used as perfumes. With enfleurage, excellent quality oils are obtained; however, being an extremely expensive method, it is used today for demonstration purposes only [38]. Compared to squeezing, enfleurage is an unusual Solid-Liquid extraction technique because there is no contact between components. Briefly, squeezing and enfleurage are two simple and ancient techniques of Solid-Liquid extraction that have few and limited applications; both techniques do not use Solid-Liquid contact and are not often applicable.

3.2. Maceration

Another simple and economic separation technique is maceration, which is carried out in steel containers that can have both small and large capacities (starting from a few liters to industrial-level amounts) and inert material both towards the solid matrix and the extracting solvent. This Solid-Liquid extraction technique is the first and the oldest that is based on diffusion and osmosis and, for this reason, is counted as the reference technique for many applications involving the extraction of active principles from officinal plants [39]. The solid to be extracted is introduced into the inert container and completely covered by the solvent. In order to obtain the most complete extraction possible, the container must be hermetically closed, and agitation of the batch is required in order to enable the diffusion of compounds extracted in the liquid and thus to avoid the equilibrium of extracted substances. The extraction process is generally quite long and requires days or even weeks to complete. In this extraction process, both diffusion and osmosis phenomena, strongly dependent on temperature, occur. The extraction process is sped up with the increase in temperature (Fick's law) or, more recently, with the use of ultrasounds or microwaves that increase the kinetic energy of the molecules that are found within the solid matrix and potentially extractables. This technique is recommended for extracting soluble or thermolabile active ingredients and for those matrices that, when hot, can lose substances of therapeutic interest (active principles). Maceration requires only occasional agitation for the diffusion of substances that are extracted in the mass of the extracting liquid. It is useful to underline that it is necessary to carry out maceration with limited quantities of solvent in several cycles and to subject the extracted material to squeezing, so as to avoid a strong loss of active ingredients. In fact, a dried and ground vegetable matrix absorbs a certain amount of solvent and, depending on its absorbent capacity, retains a more or less high portion in which the active ingredients are dissolved. A study by Ćujić et al. (2016) indicated that maceration was an effective and simple technique for the extraction of bioactive compounds from chokeberry fruit, even if it requires a long extraction time [40]. The maceration technique, with regard to extraction in aqueous phase, presents some variations, as it is not possible to use water at room temperature to extract vegetables because microbiological processes of fermentation take place more rapidly than extractive processes. To remedy this inconvenience, the infusion can be used, which can be represented as maceration for very short periods of time (1–2 min or until it has cooled), and is obtained by immersing medicinal plants or solid foods in water boiling to extract the active ingredients. In this case, the extraction is certainly faster, but the degradation of thermally labile substances also becomes faster. Briefly, maceration is a good technique of Solid-Liquid extraction and is applied in many cases for the production of extracts from officinal plants [39]. It is easy to apply, but extraction times are too long. The loss of liquid in the solid matrix is also relevant.

3.3. Decoction

Another variation of the classic maceration is decoction, which is carried out by contacting the matrix with the solvent operating at boiling temperature, for a variable time up to 30 min. At the end of extraction, the liquid is filtered, and the squeezed liquid of the extracted matrix impregnated with the solvent is added to it. This technique is therefore reserved for compact materials that have thermoresistant active principles, in accordance with which extraction requires the intervention of heat. The metabolic and antioxidant profiles of 10 herbal preparations have been assessed, and data have shown that the infusion procedure positively affected the extractability of the phenolic compounds compared to decoctions [41]. However, decoctions, like infusions, are easily altered and have limited validity; in fact, their shelf life is very short, and for this reason the extracts (generally named tisanes) must be consumed immediately after production. The last variant of classical maceration is digestion, which consists of heating the matrix in contact with the solvent from 35 to 60 °C. This technique is used when moderate heat is allowed in order to increase the extraction power of the solvent: if the solvent used is very volatile, it is necessary that the container in which the digestion is carried out is provided with a suitable refluxing refrigerant system for the recovery and recycling of the solvent itself. Digestion yields are higher than those obtained in maceration even if, with cooling and resting, cloudiness and the formation of precipitates occur [42]. Briefly, the decoction is a valid method to extract the active ingredients or aromas from the parts of medicinal plants or foods that are harder, such as roots, seeds, bark or wood, but it is not suitable for thermolabile compounds.

3.4. *Percolation*

3.4.1. Simple Percolation

In simple percolation, a particular container (cylindric percolator) filled with the matrix is used. In this matrix, the extracting liquid is recirculated by the means of a pump. Percolators can be made of glass, enameled iron, porcelain, or steel, and the shape depends on the nature of the matrix to be extracted. The percolation involves a series of fundamental operations: A good grinding is required; in fact, the degree of pulverization greatly influences the efficiency and the extraction time. Preliminary humidification is necessary, as the particles of the matrix to be extracted in contact with the solvent tend to swell. In the absence of this operation, the interstitial spaces diminish, and the regular outflow of the liquid is thus prevented. It is necessary to fill the percolator after a layer of cotton and sand is placed on the bottom, with the aim to block the solid matrix and as a filtration element. The matrix is added in a uniform and compact manner, minimizing and humidifying the contents of the chromatographic column. No spaces are left in the solid matrix. Preventive maceration occurs, and this works to soften the tissues and facilitates extraction. Finally, the extracting liquid is added to the percolator head and comes into contact with the solid matrix, and this liquid exerts a dynamic solubilization action on the matrix. In this technique, diffusion and osmosis occur as they do in maceration; the difference is in the continuous movement of extractant liquid through the solid, and this constitutes the driving force of percolation. The leachate thus forms, after filtration on cotton or sand, comes out from the percolator and is collected.

3.4.2. Continuous Percolation

In this extraction technique, a series of percolators are used, and in these the matrix to be extracted is placed and continuously fed in a counter-current by leachates that are less rich in extracted substances and that come from successive percolators, where the matrix is in a more exhausted state. Passing from one diffuser to the other, the solvent will be increasingly enriched by the extractable components while the matrix becomes increasingly impoverished in solutes. In this way, a concentration gradient is guaranteed in each diffuser. By optimizing the soaking and distribution ratios, the percolation operation can take place using a series of a few diffusers (5–10). One work by Chanda et al. (2012) reports a comparison between three different methods for extracting antioxidants from *Syzygium cumini* L. leaves: sequential cold percolation, decoction, and maceration. The results show that the sequential cold percolation extraction method is the best method of extracting leaf antioxidants from this plant [43]. In general, percolation does not require trained personnel to perform extraction operations. Furthermore, temperature and/or ultrasounds and microwaves can accelerate the extraction process; however, as for maceration, it is necessary to make the same considerations regarding thermolabile substances. Moreover, it is worth noting that, if the Solid-Liquid contact is very fast, the yield of extraction is not high, and generally about 50% (*w*/*w*) of extractable substances are lost in the matrix by the end of the process. Briefly, percolation is a very fast way to extract active principles from vegetables, but the process is not exhaustive.

3.5. Soxhlet Extraction

The Soxhlet method was introduced to determine whether it was possible to extract with the same Solid-Liquid ratio without using great quantities of extractant liquid and using "fresh" liquid [28]. Through the Soxhlet apparatus, this was possible because the liquid in contact with the solid is consistently "fresh" because of its distillation from the boiling liquid in the flask.

The Soxhlet method is used to extract compounds with high thermic stability due to a high temperature (the boiling point of the solvent). These substances are concentrated during the extractive process. The main advantage of this device is the use of a minimum quantity of solvent, thanks to its continuous purification and distillation after each passage through the matrix. The material to be extracted is placed in a porous thimble placed in the extraction chamber, which is placed on a distillation flask in which the solvent to be heated is placed. As the liquid boils, its vapors rise along a side tube up to the refrigerant mounted on the extractor. The liquid obtained from the condensation of the vapors falls into the extraction chamber passing through the material contained in the porous thimble, filling it until it reaches the elbow of the lateral siphon. At this point, due to its weight, the percolated liquid is sucked into the underlying flask, from which it is distilled again. The cycle described above is repeated several times until the extraction is considered complete: in this way, it is possible to extract all the soluble material from the matrix always using the same volume of solvent previously loaded in the boiler, renewed continuously by the distillation process. A review by De Castro and Priego-Capote (2010) describes the advantages and shortcomings of this centenary technique as well as the attempts to improve its performance and the achievements reached. In addition, currently, automation of Soxhlet procedures opened the door to the commercialization of a number of different approaches [44]. However, the Soxhlet apparatus and similar equipment cannot be used for the extraction of matrices that contain thermolabile active principles; moreover, this apparatus is not scalable to an industrial dimension. Briefly, Soxhlet extraction is a good tool for the extraction of many classes of compounds at a laboratory level; in fact, it is reported in official methods of analysis for the extraction of fats from foods, for the extraction of IPAs and PCBs from soil, etc. The limit of this type of extraction is related to the high temperature of the solvent, which does not allow for the extraction of thermo-sensible substances.

3.6. Steam Current Distillation (SCD)

SCD is a preferred method applied to the extraction of essential oils from vegetables. The solid matrix is placed in a distillation flask in which steam is forced to pass through; volatile compounds are moved in a condenser where they pass from gas to liquid form, and at the bottom of the condenser a container is placed. The principle of this technique is based on the fact that the vapor pressure of volatile substances allows them to be removed from the vegetable. For this reason, this technique of Solid-Liquid extraction is different because volatile compounds are not extracted in a liquid. However, like maceration and enfleurage, this technique is considered Solid-Liquid extraction for the final effect of extraction.

Distillation can be simple (e.g., the traditional distillation of wine waste) or in a steam current; in the latter case, the process becomes faster and the yield is higher in comparison with simple distillation. The auxiliary fluid that is used to assist distillation is generally represented by water in the form of steam, as it is very simple to generate steam (steam generators), has a high latent heat value (for this reason this process is expensive), and it is also particularly suitable for extractions of essential oils from aromatic plants. In a paper by Wei et al. (2012), steam distillation extraction and one-step high-speed counter-current chromatography were applied to separate and purify some bioactive compounds from the essential oil of *Flaveria bidentis* (L.) Kuntze, and good yields and high purity of the compounds (96.8% on average) were obtained [45]. However, despite its many advantages, steam current distillation cannot be used for all classes of organic compounds since the temperatures reached in the treatment can still be critical for the integrity of some of the molecules involved. Briefly, steam current distillation is an ancient technique for the extraction of thermally stable volatile compounds contained in vegetables, particularly for the extraction of essential oils. It is the most common technique for the production of distillates and similar products worldwide.

3.7. Microwave-Assisted Extraction (MAE)

MAE is a fast and efficient extraction technique based on the use of microwaves to heat the sample/solvent mixture in order to facilitate and speed up the extraction of the analyte. It is essentially very similar to the maceration process, but the introduction of a source of microwaves contributes to accelerating the extractive process. Unlike traditional heat sources, which act on a surface, from which heat diffuses towards the inner layers of the matrix by conduction and convection, a microwave heat source acts on the entire volume (if the medium is homogeneous) or on localized heating centers, consisting of the polar molecules present in the product. Therefore, whereas with conventional heating some time is required to heat the container before the heat is transferred to the solution, the microwaves directly heat the solution and the solid matrix, and the temperature gradient is kept to a minimum. Currently, MAE is already widely used in the laboratory for the extraction of organic pollutants from different matrices and for the isolation of natural products. It allows for a considerable reduction in the process time and in the solvent volumes used with respect to the classical extraction conducted through Soxhlet extractors. Recently, the MAE of phenolics from pomegranate peels was studied by Kaderides et al. (2019) and the extraction efficiency was compared with that of ultrasound extraction. The obtained extracts presented a high antioxidant activity [46]. However, MAE has the disadvantages that the tested samples must be thermostable, and a filtration phase is necessary, which in some cases can be very complex. Briefly, the introduction of microwaves in the maceration batch is used to perform MAE. This process is more accelerated than maceration because microwaves immediately heat both the solid and the liquid. The increasing temperature accelerates the extractive process, but at the same time if the energy of microwave is too high it is possible to damage the solid matrix and transform the active principles [47].

3.8. Ultrasonic Assisted Extraction (UAE)

The ultrasonication technique consists of passing a series of ultrasound pulses with increasing intensity through titanium probes immersed in a liquid medium. The probes convert the pulsed electrical energy applied to their heads into a vibrational impulse, which in a gaseous medium is transformed into ultrasound, while in a liquid medium, due to its incompressibility, it becomes an implosion. The waves generated by the pressure impulse in these particular vibrational conditions can cause cavitation, a phenomenon that consists in the formation of millions of small bubbles, during the negative pressure phase, which can implode in one of the subsequent compression phases. The implosion of each bubble causes a sudden change in temperature and pressure within the latter. The collapse of the cavity near a liquid-solid interface, however, differs considerably from the cavitation in a homogeneous liquid. In fact, passing through a liquid, the ultrasound expansion cycles exert a negative pressure on the liquid, with the molecules moving away from each other: if the ultrasound is sufficiently intense, the negative pressure exceeds the tensile strength of the liquid molecules generating a cavity. Cavitation bubbles form in the pre-existing weak points of the liquid and inside the solid spaces. Both were filled with gas in the powdered matter and suspended. Micro-bubbles, prior to cavitation were suspended in the irradiated liquid. Thus, there were devastating effects on the cellular structure. At high intensities, then, due to an inertial effect, a small cavity can quickly develop during the expansion half-cycle and will not have time to re-compress during the compression half-cycle. The bubble thus formed in the following cycle will suffer the same effect and increase in size, and the phenomenon will be repeated in the subsequent cycles until the bubble reaches a critical size such that it collapses with an increase in thermal energy. Instead, at lower acoustic intensities, cavity development may occur with a slower process. Under these conditions, a cavity will oscillate in size until it reaches the critical dimension, defined as a resonant dimension, where it can efficiently absorb the energy coming from the ultrasonic irradiation. The frequency range of use of the ultrasound is outside the perception limit of the human ear. Sonication is a technique that is used in many fields: the most widespread laboratory applications are in the field of biomedical and pharmaceutical research to lyse bacteria or cells in culture, in the field of environmental analysis for the extraction of various molecules, in the cosmetic and pharmaceutical industry for the preparation of creams and emulsions, and in biotechnology for the homogenization of immiscible liquids and the solubilization of difficult compounds. In the extractive field, this technique uses ultrasound frequencies to break up the cellular structure and facilitate diffusion processes. Goula et al. (2017) have carried out comparative studies between ultrasound-assisted and conventional solvent extraction in terms of processing procedure and total carotenoid content extracted from pomegranate wastes. The efficiency of the technique made it possible to produce an oil enriched with antioxidants [48]. The use of ultrasound in solvent extraction is a good remedy for the inconveniences linked to diffusion, but it is not always efficient. Due to the high energy developed inside the extraction batch, the breaking of the cellular structure results in extracts very similar to that obtained with the squeezing technique; in fact, vegetables are finely dispersed in the extractant liquid, and the resulting mixture is complex, so filtration and separation of active principles are required. Moreover, extracted compounds suffer from the direct bombing of cavitation generated by ultrasounds and can undergo transformations resulting in the loss of their beneficial activity. Briefly, for UAE, there are many variables to consider in order to obtain a good yield, so often the development of the various parameters lengthens the experimentation time. In the extraction of active ingredients from plants, for example, the results are comparable to extraction by squeezing, if not worse, due to the heating of the system for a prolonged time. The solid matrix is completely crushed, and a mixture is impossible to separate in its constituents is obtained, which makes this technique difficult to apply on an industrial level. This technique is often used in a laboratory procedure for sample preparation [49].

3.9. Supercritical Fluid Extraction (SFE)

SFE is a recent and very complex Solid-Liquid extraction technology [50]. The technique is based on the possibility of being able to use an extraction solvent that is a fluid (usually carbon dioxide) with properties that are intermediate between those of gases and liquids, named a supercritical fluid. Through modest variations in temperature and/or pressure, it is possible to modulate the properties of gases in a wide range and use their criticality to control phase behavior in extraction/separation processes. In practice, above the critical temperature, it is possible to continuously regulate the solubility of the fluid over a wide range, either with a small change in isothermal pressure or with a small isobaric temperature change. This ability to regulate the solvent power of a supercritical fluid is the main feature on which the SFE systems are based. These solvents can be used to extract and then efficiently recover the selected products. Since supercritical fluids have density, viscosity, and other properties that are intermediate between those of the substance in the gaseous state and those of the substance in the liquid state, the first and obvious advantage of this technique is that at the end of the extraction process the carbon dioxide is brought to ambient temperature and pressure, and gasification consequently leaves the substances extracted from the solid matrix. This fact makes SFE a green technique for Solid-Liquid extraction. A second advantage is represented by the best transport speeds: although the densities of supercritical fluids approximate those of conventional liquids, their transport properties are closer to those of gases. For example, the viscosity is many orders of magnitude lower than that of liquids, and the same diffusion coefficients are 100 times larger than typical ones observed in conventional liquids. The choice of CO_2 as a supercritical fluid offers the following advantages: it spreads through extractive matrices faster than typical solvents that have a larger molecular size; it is cheap and can be obtained easily; it has higher diffusion coefficients and lower viscosities than the liquid solvent; it has a strong permeability, so the extraction time can be considerably shorter than that required by extraction with a common solvent; it is odorless, non-toxic, does not burn, does not explode, and does not damage the ozone layer; the working temperature is close to room temperature (31.1 °C), particularly suitable for heat-sensitive material, which would be decomposed by heat treatment; recovery is simple and convenient and can be recycled without treatment; extraction and removal are combined in a single technique, significantly shortening processing times in a simple and convenient way; and it has a variable solvent power, depending on the selected operating conditions (pressure and temperature). The application of supercritical fluids in the extraction of bioactive compounds and their operative extraction conditions has been reported in a review by da Silva et al. (2016) [51]. Supercritical fluid technology offers features that overcome many limitations of conventional extraction methods. However, the limitation of this technique consists in the need for specialized equipment as well as a lower solubilizing capacity for water-soluble compounds, which can be solved in part by adding traces of polar liquids (methanol and acetone), and the request for specialized personnel for its use. Briefly, SFE is the best choice for the extraction of non-polar substances, such as nicotine from tobacco, caffeine from coffee, and oil from seeds, at an industrial scale; it is completely green extraction technology. Unfortunately, this technique is very expensive, requires specialized personnel, and is not versatile.

3.10. Accelerated Solvent Extraction (ASE)

ASE represents a useful and innovative approach for the extraction of a wide class of compounds from matrices of complex chemical-physical entity. The extraction of the analytes from the matrices takes place using a solvent kept in liquid phase at temperatures above the boiling temperature thanks to the application of high pressure. This means that this technique requires a stainless steel container that resists the high pressures generated by raising temperatures beyond the boiling point of the solvent. The increase in temperature, in fact, accelerates the desorption of the analytes (Fick's law) from the sample and their solubilization in the solvent, allowing for an effective extraction in a short period of time. In the use of solvents under high temperature and pressure, it is possible to influence the extraction process by modifying some chemical-physical parameters of the solvent–matrix system. The greatest effect on extraction is given by the temperature, as it influences the physical properties of the solvent and the interaction between the liquid phase and the material raising the molecular diffusion. Less significant is the effect of pressure that, even at low values, facilitates the penetration of the solvent into the pores of the sample. The essential function for ASE is that of keeping the solvent in a liquid state during the process. An advantage in the use of liquids at high temperature under pressure, with respect to supercritical fluids, is the fact that the former has a greater solvent strength and that, being used in methods that involve extractions at atmospheric pressure, no modifications or preliminary tests are required for the evaluation of their extractive efficiency. The other advantage is represented by the fact that, using liquid solvents, there are no phase changes in the return of the system to atmospheric conditions and therefore there is no need for liquid or packed restrictors or traps for the recovery of the analytes from the extract [52]. In their study, Cai et al. (2016) investigated the extraction efficiency of anthocyanins from purple sweet potatoes using conventional extraction, UAE, and ASE. The results show that extraction efficiencies were opposite for anthocyanins, and phenolics/flavonoids for the three methods [8]. On the other hand, the limits of the ASE method are represented by a partial extraction because of the static system and a possible degradation of the active ingredients due to operative conditions. Briefly, ASE is a good technique for the extraction of thermally very stable substances because raising the temperature behind the boiling point of the solvent remained in contact with the solid and the liquid at high temperatures for the entire experimental period. Due to the high pressure generated in the system, this technique is only used for sample preparation at a laboratory scale (10–20 mL).

3.11. *Rapid Solid-Liquid Extraction (RSLDE)*

Introduced in 2000 [34], RSLDE, through the use of the Naviglio extractor, represents a valid alternative to all existing Solid-Liquid extraction techniques and brings significant advantages in obtaining high quality extracts. First of all, it is not necessary to heat the extractive system, as the action performed is mechanical. Current extraction techniques (percolation, Soxhlet, steam current distillation and ultrasound), based on the principles of osmosis and diffusion, require an increase in temperature to increase the extraction efficiency. In the case of thermolabile compounds, however, the increase in temperature contributes to their degradation, as reported below. RSLDE requires a few extraction cycles, about 30 (depending on the matrix, but always within hours), to bring a large number of vegetable matrices to complete exhaustion. Compared to maceration, the official extraction method for many processes, RSLDE has been proven to be both quick and comprehensive. Moreover, it is possible to use water as an extracting liquid for many applications thanks to the reduced extraction times, while the prolonged contact of the plant solid matrices with water is unthinkable for maceration. RSLDE is an inexpensive technique and requires minimal energy expenditure when compared with SFE or ASE, both of which, among other things, require the use of high temperatures. In a recent work by Posadino et al. (2018), RSLDE was used to obtain polyphenolic antioxidants from the Cagnulari grape marc. The results indicate Naviglio extraction, as a green technology process, can be used to exploit wine waste to obtain antioxidants that can be used to produce enriched foods and nutraceuticals high in antioxidants [53].

In summary, the main advantages of RSLDE are as follows: exhaustion in a short period of time, with solid matrices containing extractable substances, at low operating temperatures (environment or sub-environment); reproducibility of the extraction since there is a real possibility of standardizing extracts for their active ingredients, with a guarantee of the production of high-quality extracts. From a careful comparison between the main characteristics of each of the Solid-Liquid extraction techniques described above, it is possible to state that, at present, no technique simultaneously provides all of the advantages offered by RSLDE, in terms of granulometry of the solid material, of the solvent type, or of the yield, time, quality, and stability of the extract (Table 1). On the other hand, due to its ease of use, its low-energy consumption, and the speed of the extraction process, RSLDE can be used as an exploratory and research technique for solid matrices that are not yet known and can be used to deal with materials that must undergo processes of washing (this means that the substances extracted from the solid are not important and for this reason will be discharged), such as polymers with clathrates, cork, etc. RSLDE has been advantageously used in processes in which it is important to fix substances inside the solid, as in leather tanning operations where both chrome solutions and solutions containing natural tannins are used (data not yet published). Moreover, RSLDE finds important applications in the beverage industry for the preparation of many alcoholic drinks; extracts from the ethyl alcohol of citrus peels (lemon, mandarin, orange, etc.) and from tonics and bitters from herbal extracts have been derived. In the perfume industry, it is possible to obtain fragrant and aromatic plant extracts by replacing maceration with RSLDE; in the same way, formulations of preparations in cosmetics and in herbal medicine have been improved compared to classical techniques, and alterations in the extracts obtained can be determined and made less active. Briefly, the application of this technology is very simple. It requires little energy for functioning, the time of extraction is low (two hours is the reference time), and the active principles are not degraded. Moreover, it is possible to apply RSLDE at temperatures below room temperature.

Figure 2 shows a pie chart showing the application percentages of each technique mentioned in the work obtained from a qualitative analysis of the literature data.

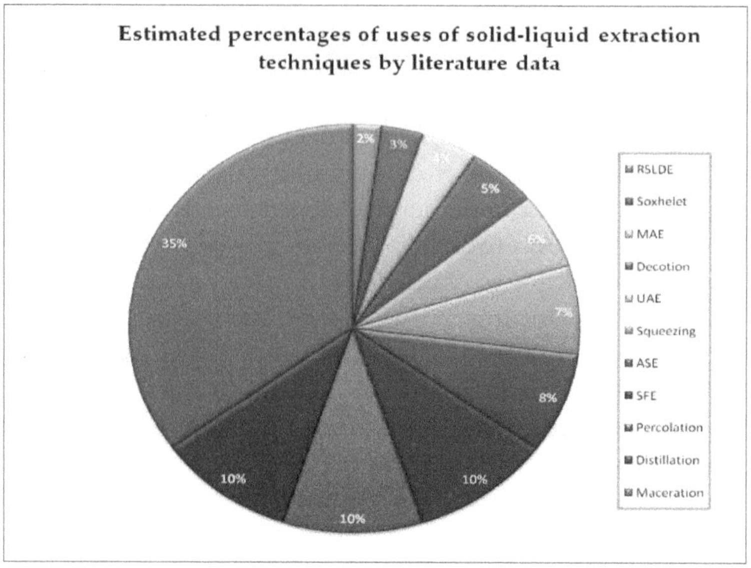

Figure 2. Pie chart showing the percentage of applications of each cited extraction technique.

4. RSLDE Applications in Various Industries

4.1. The Pharmaceutical Sector

In the pharmaceutical sector, RSLDE has been used in the preparation of high-quality standardized extracts, including medicinal plant extracts and herbal extracts, fluid extracts, mother tinctures, glycerinated extracts, glyceric macerates, liposoluble extracts, bitter medicines, etc., all of which were obtained in a much shorter period of time (4–8 h) compared to maceration, which took 21 days (maceration data provided by the Official Pharmacopoeia) [39]. The speed of the process, the extraction at low temperature, and the high efficiency guarantee the total recovery of non-degraded active ingredients contained in medicinal plants [11].

Paullinia cupana seeds, commonly called guarana, are natural sources of phenolic antioxidants and antimicrobial compounds, and the use of guarana extract is interesting for the food, pharmaceutical, and cosmetic industries, where such natural additives are required [54]. A work by Basile et al. (2005) has reported extraction from *Paullinia cupana* var. *sorbilis* Mart. (*Sapindaceae*) seeds via RSLDE. Moreover, the antibacterial and antioxidant activity of the ethanol extract was assessed towards selected bacteria and in different antioxidant models [55].

Cardiospermum halicacabum is a herbaceous plant belonging to the *Sapindaceae* family, widely used in traditional medicine for its therapeutic properties. Menichini et al. (2014) analyzed the chemical composition of extracts from aerial parts and seeds, the inhibitory properties against some enzymes, and the antioxidant effects obtained using RSLDE and the Soxhlet method. The findings suggested the potential of both seeds and aerial parts of *C. halicacabum* for the treatment of neurological disorders [56]. Moreover, RSLDE was used to extract the flowering aerial parts of *Schizogyne sericea*, a halophytic shrub that is widespread on the coastal rocks of Tenerife (Canary Islands). The extracts obtained were assayed for in vitro biological activities. Results showed that aqueous extracts, rich in phenolic acids, were endowed with relevant radical scavenging activity [57].

Therefore, among the green extraction techniques used to improve the sensitivity and the selectivity of analytical methods, RSLDE represents a sustainable alternative to classical sample-preparation procedures used in the past [58]. In a study by Cozzolino et al. (2016), the extraction of curcuminoids by RSLDE was performed from *Curcuma longa* roots, focusing the interest on curcumin, the major phenolic component of the root that has been shown to have high antioxidant activity [59]. On the other hand, some studies have shown that curcumin exerts anti-tumor effects for its ability to induce apoptosis in cancer cells without cytotoxic effects on healthy cells. Moreover, some research has demonstrated an absence of toxicity in humans when dosing this active principle for short periods of time. Therefore, for its beneficial and healing properties, curcumin obtained by the described extraction method may be used as a natural dietary supplement [60]. A comparison between three extraction processes, including traditional maceration in n-hexane and ethyl alcohol, supercritical fluid extraction (SFE), and cyclically pressurized extraction (CPE), also known as RSLDE, has been carried out for the extraction of pyrethrins, predominantly nonpolar natural compounds with insecticidal properties found in pyrethrum, an extract of certain species of chrysanthemums [61].

4.2. The Cosmetic Sector

In cosmetics and perfumery, both in production and research, it is possible to produce extracts from vegetable matrices that contain pigments and odorous substances for the production and formulation of creams and perfumes. Officinal plants are raw, cosmetic materials that have been used in numerous formulations since ancient times. Plant extraction methods are carried out to obtain active phytocomplexes, both lipo and hydrosoluble [62]. The active ingredients of plants can be obtained from the plant complex or can be taken with drugs, a term that indicates the part or parts of the plant in which the active ingredients are present. Plant drugs are essentially whole plants (fragmented or cut), parts of plants, algae, fungi, or lichens in an untreated state, generally in dried form, but sometimes fresh. Phytocosmetic plants include vasal reinforcers (e.g., root ruscus, blueberry fruits, and gingko leaves), emollients (e.g., mallow, altea, and borage), stimulants (e.g., lavender, thyme, sage, juniper, and rosemary), bioactivators (e.g., calendula and carrot). They can be used as such or through their fluid extracts, and, with the addition of natural excipients, they can be used for natural functional cosmetics. The excipients are products that support and convey active and functional plant extracts. Essential oils or essences are an important part of phytocosmetics; they are obtained by the distillation of medicinal aromatic plants, obtaining a separation of the volatile component distillable from the non-volatile. These essential oils are diluted in appropriate solvents and applied in aesthetics according to their properties [63].

A review by Barbulova et al. (2015) reported some examples of the most important applications of agricultural food by-products in cosmetics and their performance as efficacy and safety [64]. In another review, Zappelli et al. (2016) showed examples of active cosmetic ingredients developed through biotechnological systems, whose activity on the skin has been scientifically proven through in vitro and clinical studies [65]. More recently, the reasons and the characteristics as well as the challenges of plant cell culture-based productions for the cosmetic and food industries are discussed in a review by Eibl et al. (2018) [66].

In the case of RSLDE, an active action is carried out towards the substances to be extracted; in fact, the compounds not chemically bound to the solid matrix are extracted in small quantities at each extraction cycle (active process) until the matrix is completely exhausted. The advantage is that the whole process takes place in the order of hours. The important consequences of the use of this technique are the possibility of extracting vegetable matrices with water. Therefore, it is possible to extract substances at temperatures even lower than room temperature for any thermolabile compounds. Moreover, these applications can be implemented on industrial, domestic, and lab scales [11].

4.3. The Herbal Sector

In the herbal and phytotherapy sector, both in production and research, RSLDE can be used for the extraction of plants and medicinal herbs for the production of fluid extracts. Since it is not necessary to heat the extraction system, it is possible to produce teas and/or infusions at room temperature, keeping the active ingredients unaltered.

Fresh plants of *Malva silvestris* were extracted with water using RSLDE, and the effects of terpenoids and phenol isolated from this plant on the germination and growth of dicotyledon *Lactuca sativa* L. (lettuce) were studied [67].

In a study by Ferrara et al. (2014), a conventional extraction technique (UAE) and a cyclically pressurized Solid-Liquid extraction (RSLDE) were compared, in order to obtain qualitative and quantitative data related to the bioactive compounds of saffron. The results obtained showed that extracts via RSLDE had significant advantages in terms of extraction efficiency and the quality of the extract [68].

4.4. The Food and Beverage Sector

In the food sector, both in production and research, RSLDE has been applied in various ways. Lycopene, the carotenoid responsible for the red color of many fruits and vegetables, is considered fundamental for its antioxidant action. Therefore, its extraction is of great interest in various sectors. In fact, it can be used both for the formulation of functional foods and in cosmetics. In addition, lycopene can be extracted from tomato processing waste using only water as an extract liquid. The use of water as an extracting phase considerably reduces the cost of the entire process when compared with the commonly used solvent-based procedure or with the newer supercritical extraction process of lycopene from tomato waste. Lycopene, not soluble in water, was recovered in a quasi-crystalline solid form and purified by solid-phase extraction using a small amount of organic solvent [69]. Lycopene can be used as a dye and/or a natural antioxidant. Moreover, through RSLDE, it is also possible to produce limoncello, a lemon liqueur, in just two hours, avoiding the long traditional maceration that takes 7–14 days [70]. Nowadays the industrial process for the production of lemon liquor is performed via maceration, as home-made products are made, and the process requires at least 48 h of lemon peel infusion in alcohol.

In a paper by Formato et al. (2013), two Solid-Liquid extraction techniques, supercritical fluid extraction (SFE) with and without modifiers and cyclically pressurized Solid-Liquid extraction or RSLDE, were compared on the basis of the extraction of acidic compounds contained in hops. The results showed that both techniques were valid for the extraction of α and β acids from hops. By suitably varying the parameters of the two extractive procedures, it was possible to obtain extracts for use in the production of beer and of dietary supplements and drugs [71].

In order to obtain the alcoholic extracts of some herbal mixtures, the traditional maceration procedure was compared to RSLDE. Three different mixtures of various parts of plants were extracted with both methods, and results were compared. Organoleptic tests performed on alcoholic bitters obtained from different extracts have been used to determine the optimum extraction time for the two different methods used. The results showed that the bitters produced with RLSDE were more appreciated than bitters prepared by maceration [72]. In another work by Naviglio et al., (2014), the extraction process for the production of *Cinchona calisaya* elixir starting from the same vegetable mixture was performed by a conventional maceration and RSLDE. The results show that, compared to the conventional method, RSLDE allowed for extraction at room temperature using cyclical extraction pressurization. In this way, it was possible to avoid thermal stress on thermolabile substances while simultaneously reducing extraction time [73].

Based on the Naviglio extractor, a system of extraction has been devised that can be used for student laboratory experiments to illustrate RSLDE (two-syringe system). In a paper by Naviglio et al. (2015), students compare two extraction techniques for the preparation of limoncello: maceration and the two-syringe system. The development of the two-syringe system for simple manual operations reduced the risks of the procedure, allowing students to evaluate the extraction efficiency of the two methods [74]. In another work, two extraction processes for the production of limoncello—the traditional maceration of lemon peels and RSLDE—were compared. Alcoholic extracts were analyzed by gas chromatography, and alcoholic extracts were analyzed by UV spectrophotometry to identify the more abundant chemical species, while the organoleptic tests performed on the final product (limoncello) provided an indication of the taste of the final product. Results showed that the RSLDE process was 120 times faster than maceration and had a greater efficiency in a short period of time [75].

The results of a recent study by Gigliarelli et al. (2017) indicated that RSLDE is an effective technique for the extraction of piperine from fruits of *Piper longum* compared to other techniques, such as Soxhlet extraction, decotion (International Organization for Standardization), and conventional maceration [76].

Stevioside and rebaudioside A are the main diterpene glycosides present in the leaves of the *Stevia rebaudiana* plant, which is used in the production of foods and low-calorie beverages. In this context, RSLDE constitutes a valid alternative method to conventional extraction by reducing the extraction time and the consumption of toxic solvents and favoring the use of extracted metabolites as food additives and/or nutraceuticals [77,78].

Portulaca oleracea, commonly known as purslane, is a wild plant pest of orchards and gardens, but is also an edible vegetable rich in beneficial nutrients. The purpose of a work by Gallo et al. (2017) was to compare various Solid-Liquid extraction techniques to determine the most efficient technique for the extraction of biomolecules from leaves of purslane. Therefore, extraction reproducibility was tested on the same matrix in terms of weighting, and comparison experiments were performed. RSLDE showed a higher recovery and a higher quality of extract, and in no cases did it induce the alteration of heat-sensitive substances [79].

RSLDE can be used for applications other than extraction, such as the quick hydration of legumes and the de-structuring of vegetables at room temperature (heat-free cooking) for a better preservation of nutritional elements. Hydrating beans before cooking reduces cooking time, increases their tenderness and weight, and improves their appearance after cooking. Naviglio et al. (2013) described a process of cyclically pressurized soaking for the rapid hydration of cannellini beans at room temperature. The hydration process was approximately 10-fold faster than the traditional soaking procedure, and the microbial load developed by the end of this process was much lower compared to that obtained using the traditional process [80].

A large amount of food waste and by-products is produced from farm to plate and represents valuable sources for the production of compounds with a high added value. Consequently, the application of innovative approaches is necessary due to the limitation of conventional processes. In this context, RSLDE can be a useful tool to increase extract yield and quality, reducing extraction time, temperature, and toxic solvents. In a study by Bilo et al. (2018), RSLDE was used for the synthesis of a new bioplastic produced from rice straw, an agricultural waste that is generally not recovered. The results show that, depending on the environmental humidity, the material shows a dual mechanical behavior that can be exploited to obtain shrink films and sheets or to drive a shape memory effect. Therefore, rice straw bioplastics could represent a new potential eco-material for different application fields [81]. In addition, RSLDE could provide an innovative approach to increase the production of specific compounds from food waste for use as nutraceuticals or as ingredients in the design of functional foods. As it is currently known, grape pomace is a by-product of winemaking that can be conveniently reused in many different ways, including agronomic use as well as cosmetic industry applications. Moreover, the by-products can also be used in the energy field as biomass for the production of biogas or for food plants used for the production of energy. As an added value, grape pomace resulting from the production of wine also contains numerous bioactive compounds. In a very recent work by Gallo et al. (2019), to extract polyphenols, grape peels were processed via RSLDE, which does not require the use of any organic solvent, nor does it include heating or cooling processes that can cause the loss of substances of interest [82]. Still, within the framework of the guiding principles for eco-innovation, which aims at a zero waste economy, many residues have the potential to be reused as raw material for new products and applications and in other production systems, such as those in the nutraceutical and pharmaceutical sectors. Another recent work by Gallo et al. (2019) shows an alternative process for the extraction of lycopene from tomato waste through RSLDE. The high purity of lycopene obtained using this procedure make the process very attractive, and the pure product obtained could be used in various applications [83].

5. Conclusions

The efficiencies of conventional and non-conventional extraction methods mostly depend on the critical input parameters, on understanding the nature of the matrices to be extracted, on the chemistry of the compounds, and on scientific expertise. In particular, RSLDE introduces a new Solid-Liquid extractive technology that is based on the difference in pressure between the inner material and the outlet of the solid matrix, which generates a suction effect (matter transfer) that causes the compounds that are not chemically linked to the solid matrix to be extracted. Therefore, RSLDE makes it possible to optimally replace most of the current Solid-Liquid extraction techniques and brings about considerable new features and advantages in obtaining high-quality extracts. It changed the concept of Solid-Liquid extraction based only on diffusion and osmosis phenomena (passive process of extraction) using a new philosophy that bases on the generation of pressure gradient between the inner material and the outlet of the solid (active process of extraction). In addition, the areas of application of RSLDE are numerous and include the pharmaceutical, cosmetic, herbal, and food and beverage industries. Furthermore, RSLDE can be used for the extraction of food waste, a by-product of various industrial, agricultural, domestic, and other food sectors, which is currently increasing due to the increase in these activities. These by-products can be used as a potential source of bioactive and nutraceutical compounds that have important applications in the treatment of various disorders.

Author Contributions: Conceptualization, D.N. and M.G.; investigation, D.N., P.S., M.C., and M.G.; data curation, P.S., M.C.; writing—original draft preparation, M.G.; writing—review and editing, D.N. and M.G.; visualization, D.N.; supervision, M.G.

References

1. Azmir, J.; Zaidul, I.S.M.; Rahman, M.M.; Sharif, K.M.; Mohamed, A.; Sahena, F.; Omar, A.K.M. Techniques for extraction of bioactive compounds from plant materials: A review. *J. Food Eng.* **2013**, *117*, 426–436. [CrossRef]
2. Aguilera, J.M. Solid-liquid extraction. In *Extraction Optimization in Food Engineering*; CRC Press: Boca Raton, FL, USA, 2003; pp. 51–70.
3. Wang, L.; Weller, C.L. Recent advances in extraction of nutraceuticals from plants. *Trends Food Sci. Technol.* **2006**, *17*, 300–312. [CrossRef]
4. Galanakis, C.M. Recovery of high added-value components from food wastes: Conventional, emerging technologies and commercialized applications. *Trends Food Sci. Technol.* **2012**, *26*, 68–87. [CrossRef]
5. Chemat, F.; Rombaut, N.; Sicaire, A.G.; Meullemiestre, A.; Fabiano-Tixier, A.S.; Abert-Vian, M. Ultrasound assisted extraction of food and natural products. Mechanisms, techniques, combinations, protocols and applications. A review. *Ultrason. Sonochem.* **2017**, *34*, 540–560. [CrossRef] [PubMed]
6. Khaw, K.Y.; Parat, M.O.; Shaw, P.N.; Falconer, J.R. Solvent supercritical fluid technologies to extract bioactive compounds from natural sources: A review. *Molecules* **2017**, *22*, 1186. [CrossRef] [PubMed]
7. Ekezie, F.G.C.; Sun, D.W.; Cheng, J.H. Acceleration of microwave-assisted extraction processes of food components by integrating technologies and applying emerging solvents: A review of latest developments. *Trends Food Sci. Technol.* **2017**, *67*, 160–172. [CrossRef]
8. Cai, Z.; Qu, Z.; Lan, Y.; Zhao, S.; Ma, X.; Wan, Q.; Li, P. Conventional, ultrasound-assisted, and accelerated-solvent extractions of anthocyanins from purple sweet potatoes. *Food Chem.* **2016**, *197*, 266–272. [CrossRef]
9. Souza-Silva, É.A.; Jiang, R.; Rodríguez-Lafuente, A.; Gionfriddo, E.; Pawliszyn, J. A critical review of the state of the art of solid-phase microextraction of complex matrices I. Environmental analysis. *Trends Anal. Chem.* **2015**, *71*, 224–235. [CrossRef]
10. Kumar, S.J.; Kumar, G.V.; Dash, A.; Scholz, P.; Banerjee, R. Sustainable green solvents and techniques for lipid extraction from microalgae: A review. *Algal Res.* **2017**, *21*, 138–147. [CrossRef]
11. Naviglio, D. Naviglio's principle and presentation of an innovative solid–liquid extraction technology: Extractor Naviglio®. *Anal. Lett.* **2003**, *36*, 1647–1659. [CrossRef]
12. Barba, F.J.; Zhu, Z.; Koubaa, M.; Sant'Ana, A.S.; Orlien, V. Green alternative methods for the extraction of antioxidant bioactive compounds from winery wastes and by-products: A review. *Trends Food Sci. Technol.* **2016**, *49*, 96–109. [CrossRef]

13. Chemat, F.; Rombaut, N.; Meullemiestre, A.; Turk, M.; Perino, S.; Fabiano-Tixier, A.S.; Abert-Vian, M. Review of green food processing techniques. Preservation, transformation, and extraction. *Innov. Food Sci. Emerg. Technol.* **2017**, *41*, 357–377. [CrossRef]
14. Al Jitan, S.; Alkhoori, S.A.; Yousef, L.F. Phenolic acids from plants: Extraction and application to human health. In *Studies in Natural Products Chemistry*; Elsevier: Amsterdam, The Netherlands, 2018; Volume 58, pp. 389–417.
15. Azwanida, N.N. A review on the extraction methods use in medicinal plants, principle, strength and limitation. *Med. Aromat. Plants* **2015**, *4*. [CrossRef]
16. AOAC Method 43.290. *Official Methods of Analysis of the AOAC*, 15th ed.; Association of Official Analytical Chemists: Washington, DC, USA, 1990.
17. Willson, K.C.; Clifford, M.N. *Tea: Cultivation to Consumption*; Springer Science & Business Media: Berlin, Germany, 2012.
18. Liguori, L.; Russo, P.; Albanese, D.; Di Matteo, M. Production of low-alcohol beverages: Current status and perspectives. In *Food Processing for Increased Quality and Consumption*; Academic Press: Cambridge, MA, USA, 2018; pp. 347–382.
19. Aspé, E.; Fernández, K. The effect of different extraction techniques on extraction yield, total phenolic, and anti-radical capacity of extracts from *Pinus radiata* Bark. *Ind. Crop. Prod.* **2011**, *34*, 838–844. [CrossRef]
20. Božović, M.; Navarra, A.; Garzoli, S.; Pepi, F.; Ragno, R. Essential oils extraction: A 24-hour steam distillation systematic methodology. *Nat. Prod. Res.* **2017**, *31*, 2387–2396. [CrossRef] [PubMed]
21. Joana Gil-Chávez, G.; Villa, J.A.; Fernando Ayala-Zavala, J.; Basilio Heredia, J.; Sepulveda, D.; Yahia, E.M.; González-Aguilar, G.A. Technologies for extraction and production of bioactive compounds to be used as nutraceuticals and food ingredients: An overview. *Compr. Rev. Food Sci. Food Saf.* **2013**, *12*, 5–23. [CrossRef]
22. Rostagno, M.A.; Prado, J.M. *Natural Product Extraction: Principles and Applications*; No. 21; Royal Society of Chemistry: London, UK, 2013.
23. Esclapez, M.D.; García-Pérez, J.V.; Mulet, A.; Cárcel, J.A. Ultrasound-assisted extraction of natural products. *Food Eng. Rev.* **2011**, *3*, 108. [CrossRef]
24. Gallo, M.; Ferrara, L.; Naviglio, D. Application of ultrasound in food science and technology: A perspective. *Foods* **2018**, *7*, 164. [CrossRef] [PubMed]
25. Jesus, S.P.; Meireles, M.A.A. Supercritical fluid extraction: A global perspective of the fundamental concepts of this eco-friendly extraction technique. In *Alternative Solvents for Natural Products Extraction*; Springer: Berlin, Germany, 2014; pp. 39–72.
26. Sánchez-Camargo, A.D.P.; Parada-Alonso, F.; Ibáñez, E.; Cifuentes, A. Recent applications of on-line supercritical fluid extraction coupled to advanced analytical techniques for compounds extraction and identification. *J. Sep. Sci.* **2019**, *42*, 243–257. [CrossRef] [PubMed]
27. AOAC Method 963.15. *Agricultural Chemicals, Contaminants, Drugs*, 15th ed.; Association of Official Analytical Chemists: Arlington, VA, USA, 1990.
28. Jensen, W.B. The origin of the Soxhlet extractor. *J. Chem. Educ.* **2007**, *84*, 1913. [CrossRef]
29. Anderson, S. Soxtec: Its principles and applications. In *Oil Extraction and Analysis. Critical Issues and Competitive Studies*; AOCS Publishing: Champaign, IL, USA, 2004; pp. 11–25.
30. Carro, N.; Cobas, J.; García, I.; Ignacio, M.; Mouteira, A.; Silva, B. Development of a method for the determination of SCCPs (short-chain chlorinated paraffins) in bivalve mollusk using Soxtec device followed by gas chromatography-triple quadrupole tandem mass spectrometry. *J. Anal. Sci. Technol.* **2018**, *9*, 8. [CrossRef]
31. Molino, A.; Rimauro, J.; Casella, P.; Cerbone, A.; Larocca, V.; Chianese, S.; Musmarra, D. Extraction of astaxanthin from microalga *Haematococcus pluvialis* in red phase by using generally recognized as safe solvents and accelerated extraction. *J. Biotechnol.* **2018**, *283*, 51–61. [CrossRef] [PubMed]
32. He, Q.; Du, B.; Xu, B. Extraction optimization of phenolics and antioxidants from black goji berry by accelerated solvent extractor using response surface methodology. *Appl. Sci.* **2018**, *8*, 1905. [CrossRef]
33. Hilali, S.; Fabiano-Tixier, A.S.; Ruiz, K.; Hejjaj, A.; Nouh, F.A.; Idlimam, A.; Chemat, F. Green extraction of essential oils, polyphenols and pectins from orange peel employing solar energy. Towards a Zero-Waste Biorefinery. *ACS Sustain. Chem. Eng.* **2019**. [CrossRef]
34. Naviglio, D. Rapid and Dynamic Solid–Liquid Extractor Working at High Pressures and Low Temperatures for Obtaining in Short Times Solutions Containing Substances that Initially Were in Solid Matrixes Insoluble

in Extracting Liquid. Italian Patent 1,303,417, 6 November 2000.
35. Baldwin, E.A.; Bai, J.; Plotto, A.; Cameron, R.; Luzio, G.; Narciso, J.; Ford, B.L. Effect of extraction method on quality of orange juice: Hand-squeezed, commercial-fresh squeezed and processed. *J. Sci. Food Agric.* **2012**, *92*, 2029–2042. [CrossRef] [PubMed]
36. Armenta, S.; Garrigues, S.; de la Guardia, M. The role of green extraction techniques in Green Analytical Chemistry. *TrAC Trends Anal. Chem.* **2015**, *71*, 2–8. [CrossRef]
37. Vongsak, B.; Sithisarn, P.; Mangmool, S.; Thongpraditchote, S.; Wongkrajang, Y.; Gritsanapan, W. Maximizing total phenolics, total flavonoids contents and antioxidant activity of Moringa oleifera leaf extract by the appropriate extraction method. *Ind. Crop. Prod.* **2013**, *44*, 566–571. [CrossRef]
38. Stratakos, A.C.; Koidis, A. Methods for extracting essential oils. In *Essential Oils in Food Preservation, Flavor and Safety*; Preedy, V.R., Ed.; Academic Press: Cambridge, MA, USA, 2016; pp. 31–38.
39. European Pharmacopoeia Commission. *European Pharmacopoeia*, 9th ed.; European Directorate for the Quality of Medicines (EDQM): Strasbourg, France, 2014.
40. Ćujić, N.; Šavikin, K.; Janković, T.; Pljevljakušić, D.; Zdunić, G.; Ibrić, S. Optimization of polyphenols extraction from dried chokeberry using maceration as traditional technique. *Food Chem.* **2016**, *194*, 135–142. [CrossRef]
41. Fotakis, C.; Tsigrimani, D.; Tsiaka, T.; Lantzouraki, D.Z.; Strati, I.F.; Makris, C.; Zoumpoulakis, P. Metabolic and antioxidant profiles of herbal infusions and decoctions. *Food Chem.* **2016**, *211*, 963–971. [CrossRef]
42. Manousi, N.; Sarakatsianos, I.; Samanidou, V. Extraction techniques of phenolic compounds and other bioactive compounds from medicinal and aromatic plants. In *Engineering Tools in the Beverage Industry*; Woodhead Publishing: Sawston, UK, 2019; pp. 283–314.
43. Chanda, S.V.; Kaneria, M.J. Optimization of conditions for the extraction of antioxidants from leaves of *Syzygium cumini* L. using different solvents. *Food Anal. Methods* **2012**, *5*, 332–338. [CrossRef]
44. De Castro, M.L.; Priego-Capote, F. Soxhlet extraction: Past and present panacea. *J. Chromatogr. A* **2010**, *1217*, 2383–2389. [CrossRef] [PubMed]
45. Wei, Y.; Du, J.; Lu, Y. Preparative separation of bioactive compounds from essential oil of *Flaveria bidentis* (L.) Kuntze using steam distillation extraction and one step high-speed counter-current chromatography. *J. Sep. Sci.* **2012**, *35*, 2608–2614. [CrossRef] [PubMed]
46. Kaderides, K.; Papaoikonomou, L.; Serafim, M.; Goula, A.M. Microwave-assisted extraction of phenolics from pomegranate peels: Optimization, kinetics, and comparison with ultrasounds extraction. *Chem. Eng. Process. Process Intensif.* **2019**, *137*, 1–11. [CrossRef]
47. Chan, C.H.; Yusoff, R.; Ngoh, G.C.; Kung, F.W.L. Microwave-assisted extractions of active ingredients from plants. *J. Chromatogr. A* **2011**, *1218*, 6213–6225. [CrossRef] [PubMed]
48. Goula, A.M.; Ververi, M.; Adamopoulou, A.; Kaderides, K. Green ultrasound-assisted extraction of carotenoids from pomegranate wastes using vegetable oils. *Ultrason. Sonochem.* **2017**, *34*, 821–830. [CrossRef] [PubMed]
49. Tiwari, B.K. Ultrasound: A clean, green extraction technology. *Trends Anal. Chem.* **2015**, *71*, 100–109. [CrossRef]
50. Sharif, K.M.; Rahman, M.M.; Azmir, J.; Mohamed, A.; Jahurul, M.H.A.; Sahena, F.; Zaidul, I.S.M. Experimental design of supercritical fluid extraction–A review. *J. Food Eng.* **2014**, *124*, 105–116. [CrossRef]
51. Da Silva, R.P.; Rocha-Santos, T.A.; Duarte, A.C. Supercritical fluid extraction of bioactive compounds. *TrAC Trends Anal. Chem.* **2016**, *76*, 40–51. [CrossRef]
52. Nayak, B.; Dahmoune, F.; Moussi, K.; Remini, H.; Dairi, S.; Aoun, O.; Khodir, M. Comparison of microwave, ultrasound and accelerated-assisted solvent extraction for recovery of polyphenols from *Citrus sinensis* peels. *Food Chem.* **2015**, *187*, 507–516. [CrossRef]
53. Posadino, A.; Biosa, G.; Zayed, H.; Abou-Saleh, H.; Cossu, A.; Nasrallah, G.; Pintus, G. Protective effect of cyclically pressurized solid–liquid extraction polyphenols from Cagnulari grape pomace on oxidative endothelial cell death. *Molecules* **2018**, *23*, 2105. [CrossRef]
54. Santana, A.L.; Macedo, G.A. Health and technological aspects of methylxanthines and polyphenols from guarana: A review. *J. Funct. Foods* **2018**, *47*, 457–468. [CrossRef]
55. Basile, A.; Ferrara, L.; Del Pezzo, M.; Mele, G.; Sorbo, S.; Bassi, P.; Montesano, D. Antibacterial and antioxidant activities of ethanol extract from *Paullinia cupana* Mart. *J. Ethnopharmacol.* **2005**, *102*, 32–36. [CrossRef] [PubMed]
56. Menichini, F.; Losi, L.; Bonesi, M.; Pugliese, A.; Loizzo, M.R.; Tundis, R. Chemical profiling and in vitro biological effects of *Cardiospermum halicacabum* L. (*Sapindaceae*) aerial parts and seeds for applications in

neurodegenerative disorders. *J. Enzym. Inhib. Med. Chem.* **2014**, *29*, 677–685. [CrossRef] [PubMed]
57. Caprioli, G.; Iannarelli, R.; Sagratini, G.; Vittori, S.; Zorzetto, C.; Sánchez-Mateo, C.C.; Petrelli, D. Phenolic acids, antioxidant and antiproliferative activities of Naviglio® extracts from *Schizogyne sericea* (*Asteraceae*). *Nat. Prod. Res.* **2017**, *31*, 515–522. [CrossRef] [PubMed]
58. Bandar, H.; Hijazi, A.; Rammal, H.; Hachem, A.; Saad, Z.; Badran, B. Techniques for the extraction of bioactive compounds from Lebanese *Urtica Dioica*. *Am. J. Phytomed. Clin. Ther.* **2013**, *1*, 507–513.
59. Cozzolino, I.; Vitulano, M.; Conte, E.; D'Onofrio, F.; Aletta, L.; Ferrara, L.; Gallo, M. Extraction and curcuminoids activity from the roots of *Curcuma longa* by RSLDE using the Naviglio extractor. *ESJ* **2016**, 12. [CrossRef]
60. Pulido-Moran, M.; Moreno-Fernandez, J.; Ramirez-Tortosa, C.; Ramirez-Tortosa, M. Curcumin and health. *Molecules* **2016**, *21*, 264. [CrossRef]
61. Gallo, M.; Formato, A.; Ianniello, D.; Andolfi, A.; Conte, E.; Ciaravolo, M.; Naviglio, D. Supercritical fluid extraction of pyrethrins from pyrethrum flowers (*Chrysanthemum cinerariifolium*) compared to traditional maceration and cyclic pressurization extraction. *J. Supercrit. Fluids* **2017**, *119*, 104–112. [CrossRef]
62. Cabaleiro, N.; De La Calle, I.; Bendicho, C.; Lavilla, I. Current trends in liquid–liquid and solid–liquid extraction for cosmetic analysis: A review. *Anal. Methods* **2013**, *5*, 323–340. [CrossRef]
63. Ali, B.; Al-Wabel, N.A.; Shams, S.; Ahamad, A.; Khan, S.A.; Anwar, F. Essential oils used in aromatherapy: A systemic review. *Asian Pac. J. Trop. Biomed.* **2015**, *5*, 601–611. [CrossRef]
64. Barbulova, A.; Colucci, G.; Apone, F. New trends in cosmetics: By-products of plant origin and their potential use as cosmetic active ingredients. *Cosmetics* **2015**, *2*, 82–92. [CrossRef]
65. Zappelli, C.; Barbulova, A.; Apone, F.; Colucci, G. Effective active ingredients obtained through Biotechnology. *Cosmetics* **2016**, *3*, 39. [CrossRef]
66. Eibl, R.; Meier, P.; Stutz, I.; Schildberger, D.; Hühn, T.; Eibl, D. Plant cell culture technology in the cosmetics and food industries: Current state and future trends. *Appl. Microbiol. Biotechnol.* **2018**, *102*, 8661–8675. [CrossRef] [PubMed]
67. Cutillo, F.; D'Abrosca, B.; DellaGreca, M.; Fiorentino, A.; Zarrelli, A. Terpenoids and phenol derivatives from *Malva silvestris*. *Phytochemistry* **2006**, *67*, 481–485. [CrossRef] [PubMed]
68. Ferrara, L.; Naviglio, D.; Gallo, M. Extraction of bioactive compounds of saffron (*Crocus sativus* L.) by Ultrasound Assisted Extraction (UAE) and by Rapid Solid-Liquid Dynamic Extraction (RSLDE). *ESJ* **2014**, *10*. [CrossRef]
69. Naviglio, D.; Pizzolongo, F.; Ferrara, L.; Naviglio, B.; Santini, A. Extraction of pure lycopene from industrial tomato waste in water using the extractor Naviglio. *Afr. J. Food Sci.* **2008**, *2*, 037–044.
70. Naviglio, D.; Pizzolongo, F.; Romano, R.; Ferrara, L.; Naviglio, B.; Santini, A. An innovative solid-liquid extraction technology: Use of the Naviglio Extractor for the production of lemon liquor. *Afr. J. Food Sci.* **2007**, *1*, 42–50.
71. Formato, A.; Gallo, M.; Ianniello, D.; Montesano, D.; Naviglio, D. Supercritical fluid extraction of α-and β-acids from hops compared to cyclically pressurized solid–liquid extraction. *J. Supercrit. Fluids* **2013**, *84*, 113–120. [CrossRef]
72. Naviglio, D.; Ferrara, L.; Formato, A.; Gallo, M. Efficiency of conventional extraction technique compared to rapid solid-liquid dynamic extraction (RSLDE) in the preparation of bitter liquors and elixirs. *IOSR J. Pharm.* **2014**, *4*, 14–22.
73. Naviglio, D.; Formato, A.; Gallo, M. Comparison between 2 methods of solid–liquid extraction for the production of *Cinchona calisaya* elixir: An experimental kinetics and numerical modeling approach. *J. Food Sci.* **2014**, *79*, E1704–E1712. [CrossRef]
74. Naviglio, D.; Montesano, D.; Gallo, M. Laboratory production of lemon liqueur (Limoncello) by conventional maceration and a two-syringe system to illustrate rapid solid–liquid dynamic extraction. *J. Chem. Educ.* **2015**, *92*, 911–915. [CrossRef]
75. Naviglio, D.; Formato, A.; Vitulano, M.; Cozzolino, I.; Ferrara, L.; Zanoelo, E.F.; Gallo, M. Comparison between the kinetics of conventional maceration and a cyclic pressurization extraction process for the production of lemon liqueur using a numerical model. *J. Food Process Eng.* **2017**, *40*, e12350. [CrossRef]
76. Gigliarelli, G.; Pagiotti, R.; Persia, D.; Marcotullio, M.C. Optimisation of a Naviglio-assisted extraction followed by determination of piperine content in *Piper longum* extracts. *Nat. Prod. Res.* **2017**, *31*, 214–217. [CrossRef] [PubMed]

77. Gallo, M.; Vitulano, M.; Andolfi, A.; DellaGreca, M.; Conte, E.; Ciaravolo, M.; Naviglio, D. Rapid Solid-Liquid Dynamic Extraction (RSLDE): A new rapid and greener method for extracting two steviol glycosides (stevioside and rebaudioside A) from stevia leaves. *Plant Foods Hum. Nutr.* **2017**, *72*, 141–148. [CrossRef] [PubMed]
78. Gallo, M.; Formato, A.; Formato, G.; Naviglio, D. Comparison between two solid-liquid extraction methods for the recovery of steviol glycosides from dried stevia leaves applying a numerical approach. *Processes* **2018**, *6*, 105. [CrossRef]
79. Gallo, M.; Conte, E.; Naviglio, D. Analysis and comparison of the antioxidant component of *Portulaca oleracea* leaves obtained by different solid-liquid extraction techniques. *Antioxidants* **2017**, *6*, 64. [CrossRef]
80. Naviglio, D.; Formato, A.; Pucillo, G.P.; Gallo, M. A cyclically pressurised soaking process for the hydration and aromatisation of cannellini beans. *J. Food Eng.* **2013**, *116*, 765–774. [CrossRef]
81. Bilo, F.; Pandini, S.; Sartore, L.; Depero, L.E.; Gargiulo, G.; Bonassi, A.; Bontempi, E. A sustainable bioplastic obtained from rice straw. *J. Clean. Prod.* **2018**, *200*, 357–368. [CrossRef]
82. Gallo, M.; Formato, A.; Giacco, R.; Riccardi, G.; Lungo, D.; Formato, G.; Amoresano, A.; Naviglio, D. Mathematical optimization of the green extraction of polyphenols from grape peels through a cyclic pressurization process. *Heliyon* **2019**, e01526. [CrossRef]
83. Gallo, M.; Formato, A.; Ciaravolo, M.; Langella, C.; Cataldo, R.; Naviglio, D. A water extraction process for lycopene from tomato waste using a pressurized method: An application of a numerical simulation. *Eur. Food Res. Technol.* **2019**. [CrossRef]

11

The Food and Beverage Occurrence of Furfuryl Alcohol and Myrcene—Two Emerging Potential Human Carcinogens?

Alex O. Okaru [1,2] and Dirk W. Lachenmeier [2,*]

1 Department of Pharmaceutical Chemistry, University of Nairobi, P.O. Box 19676-00202, Nairobi, Kenya; alex.okaru@gmail.com
2 Chemisches und Veterinäruntersuchungsamt (CVUA) Karlsruhe, Weissenburger Strasse 3, 76187 Karlsruhe, Germany
* Correspondence: lachenmeier@web.de

Academic Editor: David Bellinger

Abstract: For decades, compounds present in foods and beverages have been implicated in the etiology of human cancers. The World Health Organization (WHO) International Agency for Research on Cancer (IARC) continues to classify such agents regarding their potential carcinogenicity in humans based on new evidence from animal and human studies. Furfuryl alcohol and β-myrcene are potential human carcinogens due to be evaluated. The major source of furfuryl alcohol in foods is thermal processing and ageing of alcoholic beverages, while β-myrcene occurs naturally as a constituent of the essential oils of plants such as hops, lemongrass, and derived products. This study aimed to summarize the occurrence of furfuryl alcohol and β-myrcene in foods and beverages using literature review data. Additionally, results of furfuryl alcohol occurrence from our own nuclear magnetic resonance (NMR) analysis are included. The highest content of furfuryl alcohol was found in coffee beans (>100 mg/kg) and in some fish products (about 10 mg/kg), while among beverages, wines contained between 1 and 10 mg/L, with 8 mg/L in pineapple juice. The content of β-myrcene was highest in hops. In conclusion, the data about the occurrence of the two agents is currently judged as insufficient for exposure and risk assessment. The results of this study point out the food and beverage groups that may be considered for future monitoring of furfuryl alcohol and β-myrcene.

Keywords: furfuryl alcohol; β-myrcene; carcinogens; occurrence

---

1. Introduction

The production and processing of foods and beverages may invariably lead to significant changes in the chemical composition of the products. The Maillard reaction—which yields furanic compounds such as furfural and 5-hydroxymethylfurfural (HMF) and furfuryl alcohol, among other products—is common during processes that involve heating or roasting [1–5]. Furfuryl alcohol is a food contaminant which occurs in significant amounts in thermally processed foods such as coffee, fruit juices, baked foods; in cask-stored alcoholic beverages such as wines, wine-derived spirits such as brandy, and whiskies as a result of enzymatic or chemical reduction of furfural [6–8]; and in butter and butterscotch when furfuryl alcohol is used as a flavouring agent [9]. Furfuryl alcohol may also be formed from quinic acid or 1,2-enediols as precursors during the heating of foods such as coffee beans [5] . In acidic conditions, furfuryl alcohol polymerizes to aliphatic polymers that give a brown colouration to foods [5].

Myrcene is a terpenoid compound that exists in two forms—β and α, with the former occurring naturally in essential oils of plants such as hops, bay, lemongrass [10,11], and orange juice [12], and is permitted for use as a flavouring additive of food both by the United States Food and Drug Administration (FDA) since 1965 and by the European Council since 1974. β-Myrcene is also an ingredient in the preparation of olefinic scents such as menthol, and the alcohols linalool, nerol, and geraniol [13], found in household items.

Analysis of furfuryl alcohol can be done by either gas and liquid chromatography with UV, biosensor, or fluorescence detection [5,6,14–17], while β-myrcene is typically determined using gas chromatography with mass spectrometry or flame ionization detection [18–21].

Diet is considered to be the greatest source of human exposure to furfuryl alcohol and β-myrcene. However, unlike the furanic compounds furan, 5-hydroxymethylfurfural (HMF), and furfural, and other food and beverage constituents such as ethanol, ethyl carbamate, or polycyclic aromatic hydrocarbons for which extensive occurrence data is available [22–26], there is a paucity of information on human dietary exposure to furfuryl alcohol and β-myrcene. The two agents are due for assessment as to their carcinogenicity by the International Agency for Research in Cancer (IARC) expert working group in their meeting to be held in June 2017. This study aims to provide an overview of the occurrence of furfuryl alcohol and β-myrcene in foods and beverages.

2. Materials and Methods

Occurrence data on furfuryl alcohol and β-myrcene were searched in the following databases: PubMed, Toxnet and ChemIDplus (U.S. National Library of Medicine, Bethesda, MD, USA), Web of Science (Clarivate Analytics, Philadelphia, PA, USA), and IPCS/INCHEM (International Programme on Chemical Safety/Chemical Safety Information from Intergovernmental Organizations, WHO, Geneva, Switzerland). Reference lists of all articles were hand-searched for relevant studies not included in the original search results. The literature sources (including abstracts) were evaluated using Mendeley (Mendeley Inc., New York, NY, USA). By manual screening, relevant articles were identified and ordered in full-text. No unpublished study was identified.

Additional data on the occurrence of furfuryl alcohol was also obtained from in-house analysis of 30 coffee (roasted coffee as beans, powder, or pods), 15 bread, 20 wine, and 50 alcoholic spirit samples (whiskey, brandy, and rum) submitted to our laboratory in the context of official control using nuclear magnetic resonance spectroscopy (NMR) [27]. For this, spectra previously acquired for other purposes were re-quantified for furfuryl alcohol. The coffee samples were analysed according to Monakhova et al. [28]. Quantification was conducted using the integral of the CH group at the C5 resonance of furfuryl alcohol (δ 7.47–7.35 ppm) in relation to the internal standard 1,2,4,5-tetrachloro-3-nitrobenzene (δ 7.75–7.72 ppm). Quantification was conducted using TopSpin 3.2 (BrukerBioSpin GmbH, Rheinstetten, Germany) and Mestrenova V. 11.0.2 (Mestrelab Research, Santiago de Compostela, Spain) [29]. For evaluation of spirits, the NMR method of Monakhova et al. [27] was applied. The NMR methods achieved a limit of detection (LOD) of 3.2 mg/L and limit of quantification (LOQ) of 8.6 mg/L. The results of NMR must be interpreted as semi-quantitative, because only one single non-overlapped signal of furfuryl alcohol was available for quantification. Identity was confirmed by spiking with pure furfuryl alcohol to authentic samples, but co-occurrence of compounds with a similar chemical shift cannot be completely excluded. The statistical parameters of mean, median, and percentiles (90th, 95th, 97.5th, and 99th) were used to describe the occurrence data. Similar NMR analysis of β-myrcene (e.g., in hops) was not possible due to considerable matrix interferences of all relevant signals. The concentration of β-myrcene in compounded products such as beer was below the detection limit of NMR.

3. Results

This study summarizes the occurrence of furfuryl alcohol and β-myrcene in various foods and beverages. Limited studies on β-myrcene (7) were observed compared to 19 studies for furfuryl alcohol. Meta-analysis was not possible due to the sparsity of studies for each type of food and beverage. The occurrence of furfuryl alcohol was recorded in many foods and beverages that had been subjected to thermal processing. The literature studies summarized in Table 1 were extended by inclusion of original results from our own analyses on furfuryl alcohol in 30 coffee, 15 bread, 20 wine, and 50 aged alcoholic spirit samples. From these, only coffee samples were positive (average furfuryl alcohol content of 251 mg/kg), while all other samples were below the detection limit of the method. A typical spectrum of a coffee sample is shown in Figure 1.

Out of the seven studies on β-myrcene, four were in hops and related products, while two were in beer, and the final reference reported about general use levels in various foods/beverages. Chewing gum, gelatin, beer, and hops were suggested as products with high concentration of β-myrcene. The studies are summarized in Table 2.

Table 1. Furfuryl alcohol content in various foods and beverages.

Category [Reference]	N	Furfuryl Alcohol Concentration							Units [a]
		Mean	Median	P90	P95	P97.5	P99	Maximum	
Roasted coffee/This study	30	251	243	342	392	402	406	408	mg/kg
Bread/This study	15	<LOD [b]	-	-	-	-	-	-	mg/kg
Wine/This study	20	<LOD [b]	-	-	-	-	-	-	mg/L
Spirits/This study	50	<LOD [b]	-	-	-	-	-	-	mg/L
Sweet potatoes [4]	1	0.014	-	-	-	-	-	-	mg/kg
Wine [7]	8	3.4	2.9	7.3	8.5	9.0	9.4	9.6	mg/L
Baked goods [9] [d]	-	110	-	-	-	-	-	-	ppm
Spirits [9] [d]	-	10	-	-	-	-	-	-	ppm
Candy [9] [d]	-	59	-	-	-	-	-	-	ppm
Ice cream/ices [9] [d]	-	88	-	-	-	-	-	-	ppm
Beverages [9] [d]	-	19	-	-	-	-	-	-	ppm
Honey [14]	1	1.6	-	-	-	-	-	-	mg/kg
Popcorns [15]	6	0.064	0.067	0.081	0.081	0.082	0.082	0.082	mg/kg
Fried fish [16]	1	10.5	-	-	-	-	-	-	mg/kg
Breaded fish products [17]	4	10.3	8.8	16	18	18	19	19	mg/kg
Wine [30]	6	1.51	0.89	1.57	1.60	1.62	1.63	1.64	mg/L
Vinegar [31] [c]	27	0.35	0.28	0.58	0.59	0.59	0.59	0.59	mg/L
Vinegar [32] [c]	9	0.34	0.28	0.58	0.59	0.59	0.59	0.59	mg/L
Coffee [33]	7	49	49	64	67	68	69	70	mg/kg
Instant coffee [34]	1	267	-	-	-	-	-	-	mg/kg
Roasted coffee [34]	1	564	-	-	-	-	-	-	mg/kg
Pineapple juice [34]	1	8.3	-	-	-	-	-	-	mg/L
Rice cakes [35]	2	2, 2.3	-	-	-	-	-	2.3	mg/kg
Bread [36]	1	187	-	-	-	-	-	-	mg/kg
Toasted almonds [37]	3	6.4	6.0	8.3	8.6	8.7	8.8	8.9	mg/kg
Non-fat dried milk [38]	1	15	-	-	-	-	-	-	mg/kg
Corn tortilla chips [39]	1	0.54	-	-	-	-	-	-	mg/kg
Cocoa powder [40]	1	0.02	-	-	-	-	-	-	mg/kg
Palm sugar [41]	1	0.14, 0.52	-	-	-	-	-	-	mg/kg

[a] The ambiguous unit ppm was interpreted as mg/L for liquids/beverages and as mg/kg for solid foods. [b] All samples evaluated (spirits types whiskey, rum, brandy as well as various wines and breads) were below the limit of detection (LOD; 3.2 mg/L). [c] Studies from the same research group with probably overlapping data. [d] Number of samples not provided. The data are suggested as being "usual concentrations" found in these food/beverage types.

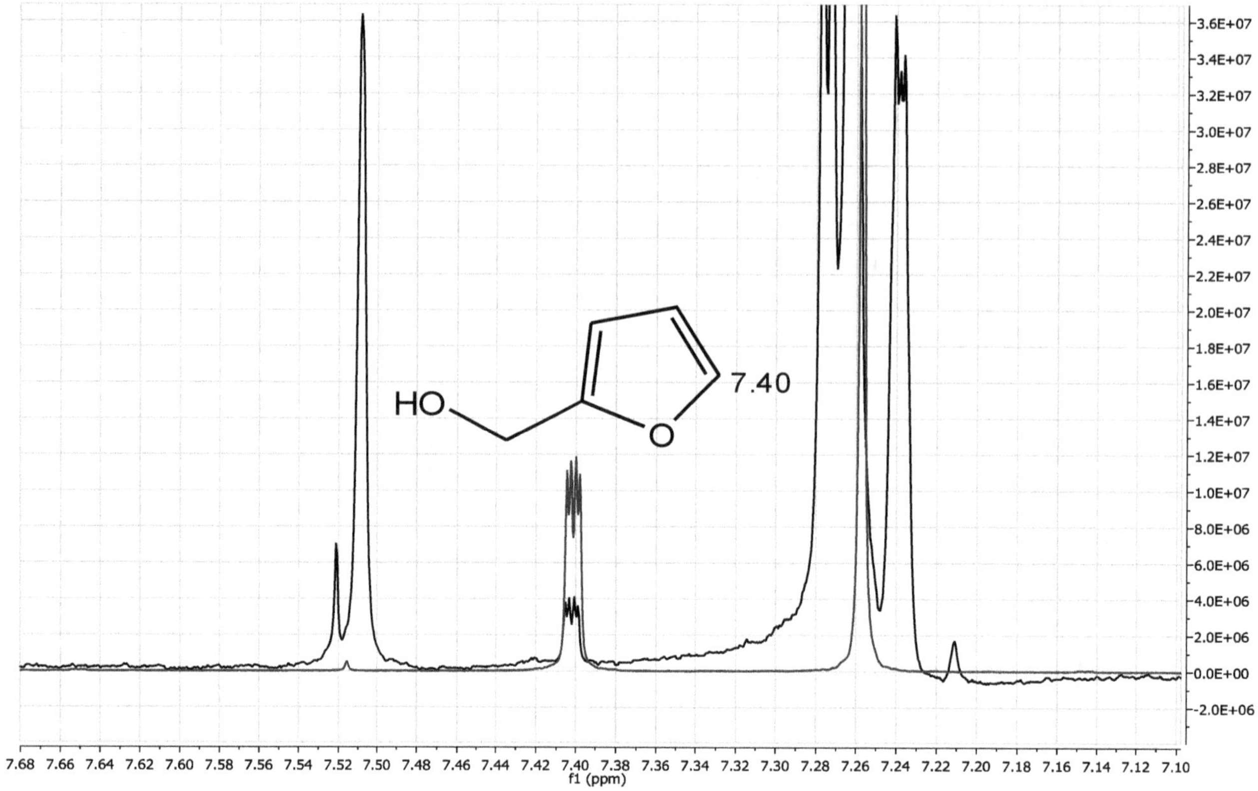

Figure 1. NMR spectra of an authentic coffee sample (**blue line**) containing 408 mg/kg of furfuryl alcohol compared to the reference standard (**red line**).

Table 2. β-Myrcene content in various matrices.

Matrix [Reference]	N	Concentration							Units [a]
		Mean	Median	P90	P95	P97.5	P99	Maximum	
Hops oil [18]	4	479	424	776	852	890	912	927	mg/L
Hops [42]	12	5489	4804	8580	9450	9972	10,285	10,494	mg/kg
Hops [43]	8	15	14	28	29	29	29	29	µg/L
Hops [44]	12	1082	705	2369	2795	3043	3191	3290	mg/kg
Pilsner beer [45]	2	46, 79	-	-	-	-	-	79	µg/L
Beer [46]	2	0.5, 0.6	-	-	-	-	-	0.6	µg/L
Alcoholic beverages [47] [b]	-	1.1	-	-	-	-	-	-	mg/L
Baked goods [47] [b]	-	10	-	-	-	-	-	-	mg/kg
Chewing gum [47] [b]	-	116	-	-	-	-	-	-	mg/kg
Condiment [47] [b]	-	5	-	-	-	-	-	-	mg/kg
Frozen dairy [47] [b]	-	12	-	-	-	-	-	-	mg/kg
Gelatin, pudding [47] [b]	-	20	-	-	-	-	-	-	mg/kg
Meat products [47] [b]	-	5	-	-	-	-	-	-	mg/kg
Non-alcoholic beverages [47] [b]	-	8	-	-	-	-	-	-	mg/L
Soft candy [47] [b]	-	6	-	-	-	-	-	-	mg/kg

[a] The ambiguous unit ppm was interpreted as mg/L for liquids/beverages and as mg/kg for solid food. [b] Number of samples not provided. The data are suggested as being "usual concentrations" found in these food/beverage types.

4. Discussion

4.1. Occurrence of Furfuryl Alcohol

The concentration of furfuryl alcohol was highest in coffee (beans 564 mg/kg and 267 mg/kg in instant coffee powder). Our new data on coffee with an average of 251 mg/kg and a maximum of 408 mg/kg corresponds well to the previous data. Green coffee is free of furfuryl alcohol (confirmed in eight samples with non-detectable levels), so the occurrence of furfuryl alcohol in coffee has been confirmed as being attributable to the roasting process [5]. This observation parallels the high content of furan found in coffee compared to other foods [24]. Other thermally processed foods, such as bread (187 mg/kg), baked goods (110 mg/kg), ice cream/ices (88 mg/kg), and fried fish (about 10 mg/kg) were also found to contain detectable amounts of furfuryl alcohol. Among beverages, higher concentrations of furfuryl alcohol arising from aging in oak barrels [30] were found in spirits (10 mg/L) than in wine (1.5–3.4 mg/L). However, the content was lower compared to bread, baked goods, fish, and coffee. Relatively lower concentrations (less than 1 mg/kg) were observed in palm sugar, chips, popcorns, sweet potatoes, and vinegar. The variation in the concentration of furfuryl alcohol in the foods/beverages may be related to the type of raw materials and processing conditions used. The Joint FAO/WHO Expert Committee on Food Additives (JECFA) set a group acceptable daily intake (ADI) of 0–0.5 mg/kg body weight for furfuryl alcohol, and suggested the compound as being of no safety concern at current levels of intake when used as a flavouring agent [48]. Despite the concentrations reported here being low for a majority of individual foods and beverages, a cumulative amount of furfuryl alcohol may be ingested from consuming a combination of different foods and beverages. According to the National Toxicology Program (NTP) report [49], exposure of male mice to 32 ppm (equivalent to 60 mg/kg bw/day [50]) of furfuryl alcohol was found to induce tumours of renal tubular epithelium. The postulated mechanism of carcinogenicity of furfuryl alcohol is through activation by sulfotransferases resulting in the formation of a 2-methylfuranyl-DNA adduct [50,51]. According to estimation from the typical intake levels of the food items listed in Table 1, concentrations of toxicological concern are probably not reached. However, food legislation demands to reduce food contaminants as low as reasonably achievable (ALARA principle). More data are clearly necessary to provide exposure estimations and risk assessment for this compound.

4.2. Occurrence of β-Myrcene

A majority of the studies on β-myrcene are qualitative, and the few quantitative data were focusing on hops and beers, despite the widespread occurrence of myrcene in many plants that are used in foods and beverages. Hop oil and chewing gum were found to contain the highest content of β-myrcene compared to other products. The low concentration of β-myrcene in beers is plausible, since there is a very variable extraction of β-myrcene from hops to beer postulated to be in the range of 0.5%–5.6% from cones and 8.4%–25.8% from pellets [52], and hops contain other volatile components such linanool, humulene, and α-terpineol in higher proportions than β-myrcene. Additionally, β-myrcene may be destroyed during the heating processes, and thus a low level is expected in the final beer. The NTP report links β-myrcene with neoplasms of the kidney in male rats and liver cancer in male mice [53]. The daily per capita intake (eaters only) for β-myrcene was estimated as being 164 μg corresponding to 3 μg/kg bw [54].

5. Conclusions

Consistent with the relatively high amounts of furfuryl alcohol (above 10 mg/kg) observed in coffee, baked goods, bread, fish, and some spirit drinks, monitoring these items for furfuryl alcohol is advisable for comprehensive estimation of exposures and the risk of these foods, while more research on the occurrence of β-myrcene in foods and beverages in general is required for meaningful risk assessment.

Acknowledgments: Andreas Scharinger is thanked for excellent technical assistance for the NMR work. Alex Okaru is indebted to the DAAD (German Academic Exchange Service) programme funding No. 5250960 for the scholarship award.

Author Contributions: All authors contributed equally to this study.

References

1. Monakhova, Y.B.; Lachenmeier, D.W. The Margin of Exposure of 5-Hydroxymethylfurfural (HMF) in alcoholic beverages. *Environ. Health Toxicol.* **2012**, *27*, e2012016. [CrossRef] [PubMed]

2. Martins, S.I.; Jongen, W.M.; Boekel, M.A. A review of Maillard reaction in food and implications to kinetic modelling. *Trends Food Sci. Technol.* **2001**, *11*, 364–373. [CrossRef]
3. Vanderhaegen, B.; Delvaux, F.; Daenen, L.; Verachtert, H.; Delvaux, F.R. Aging characteristics of different beer types. *Food Chem.* **2007**, *103*, 404–412. [CrossRef]
4. Wang, Y.; Kays, S.J. Contribution of volatile compounds to the characteristic aroma of baked "Jewel" sweetpotatoes. *J. Amer. Soc. Hort. Sci.* **2000**, *125*, 638–643.
5. Swasti, Y.R.; Murkovic, M. Characterization of the polymerization of furfuryl alcohol during roasting of coffee. *Food Funct.* **2012**, *3*, 965–969. [CrossRef] [PubMed]
6. Pérez-Prieto, L.J.; López-Roca, J.M.; Martínez-Cutillas, A.; Pardo-Mínguez, F.; Gómez-Plaza, E. Extraction and formation dynamic of oak-related volatile compounds from different volume barrels to wine and their behavior during bottle storage. *J. Agric. Food Chem.* **2003**, *51*, 5444–5449. [CrossRef] [PubMed]
7. Spillman, P.J.; Pollnitz, A.P.; Liacopoulos, D.; Pardon, K.H.; Sefton, M.A. Formation and degradation of furfuryl alcohol, 5-Methylfurfuryl alcohol, vanillyl alcohol, and their ethyl ethers in barrel-aged wines. *J. Agric. Food Chem.* **1998**, *46*, 657–663. [CrossRef] [PubMed]
8. Mochizuki, N.; Kitabatake, K. Analysis of 1-(2-furyl)propane-1,2-diol, a furfural metabolite in beer. *J. Ferment. Bioeng.* **1997**, *83*, 401–403. [CrossRef]
9. National Research Council (U.S.). *Chemicals Used in Food Processing;* National Academy of Sciences: Washington DC, USA, 1965.
10. Paumgartten, F.J.; De-Carvalho, R.R.; Souza, C.A.; Madi, K.; Chahoud, I. Study of the effects of beta-myrcene on rat fertility and general reproductive performance. *Braz. J. Med. Biol. Res.* **1998**, *31*, 955–965. [CrossRef] [PubMed]
11. Mohamed Hanaa, A.R.; Sallam, Y.I.; El-Leithy, A.S.; Aly, S.E. Lemongrass (Cymbopogon citratus) essential oil as affected by drying methods. *Ann. Agric. Sci.* **2012**, *57*, 113–116. [CrossRef]
12. Lachenmeier, K.; Musshoff, F.; Madea, B.; Lachenmeier, D.W. Application of experimental design to optimise solid-phase microextraction of orange juice flavour. Electron. *J. Environ. Agric. Food Chem.* **2006**, *5*, 1380–1388.
13. Behr, A.; Johnen, L. Myrcene as a natural base chemical in sustainable chemistry: A critical review. *ChemSusChem* **2009**, *2*, 1072–1095. [CrossRef] [PubMed]
14. Vázquez, L.; Verdú, A.; Miquel, A.; Burló, F.; Carbonell-Barrachina, A.A. Changes in physico-chemical properties, hydroxymethylfurfural and volatile compounds during concentration of honey and sugars in Alicante and Jijona turron. *Eur. Food Res. Technol.* **2007**, *225*, 757–767. [CrossRef]
15. Park, D.; Maga, J.A. Identification of key volatiles responsible for odour quality differences in popped popcorn of selected hybrids. *Food Chem.* **2006**, *99*, 538–545. [CrossRef]
16. Pérez-Palacios, T.; Petisca, C.; Melo, A.; Ferreira, I.M. Quantification of furanic compounds in coated deep-fried products simulating normal preparation and consumption: Optimisation of HS-SPME analytical conditions by response surface methodology. *Food Chem.* **2012**, *135*, 1337–1343. [CrossRef] [PubMed]
17. Pérez-Palacios, T.; Petisca, C.; Henriques, R.; Ferreira, I.M. Impact of cooking and handling conditions on furanic compounds in breaded fish products. *Food Chem. Toxicol.* **2013**, *55*, 222–228. [CrossRef] [PubMed]
18. Van Opstaele, F.; De Causmaecker, B.; Aerts, G.; De Cooman, L. Characterization of novel varietal floral hop aromas by headspace solid phase microextraction and gas chromatography-mass spectrometry/olfactometry. *J. Agric. Food Chem.* **2012**, *60*, 12270–12281. [CrossRef] [PubMed]
19. Gonçalves, J.; Figueira, J.; Rodrigues, F.; Câmara, J.S. Headspace solid-phase microextraction combined with mass spectrometry as a powerful analytical tool for profiling the terpenoid metabolomic pattern of hop-essential oil derived from Saaz variety. *J. Sep. Sci.* **2012**, *35*, 2282–2296. [CrossRef] [PubMed]
20. Roberts, M.T.; Dufour, J.P.; Lewis, A.C. Application of comprehensive multidimensional gas chromatography combined with time-of-flight mass spectrometry (GC x GC-TOFMS) for high resolution analysis of hop essential oil. *J. Sep. Sci.* **2004**, *27*, 473–478. [CrossRef] [PubMed]
21. Kishimoto, T.; Wanikawa, A.; Kono, K.; Shibata, K. Comparison of the odor-active compounds in unhopped beer and beers hopped with different hop varieties. *J. Agric. Food Chem.* **2006**, *54*, 8855–8861. [CrossRef] [PubMed]
22. Lachenmeier, D.W.; Przybylski, M.C.; Rehm, J. Comparative risk assessment of carcinogens in alcoholic beverages using the margin of exposure approach. *Int. J. Cancer* **2012**, *131*, E995–E1003. [CrossRef] [PubMed]
23. Lachenmeier, D.W. Carcinogens in Food: Opportunities and Challenges for Regulatory Toxicology. *Open Toxicol. J.* **2009** *3*, 30–34. [CrossRef]

24. Waizenegger, J.; Winkler, G.; Kuballa, T.; Ruge, W.; Kersting, M.; Alexy, U.; Lachenmeier, D.W. Analysis and risk assessment of furan in coffee products targeted to adolescents. *Food Addit. Contam.* **2012**, *29*, 19–28. [CrossRef] [PubMed]
25. European Food Safety Authority. Polycyclic Aromatic Hydrocarbons in Food. *EFSA J.* **2008**, *724*, 1–114.
26. Singh, L.; Varshney, J.G.; Agarwal, T. Polycyclic aromatic hydrocarbons' formation and occurrence in processed food. *Food Chem.* **2016**, *199*, 768–781. [CrossRef] [PubMed]
27. Monakhova, Y.B.; Schäfer, H.; Humpfer, E.; Spraul, M.; Kuballa, T.; Lachenmeier, D.W. Application of automated eightfold suppression of water and ethanol signals in 1H NMR to provide sensitivity for analyzing alcoholic beverages. *Magn. Reson. Chem.* **2011**, *49*, 734–739. [CrossRef] [PubMed]
28. Monakhova, Y.B.; Ruge, W.; Kuballa, T.; Ilse, M.; Winkelmann, O.; Diehl, B.; Thomas, F.; Lachenmeier, D.W. Rapid approach to identify the presence of Arabica and Robusta species in coffee using 1H NMR spectroscopy. *Food Chem.* **2015**, *182*, 178–184. [CrossRef] [PubMed]
29. Bernstein, M.A.; Sýkora, S.; Peng, C.; Barba, A.; Cobas, C. Optimization and automation of quantitative NMR data extraction. *Anal. Chem.* **2013**, *85*, 5778–5786. [CrossRef] [PubMed]
30. Carrillo, J.D.; Garrido-López, Á.; Tena, M.T. Determination of volatile oak compounds in wine by headspace solid-phase microextraction and gas chromatography-mass spectrometry. *J. Chromatogr. A* **2006**, *1102*, 25–36. [CrossRef] [PubMed]
31. Morales, M.L.; Benitez, B.; Troncoso, A.M. Accelerated aging of wine vinegars with oak chips: Evaluation of wood flavour compounds. *Food Chem.* **2004**, *88*, 305–315. [CrossRef]
32. Tesfaye, W.; Morales, M.L.; Benítez, B.; García-Parrilla, M.C.; Troncoso, A.M. Evolution of wine vinegar composition during accelerated aging with oak chips. *Anal. Chim. Acta* **2004**, *513*, 239–245. [CrossRef]
33. Petisca, C.I. Furanic Compounds in Food Products: Assessment and Mitigation Strategies. Ph.D. Thesis, Universidade Do Porto, Porto, Portugal, 2013.
34. Golubkova, T. Bildung von Potentiell Toxischen Furanderivaten in Lebensmitteln. Masters Thesis, TU Graz, Austria, 2011.
35. Moon, J.K.; Shibamoto, T. Role of roasting conditions in the profile of volatile flavor chemicals formed from coffee beans. *J. Agric. Food Chem.* **2009**, *57*, 5823–5831. [CrossRef] [PubMed]
36. Jensen, S.; Ostdal, H.; Skibsted, L.H.; Thybo, A.K. Antioxidants and shelf life of whole wheat bread. *J. Cereal Sci.* **2011**, *53*, 291–297. [CrossRef]
37. Vázquez-Araújo, L.; Enguix, L.; Verdú, A.; García-García, E.; Carbonell-Barrachina, A.A. Investigation of aromatic compounds in toasted almonds used for the manufacture of turrón. *Eur. Food Res. Technol.* **2008**, *227*, 243–254. [CrossRef]
38. Karagül-Yüceer, Y.; Cadwallader, K.R.; Drake, M.A. Volatile flavor components of stored nonfat dry milk. *J. Agric. Food Chem.* **2002**, *50*, 305–312. [CrossRef]
39. Buttery, R.G.; Ling, L.C. Additional Studies on Flavor Components of Corn Tortilla Chips. *J. Agric. Food Chem.* **1998**, *46*, 2764–2769. [CrossRef]
40. Bonvehí, J.S. Investigation of aromatic compounds in roasted cocoa powder. *Eur. Food Res. Technol.* **2005**, *221*, 19–29. [CrossRef]
41. Ho, C.W.; Aida, W.M.W.; Maskat, M.Y.; Osman, H. Changes in volatile compounds of palm sap (Arenga pinnata) during the heating process for production of palm sugar. *Food Chem.* **2007**, *102*, 1156–1162. [CrossRef]
42. Aberl, A.; Coelhan, M. Determination of volatile compounds in different hop varieties by headspace-trap GC/MS—In comparison with conventional hop essential oil analysis. *J. Agric. Food Chem.* **2012**, *60*, 2785–2792. [CrossRef]
43. Mitter, W.; Cocuzza, S. Dry Hopping—A Study of Various Parameters Consequences of the Applied Dosing Method. Available online: http://hopsteiner.com/wp-content/uploads/2016/03/3_Dry-Hopping-A-Study-of-Various-Parameters.pdf (accessed on 10 September 2016).
44. Peacock, V.E.; Deinzer, M.L. Chemistry of hop aroma in beer. *J. Am. Soc. Brew. Chem* **1979**, *39*, 136–141.
45. Schmidt, C.; Biendl, M. Headspace Trap GC-MS analysis of hop aroma compounds in beer. *BrewingScience* **2016**, *69*, 9–15.
46. Mikyška, A.; Olšovská, J. Czech Research and Development in the Field of Brewing Raw Materials. Available online: http://www.hmelj-giz.si/ihgc/doc/LdVBS-RIBM_raw_materials_research.pdf (accessed on 11 September 2016).
47. Burdock, G.A. *Fenaroli's Handbook of Flavor Ingredients*; CRC Press: Boca Raton, FL, USA, 2004.

48. World Health Organization (WHO). *Evaluation of Certain Food Additives and Contaminants*; 55th report of the Joint FAO/WHO Expert Committee on Food Additives; WHO: Geneva, Switzerland, 2001.
49. National Toxicology Program (NTP). *Technical Report on the Toxicology and Carcinogenesis Studies of Furfuryl Alcohol (CAS NO. 98-00-0) in F344/N Rats and B6C3F1 Mice (Inhalational Studies). NTP TR 482. NIH Publication No. 99-3972*; National Toxicology Program: Research Triangle Park, NC, USA, 1999.
50. Sachse, B.; Meinl, W.; Sommer, Y.; Glatt, H.; Seidel, A.; Monien, B.H. Bioactivation of food genotoxicants 5-hydroxymethylfurfural and furfuryl alcohol by sulfotransferases from human, mouse and rat: A comparative study. *Arch. Toxicol.* **2016**, *90*, 137–148. [CrossRef] [PubMed]
51. Sachse, B.; Meinl, W.; Glatt, H.; Monien, B.H. The effect of knockout of sulfotransferases 1a1 and 1d1 and of transgenic human sulfotransferases 1A1/1A2 on the formation of DNA adducts from furfuryl alcohol in mouse models. *Carcinogenesis* **2014**, *35*, 2339–2345. [CrossRef] [PubMed]
52. Wolfe, P.H. A Study of Factors Affecting the Extraction of Flavor When Dry Hopping Beer. Masters Thesis, Oregon State University, Corvallis, OR, USA, 2012.
53. National Toxicology Program (NTP). *Technical Report on the Toxicology and Carcinogenesis Studies of β-Myrcene (CAS NO. 123-35-3) in F344/N Rats and B6C3F1 Mice. NTP TR 557. NIH Publication No. 09-5898*; National Toxicology Program: Reseach Triangle Park, NC, USA, 2011.
54. Adams, T.B.; Gavin, C.L.; McGowen, M.M.; Waddell, W.J.; Cohen, S.M.; Feron, V.J.; Marnett, L.J.; Munro, I.C.; Portoghese, P.S.; Rietjens, I.M.C.M.; Smith, R.L. The FEMA GRAS assessment of aliphatic and aromatic terpene hydrocarbons used as flavor ingredients. *Food Chem. Toxicol.* **2011**, *49*, 2471–2494. [CrossRef] [PubMed]

12

ELISA and Chemiluminescent Enzyme Immunoassay for Sensitive and Specific Determination of Lead (II) in Water, Food and Feed Samples

Long Xu [1,2,†], Xiao-yi Suo [3,†], Qi Zhang [1,3], Xin-ping Li [3], Chen Chen [1] and Xiao-ying Zhang [1,2,3,*]

1. College of Biological Science and Engineering, Shaanxi University of Technology, Hanzhong 723000, China; xu_lon@163.com (L.X.); 13259850610@163.com (Q.Z.); cchen2008@yahoo.com (C.C.)
2. Centre of Molecular and Environmental Biology, University of Minho, Department of Biology, Campus de Gualtar, 4710-057 Braga, Portugal
3. College of Veterinary Medicine, Northwest A&F University, Yangling 712100, China; suoxiaoyi@nwafu.edu.cn (X.-y.S.); lxp67cqu@163.com (X.-p.L.)

* Correspondence: zhang@bio.uminho.pt

† The first two authors contributed equally to this work.

Abstract: Lead is a heavy metal with increasing public health concerns on its accumulation in the food chain and environment. Immunoassays for the quantitative measurement of environmental heavy metals offer numerous advantages over other traditional methods. ELISA and chemiluminescent enzyme immunoassay (CLEIA), based on the mAb we generated, were developed for the detection of lead (II). In total, 50% inhibitory concentrations (IC_{50}) of lead (II) were 9.4 ng/mL (ELISA) and 1.4 ng/mL (CLEIA); the limits of detection (LOD) were 0.7 ng/mL (ic-ELISA) and 0.1 ng/mL (ic-CLEIA), respectively. Cross-reactivities of the mAb toward other metal ions were less than 0.943%, indicating that the obtained mAb has high sensitivity and specificity. The recovery rates were 82.1%–108.3% (ic-ELISA) and 80.1%–98.8% (ic-CLEIA), respectively. The developed methods are feasible for the determination of trace lead (II) in various samples with high sensitivity, specificity, fastness, simplicity and accuracy.

Keywords: lead (II); ELISA; monoclonal antibody (mAb); isothiocyanobenzyl-EDTA (ITCBE); chemiluminescent enzyme immunoassay (CLEIA)

1. Introduction

Environmental pollution from heavy metals is a worldwide issue. Lead has been widely used in the nuclear industry, glass manufacturing, battery industry, pipe industry, cosmetics industry, toy industry and paint industry [1]. Lead can be accumulated in the environment, as it cannot be rendered harmless through a chemical or bioremediation process. Plant leaves and roots are prone to accumulate toxic metals and can therefore be used for environmental monitoring, as a tool for assessing soil-contamination levels [2].

The major sources of lead exposure include piped drinking water, soldering from canned foods, beverages and traditional medicines. When indirectly ingested through contaminated food or inhalation, lead enters the food chain from the soil, water, deposition from the air, containers or dishes, and/or from food-processing equipment. Lead primarily accumulates in blood, soft tissues, bone and neurons, and this accumulation may cause behavioral changes, cognitive obstacles, blindness, encephalopathy, kidney failure and death. Children are more susceptible and vulnerable to lead due to its impact on the nervous system, as well as on development and behavioral performance [3]. Nowadays, lead pollution has become increasingly serious because of its excessive usage. The recent water contamination of lead in Flint, Michigan, remains a topical issue in public health. The decreased intelligence of children is directly positively correlated with blood lead and bone lead levels [4]. Although regulatory authorities have established safe levels of lead in foods (Table 1), the consensus is that there is no safe level of lead.

The spectroscopy methods for the detection of lead (II) included atomic absorption spectrometry (AAS) [5], atomic fluorescence spectrometry (AFS) [6] and multiple collectors inductively coupled plasma mass spectrometry (MC-ICP-MS) [7]. An AAS was used to detect Pb^{2+} in food with detection limit of 6 ng/mL [8]. These methods are sensitive and accurate, but costly and require intricate equipment and highly qualified technicians, making them unsuitable for onsite detection [9]. Recently, several sensors based on fluorophores, organic molecules and gold nanoparticles [10] have been reported to detect lead ions. A biomimetic sensor was applied in detecting Pb^{2+} in water, with a limit of detection of 9.9 ng/mL [11]. Lead (II) fluorescent sensors detections show high sensitivity, but require fluorophores and/or quenchers. Furthermore, the background signal could lead to serious interference due to its high fluorescence intensity [12]. Electrochemical sensors need tailor-made tactical materials and biological molecules, require skillful design and lengthy sample preparation and lack sufficient specificity. Therefore, the prospects of applying these sensors are limited [13,14].

Immunoassays have been applied for heavy-metal detection (e.g., cadmium, lead, chromium, uranium and mercury) [15], as they are quick, inexpensive, easy to perform, and highly sensitive and selective. ELISA and gold immunochromatographic assay (GICA) have been applied for detection of lead ions in water samples [16,17]. Chemiluminescent enzyme immunoassay (CLEIA), which has been widely used in pesticide and veterinary drug residue analysis, uses the energy generated by chemical reactions to excite luminescence, eliminating the need for external light sources. As a pilot attempt, this study aimed to develop CLEIA and the most commonly used ELISA for Pb^{2+} analysis in water, food and feed samples, to better address the current rapid and sensitive need on Pb^{2+} detection in environment and food contamination.

Table 1. Permitted maximum amount of lead.

	Fresh Vegetable (mg/kg)	Cereals (mg/kg)	Fresh Fruits (mg/kg)	Mushroom (mg/kg)	Beans (mg/kg)	Livestock and Poultry Meat (mg/kg)	Livestock and Poultry Gut (mg/kg)	Fish (mg/kg)	Salt (mg/kg)	Drinking Water (mg/L)
CAC [18]	0.1	0.2	0.1	-	0.2	0.1	0.5	0.3	2	0.01
EFSA [19]	0.1	0.2	0.1	0.3	0.2	0.1	0.5	0.3	-	0.01
CFDA [20]	0.1	0.2	0.1	1	0.2	0.2	0.5	0.5	2	0.01
FSANZ [21]	0.1	0.2	0.1	-	0.2	0.1	0.5	0.5	-	0.05

Notes: CAC: Codex Alimentarius Commission; CFDA; Chinese Food and Drug Administration; EFSA: European Food Safety Authority; FSANZ: Food Standards Australia New Zealand.

2. Materials and Methods

2.1. Ethics Statement

All experimental animal protocols were reviewed and approved by the Ethics Committee of Shaanxi University of Technology for the Use of Laboratory Animals.

2.2. Chemicals and Reagents

Isothiocyanobenzyl-EDTA (ITCBE) was purchased from Dojindo (Kyushu, Japan). N, N′-dicyclohexylcarbodiimide (DCC), N-hydroxysuccinimide (NHS), dimethyl formamide (DMF), 3, 3′, 5, 5′-tetramethylbenzidine (TMB) and luminol were purchased from Solarbio (Beijing, China). HAT medium, keyhole hemocyanin (KLH) and bovine serum albumin (BSA) were purchased from Sigma (St. Louis, MO, USA). Goat anti-mouse IgG-HRP was purchased from Thermo (Waltham, MA, USA). $Pb(NO_3)_2$, $HgSO_4$, $3CdSO_4{\cdot}8H_2O$, $Cr_2(SO_4)_3{\cdot}6H_2O$, $CuSO_4$, $CoCl_2{\cdot}6(H_2O)$, $NiSO_4.6H_2O$, $ZnSO_4{\cdot}7H_2O$ and $FeSO_4{\cdot}7H_2O$ were purchased from Sinopharma chemical reagent (Shanghai, China). OriginPro 8.1 (OriginLab, Northampton, MA, USA) was used for processing the analytical data.

2.3. Synthesis of Artificial Antigens of Lead

The ITCBE was conjugated to lead ions, BSA or KLH, using the DCC/NHS ester method. Briefly, equimolar amounts (0.06 mmol) of ITCBE, NHS and DCC were dissolved in 200 μL of DMF, and the same amount of lead nitrate was added to the mixture and stirred overnight. After centrifugation of the solution at 13,400× *g* for 10 min, the supernatant was added dropwise to 40 mg of BSA or KLH dissolved in 3 mL of 0.13 M $NaHCO_3$ (pH 8.3), under stirring. After reaction for 4 h and centrifugation, the supernatant was dialyzed in phosphate buffered saline (PBS; 0.01 M; pH 7.4) for 4 days, with daily change of buffer.

UV spectra of lead (II)-ITCBE, BSA and lead (II)-ITCBE-BSA were tested at a wavelength ranging from 200 to 400 nm.

2.4. Production of Monoclonal Antibody

Four female BALB/C mice were immunized subcutaneously with 100 μg of lead (II)-ITCBE-KLH emulsified with an equal volume of Freund's complete adjuvant. In the next two sequential booster immunizations, 50 μg of immunogen emulsified with the same volume of incomplete Freund's adjuvant was given to each mouse, in the same way, at 2-week intervals. The fourth injection was administered intraperitoneally without adjuvant. Three days after the final booster injection, the mice were killed. Their spleen cells were removed and fused with mouse SP2/0 myeloma cells, using 50% PEG 4000 (*w/v*) as fusion agent. The mixture was spread in 96-well culture plates supplemented with hypoxanthine–aminopterin–thymidine (HAT) medium containing 20% fetal calf serum and peritoneal macrophages as feeder cells from BALB/C mice. The plates were incubated at 37 °C, with 5% CO_2. After about 2 weeks, the supernatants were screened by an indirect competitive ELISA, using lead (II)-ITCBE-BSA as coating antigen. ITCBE, lead ions and lead (II)-ITCBE were tested as competitors. The hybridomas which were positive to lead (II)-ITCBE-BSA and negative to ITCBE-BSA were subcloned three times, using the limiting dilution method. Stable antibody-producing clones were expanded and cryopreserved in liquid nitrogen. Antibodies were collected and subjected to purification by ammonium sulfate precipitation. The purified mAb was stored at −20 °C, in the presence of 50% glycerol.

2.5. Indirect Competitive ELISA

The 96-well microtiter plates were coated with lead (II)-ITCBE-BSA conjugation (1 μg/mL, 100 μL/well) in carbonate buffer (CBS, 0.05 M, pH 9.6), and then incubated overnight at 4 °C. The plates were washed three times with PBST (PBS containing 0.05% Tween-20), using an automated plate washer, and blocked with blocking buffer (2% BSA in PBS, 200 μL/well) for 2 h, at 37 °C. After washing, diluted mAbs (stock concentration: 3.5 mg/mL, 1:32 000 dilution, 50 μL/well) were added to lead ions standard solutions (0.2, 1, 2, 5, 10, 20, 50, 100 and 200 ng/mL) or samples (50 μL/well) and incubated for 40 min, at 37 °C. After washing three times, the plates were incubated with goat anti-mouse IgG-HRP (stock concentration: 1.5 mg/mL, 1:8000, 100 μL/well), at 37 °C, for 40 min. Then, the washed plates were added with the substrate solution (TMB+H_2O_2, 100 μL/well). After 10 min of incubation, H_2SO_4 (2 M, 50 μL/well) was added, and the absorbance was measured at 450 nm. Normalized calibration curves were constructed in the form of $(B/B_0) \times 100(\%)$ vs. log C (lead ions) (where B and B_0 were the absorbance of the analyte at the standard point and at zero concentration of the analyte, respectively.

2.6. Cross-Reactivity

The specificity of the mAb was investigated by cross-reactivity (CR). Different metal ions, including Hg^{2+}, Cu^{2+}, Ni^{2+}, Zn^{2+}, Cd^{2+}, Fe^{2+}, Co^{2+}, Mg^{2+} and Ca^{2+} (in the form of their soluble chloride, nitrate, carbonate or sulfate salts), were analyzed. The standard solutions of cross-reacting chemicals were prepared in the concentration range of 0.001–1000 ng/mL. CR (%) = [IC_{50} for lead ions]/ [IC_{50} for competing chemical] × 100 (%).

2.7. Indirect Competitive CLEIA

The optimal concentrations of lead (II)-ITCBE-BSA and anti-lead antibody were selected, using ELISA, by checkerboard titration. The indirect competitive CLEIA was described as follows: 100 μL/well of lead (II)-ITCBE-BSA (1 μg/mL) in 0.05 M CBS (pH 9.6) was coated on the 96-well polystyrene microtiter plates and incubated at 4 °C overnight. The following day, the plate was washed three times, using PBST, and blocked with 2% BSA in PBS (200 μL per well), at 37 °C, for 2 h. After a further washing step, 50 μL of diluted mAb (stock concentration: 3.5 mg/mL, 1:32 000 dilution) and 50 μL of lead ions standard solution were added to each well and incubated at 37 °C, for 40 min. Lead ions standard solution was prepared by diluting with PBS at a series of concentrations (0.2, 0.5, 1, 2, 5, 10, 20, 50, 100 and 200 ng/mL). After washing with PBST, the plates were incubated, and goat anti-mouse IgG-HRP (stock concentration: 1.5 mg/mL, 1:8000, 100 μL per well) was added and incubated at 37 °C, for 40 min. Finally, 100 μL of substrate solution prepared freshly was added into each well and incubated for 5 min, in the dark. Then chemiluminescence intensity was monitored on Synergy H1. The standard curve was evaluated by plotting chemiluminescence intensity against the logarithm of each concentration and fit to a logistic equation, using OriginLab 8.1 program.

2.8. Graphite Furnace Atomic Absorption Spectrometry (GFAAS)

The operating parameters of the GFAAS system were as follows: lead hollow lamp current 30 mA, wavelength 283.3 nm, shielding gas (Ar) flow rate 1500 mL/min, carrier gas (Ar) flow rate 500 mL/min, and ashing temperature and time were 450 °C and 9 s. The atomization temperature, heating rate and heating time were 2250 °C, 2200 °C/s and 3 s, respectively. The carrier solution was HNO_3 (5.0%, *v/v*). The calibration curve for lead ions was constructed with standards of 0, 0.1, 0.2, 0.4, 0.6, 0.8, 1.0, 1.4, 1.8, 2.4 and 3.0 μg/L.

2.9. Sample Preparation and Spiked Experiment

Spiked samples were used to examine the assay accuracy and precision.

Water samples, including ultrapure water, tap water and river water, were collected from different sites in Yangling, Shaanxi province, China. Water samples (100 mL) were added with Pb standard solution (1 mg/mL) at the final concentration of 100, 200 and 500 ng/mL. Ultrapure water and tap water were analyzed without any dilution and sample preparation. The river water was filtrated with a 0.45 μm nylon membrane filter and adjusted to pH 7.0 before analysis.

Milk samples were collected from the local market. Milk samples (100 mL) were added with Pb standard solution (1 mg/mL) at the final concentration of 100, 200 and 500 ng/mL. Then the samples were boiled to remove the denatured protein and fat, and then an equal volume of acetate buffer solution (0.1 M, pH 5.7) was added for precipitation. After being maintained at room temperature for 2 h, the mixture was centrifuged at 13,400× *g* for 10 min. The pH of the supernatants was adjusted to 7.0 with 1 M NaOH and diluted with pure water for analysis.

Chicken, rice and feed samples (1.0 g) were homogenized and added with Pb standard solution (1 mg/mL) at the final amounts of 100, 200 and 500 ng. Then the samples were extracted by acid leach method. The samples were soaked with 20% HNO_3, overnight, at room temperature, followed by boiling until fully dissolved. After cooling, the solution was centrifuged, and the supernatant was adjusted to a pH value of 7.0 with 1 M NaOH and diluted with pure water for further analysis.

2.10. Pretreatment of Samples for GFAAS

Water samples (10 mL) were added with Pb standard solution (1 mg/mL) at the final amounts of 1, 2 and 5 μg. Then the samples were mixed with 50% HCl (1 mL), 0.8 mL of a solution containing $KBrO_3$ (0.1 M) and KBr (0.084 M). After reaction for 15 min, an appropriate amount of hydroxylamine hydrochloride/sodium chloride (both at a concentration of 120 g/L) solution was added until the yellow color disappeared. The solution was further diluted with pure water, to 200 mL, and determined by GFAAS.

Chicken, rice and feed samples were pretreated, using a microwave-assisted acid-digestion procedure. Samples (1.0 g) were homogenized and added with Pb standard solution (1 mg/mL) at the final amounts of 100, 200 and 500 ng. Then the samples were transferred into polytetrafluoroethylene (PTFE) flasks, and then HNO_3 (8 mL) and H_2O_2 (2 mL) were added to each flask and kept for 15 min, at room temperature. The flasks were sealed and subjected to microwave digestion. Finally, the samples were diluted with pure water, to 200 mL, for GFAAS detection.

3. Results

3.1. Characterization of the Artificial Antigen and the Monoclonal Antibody

Lead (II)-ITCBE, BSA and Lead (II)-ITCBE-BSA spectra were recorded from 200 to 400 nm. BSA exhibits a characteristic ultraviolet absorption peak at 229 and 278 nm, and lead (II)-ITCBE-BSA exhibits a characteristic ultraviolet absorption peak at 215 nm. The shift of the ultraviolet absorption peak proved that the artificial antigen synthesis was successful (see Figure 1).

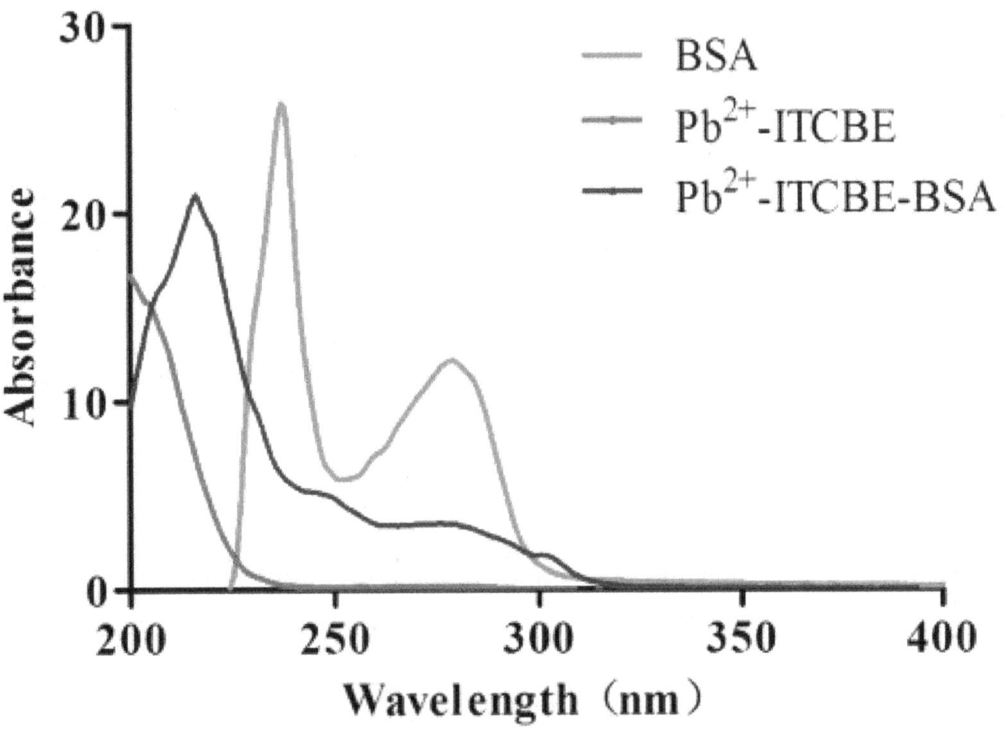

Figure 1. UV absorbance spectra of lead (II)-ITCBE, BSA and lead (II)-ITCBE-BSA.

The anti-lead mAb was purified from mice ascites, using ammonium sulfate precipitation and protein G column affinity chromatography with an obtained concentration of 3.5 mg/mL. The isotype of mAb was IgG1 with a kappa light chain.

3.2. Development of ic-ELISA

Sensitivity of ELISA was determined under optimal conditions. In the representative competitive inhibition curve for lead ions (see Figure 2), the regression curve equation of the anti-lead mAb was $Y = -0.352X + 1.195$ ($R^2 = 0.990$, $n = 3$), with an IC_{50} value of 9.4 ng/mL and limit of detection (IC_{10} value) of 0.7 ng/mL. The ELISA could be used for Pb^{2+} detection with a linear range from 1 to 100 ng/mL.

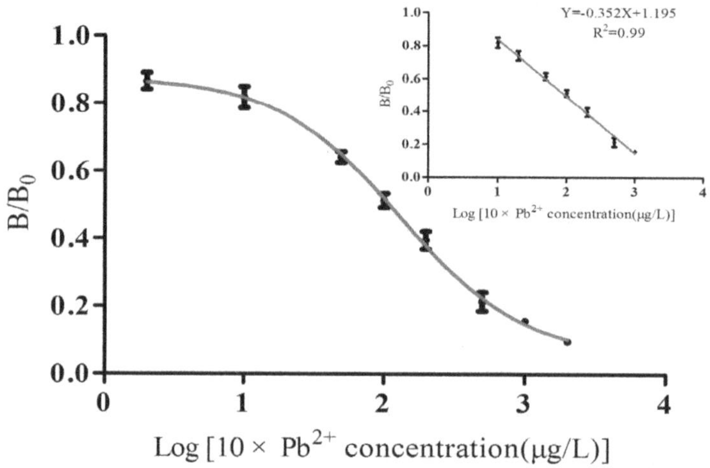

Figure 2. Standard curve of the competitive ELISA for lead ions.

3.3. Cross-Reactivity

The obtained mAb did not recognize the other eight common metal ions (see Table 2).

Table 2. Cross-reactivity of anti-lead IgG with other metal ions ($n = 3$).

Compounds	IC_{50} (μg/L)	Cross-Reactivity (%)
Pb^{2+} -ITCBE	9.4	100
Hg^{2+} -ITCBE	>1×10^3	<0.943
Cd^{2+} -ITCBE	>1×10^3	<0.943
Cr^{3+} -ITCBE	>1×10^3	<0.943
Cu^{2+} -ITCBE	>1×10^3	<0.943
Co^{2+} -ITCBE	>1×10^3	<0.943
Ni^{2+} -ITCBE	>1×10^3	<0.943
Zn^{2+} -ITCBE	>1×10^3	<0.943
Fe^{2+} -ITCBE	>1×10^3	<0.943

3.4. Chemiluminescence Immunoassay

The sensitivity of ic-CLEIA was determined under optimal conditions. The representative competitive inhibition curve (see Figure 3) revealed the regression curve equation of $Y = -0.319X + 0.862$ ($R^2 = 0.992$, $n = 3$), with IC_{50} value of 1.4 ng/mL, the limit of detection (IC_{10} value) of 0.1 ng/mL and the linear range from 0.2 to 50 ng/mL.

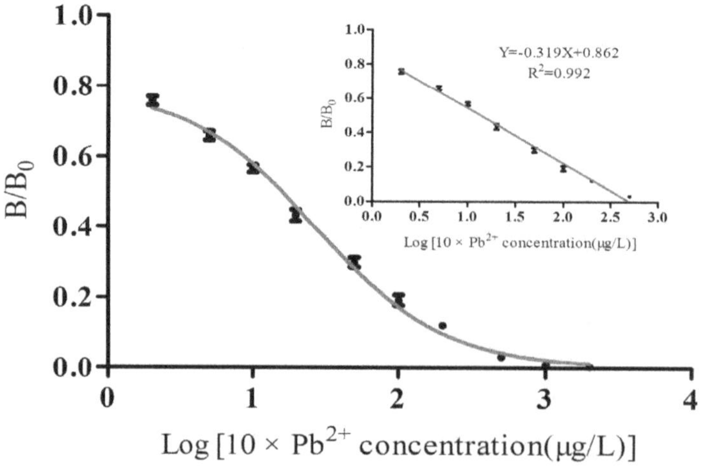

Figure 3. Standard curve of the competitive CLEIA for lead ions.

3.5. GFAAS Analysis of Pb^{2+}

The sensitivity of GFAAS was determined under optimal conditions. The regression curve equation was Y = 2.857X − 0.020 (R^2 = 0.999, n = 3; see Figure 4). The linearity ranged from 0 to 3.0 μg/L. The limit of quantification was 0.86 μg/L.

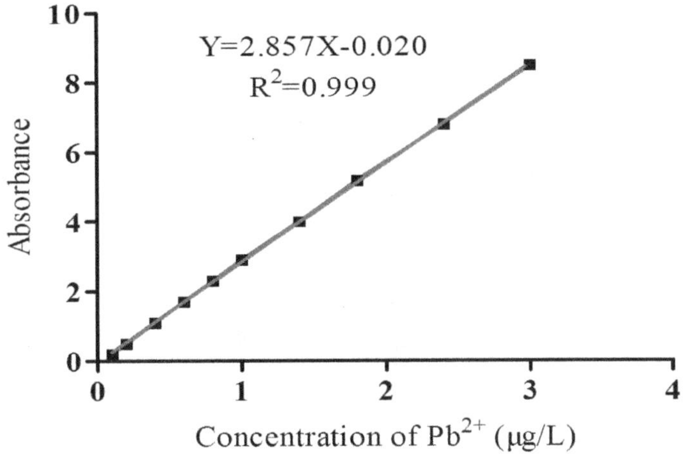

Figure 4. Standard curve of the GFAAS for lead ions.

3.6. Precision and Recovery in Sample Test

The spiked chicken, rice, chicken feed, rat feed, milk and tap water samples containing different concentrations of lead ions (100, 200 and 500 ng/g, respectively) were detected by using the proposed ic-ELISA and ic-CLEIA, respectively, and both methods showed high recoveries and low coefficients of variation (see Table 3). The recovery of the spiked samples suggested that the CLEIA is suitable as a rapid and reliable method to detect lead ions in several matrices.

Table 3. Recovery ratio of Pb^{2+} from different samples (n = 4).

Samples	Spiked (ng/g)	ELISA Mea-sured (ng/g)	ELISA Recovery (%)	ELISA RSD (%)	CLEIA Mea-sured (ng/g)	CLEIA Recovery (%)	CLEIA RSD (%)	AAS Mea-sured (ng/g)	AAS Recovery (%)	AAS RSD (%)
Chicken	0	<LOD	–	—	<LOD	–	—	<LOD	–	—
	100	91.0	91.0	10.1	88.2	88.2	16.0	92.1	92.1	8.9
	200	197.5	98.7	11.4	182.2	91.1	2.5	196.4	98.2	0.2
	500	442.3	88.5	15.0	487.4	97.5	4.3	482.5	96.5	0.3
Rice	0	<LOD	–	—	<LOD	–	—	<LOD	–	—
	100	82.1	82.1	5.5	98.8	98.8	12.1	90.2	90.2	9.1
	200	189.1	94.6	8.4	176.4	88.2	9.3	204.0	102.0	4.7
	500	405.2	81.0	1.7	419.3	83.9	11.7	521.2	104.2	0.2
Chicken feed	0	<LOD	–	—	<LOD	–	—	<LOD	–	—
	100	83.8	83.8	6.8	91.3	91.3	6.8	83.3	83.3	8.7
	200	216.5	108.3	12.1	164.5	82.2	12.0	209.2	104.6	0.3
	500	430.8	86.1	6.8	429.9	85.9	10.6	495.7	99.1	0.2
Rat feed	0	<LOD	–	—	<LOD	–	—	<LOD	–	—
	100	83.0	83.0	9.6	82.0	82.0	11.0	80.3	80.3	1.4
	200	171.9	85.9	8.5	175.2	87.6	8.2	195.5	97.7	0.3
	500	410.5	82.1	9.6	453.5	90.7	10.9	522.5	104.5	1.6
Milk	0	<LOD	–	—	<LOD	–	—	<LOD	–	—
	100	94.5	94.5	12.0	85.6	85.6	12.3	88.01	88.01	8.0
	200	175.0	87.5	10.5	160.3	80.1	2.3	194.6	97.3	0.6
	500	486.7	97.3	14.1	463.2	92.6	8.4	456.5	91.3	0.3
Tap water	0	<LOD	–	—	<LOD	–	—	<LOD	–	—
	100	107.7	107.7	10.2	90.9	90.9	8.7	94.2	94.2	9.3
	200	188.3	94.1	9.8	172.8	86.4	5.0	187.4	93.7	0.5
	500	513.0	102.6	3.6	426.3	85.3	4.2	524.5	104.9	0.2

3.7. Comparison of ELISA, CLEIA and GFAAS Results for Lead (II) in Samples

The linear regression curves of ELISA (see Figure 5a) and CLEIA (see Figure 5b) showed good correlation coefficients square of 0.962 and 0.972, respectively, as compared to GFAAS, indicating that the two methods developed could achieve reliable and accurate determination of lead (II) ions in samples.

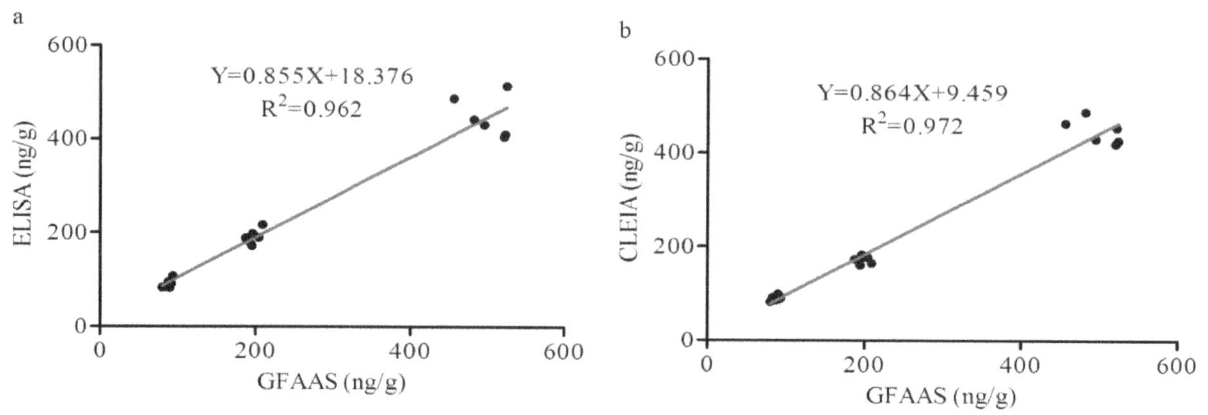

Figure 5. Correlation of ELISA and CLEIA to GFAAS on lead ions analysis.

4. Discussion

The small size and simple structure of heavy metal ions result in poor immunogenicity; as such, they are classified as incomplete antigen. To generate complete antigens for immunological assays, a highly effective bifunctional chelating agent (ITCBE) was selected to connect the lead ion and the carrier protein, which has a large relative molecular mass, reduced toxicity and enhanced immunogenicity [22,23]. The mAb we obtained is superior in sensitivity and specificity as compared to the mAb generated by using the conjugation of lead and S-2-(4-aminobenzyl) diethylenetriamine penta-acetic acid (DTPA) as immunogen, which was applied in ELISA with a limit of detection of 11.6 ng/mL and cross-activity less than 3% [24].

Sample pretreatment is a primary factor for enriching heavy metals and minimizing matrix interference in practical application, as in real detection conditions, lead ions often bind tightly to larger molecules, such as proteins, carbohydrates and colloids [17]. Several methods have been used for heavy-metal sample pretreatment. Microwave digestion method was often used to extract heavy metals from solid samples, including chicken, fish, feces and soil, with high accuracy and recovery rate, but it has limitations in real-time and high throughput detection [25]. A recent study on extracting lead ions in skin-whitening cosmetics, using microwave digestion coupled with plasma atomic emission spectrometry, showed a detection limit of 3.8 μg/kg [26]. The dry ash method is usually used to enrich heavy metals from food and plant samples; however, it demonstrated low accuracy, low recovery rate and high blank value, and it is not suitable for food containing highly volatile inorganic salt [27]. Dry ash extraction has been used for GFAAS-based lead measurement from green vegetables with obtained recovery ranging from 67% to 103% [28]. To better separate the lead ions, we used the acid leach method to enriched lead ions in samples, and have achieved high recovery and a good variable coefficient (Table 3). Furthermore, the acid leach method is easy to operate and has no loss of element, as compared to the other methods, such as microwave digestion, wet digestion and dry ash.

5. Conclusions

In this study, a monoclonal antibody against lead (II) was raised by immunizing Balb/c mouse and hybridoma technique. The LOD of ic-ELISA and ic-CLEIA were 0.7 and 0.1 ng/mL, respectively. The ic-ELISA and ic-CLEIA demonstrated low coefficient of variation. Compared to GFAAS, the two developed methods showed a wide detection range, and the ic-CLEIA showed even more sensitivity compared to the ic-ELISA.

Collectively, ic-ELISA and ic-CLEIA were developed for handy, sensitive and specific detection of lead (II) ions in water, food and feed.

Author Contributions: Conceptualization, L.X., X.-y.S. and X.-y.Z.; data curation, C.C.; formal analysis, Q.Z.; investigation, L.X. and X.Y.S.; methodology, L.X. and X.Y.S.; resources, X.-p.L. and X.-y.Z.; validation, L.X. and X.-y.S.; writing—original draft, L.X. and X.-y.S.; writing—review and editing, X.-y.Z. All authors have read and agreed to the published version of the manuscript.

References

1. Molera, J.; Pradell, T.; Salvadó, N.; Vendrell-Saz, M. Interactions between Clay Bodies and Lead Glazes. *J. Am. Ceram. Soc.* **2001**, *84*, 1120–1128. [CrossRef]
2. Szczyglowska, M.; Bodnar, M.; Namiesnik, J.; Konieczka, P. The Use of Vegetables in the Biomonitoring of Cadmium and Lead Pollution in the Environment. *Crit. Rev. Anal. Chem.* **2014**, *44*, 2–15. [CrossRef] [PubMed]
3. Sang Yong, E.; Young-Sub, L.; Seul-Gi, L.; Mi-Na, S.; Byung-Sun, C.; Yong-Dae, K.; Ji-Ae, L.; Myung-Sil, H.; Ho-Jang, K.; Yu-Mi, K. Lead, Mercury, and Cadmium Exposure in the Korean General Population. *J Korean Med. Sci.* **2018**, *33*, e9.
4. Wasserman, G.A.; Factor-Litvak, P.; Liu, X.; Todd, A.C.; Graziano, J.H. The Relationship Between Blood Lead, Bone Lead and Child Intelligence. *Child Neuropsychol.* **2003**, *9*, 22–34. [CrossRef]
5. Oliveira de, T.M.; Peres, J.A.; Felsner, M.L.; Justi, K.C. Direct determination of Pb in raw milk by graphite furnace atomic absorption spectrometry (GF AAS) with electrothermal atomization sampling from slurries. *Food Chem.* **2017**, *229*, 721–725. [CrossRef]
6. da Silva, D.L.F.; da Costa, M.A.P.; Silva, L.O.B.; dos Santos, W.N.L. Simultaneous determination of mercury and selenium in fish by CVG AFS. *Food Chem.* **2019**, *273*, 24–30. [CrossRef]
7. Chen, K.y.; Fan, C.; Yuan, H.l.; Bao, Z.a.; Zong, C.l.; Dai, M.n.; Ling, X.; Yang, Y. High-Precision In Situ Analysis of the Lead Isotopic Composition in Copper Using Femtosecond Laser Ablation MC-ICP-MS and the Application in Ancient Coins. *Spectrosc. Spect. Anal.* **2013**, *33*, 1342–1349.
8. Da Col, J.A.; Domene, S.M.A.; Pereira-Filho, E.R. Fast Determination of Cd, Fe, Pb, and Zn in Food using AAS. *Food Anal. Methods* **2009**, *2*, 110–115. [CrossRef]
9. Wei, C.; Shunbi, X.; Jin, Z.; Dianyong, T.; Ying, T. Immobilized-free miniaturized electrochemical sensing system for Pb^{2+} detection based on dual Pb^{2+} -DNAzyme assistant feedback amplification strategy. *Biosens. Bioelectron.* **2018**, *117*, 312–318.
10. Wang, X.Y.; Niu, C.G.; Guo, L.J.; Hu, L.Y.; Wu, S.Q.; Zeng, G.M.; Li, F. A Fluorescence Sensor for Lead (II) Ions Determination Based on Label-Free Gold Nanoparticles (GNPs)-DNAzyme Using Time-Gated Mode in Aqueous Solution. *J. Fluoresc.* **2017**, *27*, 643–649. [CrossRef]
11. Chu, W.; Zhang, Y.; Li, D.; Barrow, C.J.; Wang, H.; Yang, W. A biomimetic sensor for the detection of lead in water. *Biosens. Bioelectron.* **2015**, *67*, 621–624. [CrossRef] [PubMed]
12. Liu, C.W.; Huang, C.C.; Chang, H.T. Highly Selective DNA-Based Sensor for Lead(II) and Mercury(II) Ions. *Anal. Chem.* **2009**, *81*, 2383–2387. [CrossRef] [PubMed]
13. Tang, S.r.; Lu, W.; Gu, F.; Tong, P.; Yan, Z.; Zhang, L. A novel electrochemical sensor for lead ion based on cascade DNA and quantum dots amplification. *Electrochim. Acta* **2014**, *134*, 1–7. [CrossRef]
14. Zhang, H.; Jiang, B.; Xiang, Y.; Su, J.; Chai, Y.; Yuan, R. DNAzyme-based highly sensitive electronic detection of lead via quantum dot-assembled amplification labels. *Biosens. Bioelectron.* **2011**, *28*, 135–138. [CrossRef] [PubMed]
15. Xiang, J.J.; Zhai, Y.f.; Tang, Y.; Wang, H.; Liu, B.; guo, C.W. A competitive indirect enzyme-linked immunoassay for lead ion measurement using mAbs against the lead-DTPA complex. *Environ. Pollut.* **2010**, *158*, 1376–1380. [CrossRef] [PubMed]
16. Tang, Y.; Zhai, Y.F.; Xiang, J.J.; Wang, H.; Guo, C.W. Colloidal gold probe-based immunochromatographic assay for the rapid detection of lead ions in water samples. *Biosens. Bioelectron.* **2010**, *158*, 2074–2077. [CrossRef]
17. Mandappa, I.M.; Ranjini, A.; Haware, D.J.; Manonmani, H.K. Immunoassay for lead ions: analysis of spiked food samples. *J. Immunoassay* **2014**, *35*, 1–11. [CrossRef]
18. Codex Alimentarius Commission (CSC). Codex general standard for contaminants and toxins in food and

feed. *Codex stan* **1995**, *193*, 229–234.

19. The commission of the European Community. Commission regulation (EC) NO 1881/2006 setting maximum levels for certain contaminants in foodstuffs. *OJEC* **2006**, *364*, 5–24.
20. National food safety standard for maximum levels of contaminants in foods. Available online: http://www.nhc.gov.cn/ewebeditor/uploadfile/2013/01/20130128114248937.pdf. (accessed on 29 January 2013).
21. Australia New Zealand Standard 1.4.1 contaminants and natural toxicants. Available online: https://www.foodstandards.gov.au/code/Documents/Sched%2019%20Contaminant%20MLs%20v157.pdf (accessed on 1 March 2016).
22. Perrin, C.L.; Kim, Y.J. Symmetry of Metal Chelates. *Inorg. Chem.* **2000**, *39*, 3902–3910. [CrossRef]
23. Love, R.A.; Villafranca, J.E.; Aust, R.M.; Nakamura, K.K.; Jue, R.A.; Major, J.G.; Radhakrishnan, R.; Butler, W.F. How the anti-(metal chelate) antibody CHA255 is specific for the metal ion of its antigen: X-ray structures for two Fab'/hapten complexes with different metals in the chelate. *Biochemistry* **1993**, *32*, 10950–10959. [CrossRef] [PubMed]
24. Zhu, X.; Hu, B.; Lou, Y.; Xu, L.; Yang, F.; Yu, H.; Blake, D.A.; Liu, F. Characterization of monoclonal antibodies for lead chelate complexes: applications in antibody-based assays. *J. Agric. Food Chem.* **2007**, *55*, 4993–4998. [CrossRef] [PubMed]
25. Safari, Y.; Karimaei, M.; Sharafi, K.; Arfaeinia, H.; Moradi, M.; Fattahi, N. Persistent sample circulation microextraction combined with graphite furnace atomic absorption spectroscopy for trace determination of heavy metals in fish species marketed in Kermanshah, Iran and human health risk assessment. *J. Sci. Food Agric.* **2017**, *98*, 2915–2924. [CrossRef] [PubMed]
26. Alqadami, A.A.; Mu, N.; Abdalla, M.A.; Khan, M.R.; Alothman, Z.A.; Wabaidur, S.M.; Ghfar, A.A. Determination of heavy metals in skin-whitening cosmetics using microwave digestion and inductively coupled plasma atomic emission spectrometry. *IET Nanobiotechnol.* **2017**, *11*, 597–603. [CrossRef]
27. Hadiani, M.R.; Farhangi, R.; Soleimani, H.; Rastegar, H.; Cheraghali, A.M. Evaluation of heavy metals contamination in Iranian foodstuffs: Canned tomato paste and tomato sauce (ketchup). *Food Addit. Contam. Part B* **2014**, *7*, 74–78. [CrossRef]
28. Baxter, M.J.; Burrell, J.A.; Crews, H.M.; Massey, R.C.; McWeeny, D.J. A procedure for the determination of lead in green vegetables at concentrations down to 1 μg/kg. *Food Addit. Contam.* **1989**, *6*, 341–349. [CrossRef]

13

Bioaccumulation and Distribution of Indospicine and its Foregut Metabolites in Camels Fed *Indigofera spicata*

Gabriele Netzel [1], Eddie T. T. Tan [1,2], Mukan Yin [1], Cindy Giles [3], Ken W. L. Yong [3], Rafat Al Jassim [1] and Mary T. Fletcher [1,*]

[1] Queensland Alliance for Agriculture and Food Innovation (QAAFI), The University of Queensland, Health and Food Sciences Precinct, Coopers Plains, QLD 4108, Australia; g.netzel@uq.edu.au (G.N.); eddietan@ns.uitm.edu.my (E.T.T.T.); mukan.yin@uq.net.au (M.Y.); r.aljassim@uq.edu.au (R.A.J.)

[2] Alliance of Research and Innovation for Food (ARIF), Faculty of Applied Sciences, Universiti Teknologi MARA, Cawangan Negeri Sembilan, Kuala Pilah Campus, Negeri Sembilan 72000, Malaysia

[3] Department of Agriculture and Fisheries, Health and Food Sciences Precinct, Coopers Plains, QLD 4108, Australia; cindy.giles@daf.qld.gov.au (C.G.); ken.yong@daf.qld.gov.au (K.W.L.Y.)

* Correspondence: mary.fletcher@uq.edu.au

Abstract: In vitro experiments have demonstrated that camel foregut-fluid has the capacity to metabolize indospicine, a natural toxin which causes hepatotoxicosis, but such metabolism is in competition with absorption and outflow of indospicine from the different segments of the digestive system. Six young camels were fed *Indigofera spicata* (337 μg indospicine/kg BW/day) for 32 days, at which time three camels were euthanized. The remaining camels were monitored for a further 100 days after cessation of this indospicine diet. In a retrospective investigation, relative levels of indospicine foregut-metabolism products were examined by UHPLC-MS/MS in plasma, collected during both accumulation and depletion stages of this experiment. The metabolite 2-aminopimelamic acid could be detected at low levels in almost all plasma samples, whereas 2-aminopimelic acid could not be detected. In the euthanized camels, 2-aminopimelamic acid could be found in all tissues except muscle, whereas 2-aminopimelic acid was only found in the kidney, pancreas, and liver tissues. The clearance rate for these metabolites was considerably greater than for indospicine, which was still present in plasma of the remaining camels 100 days after cessation of *Indigofera* consumption.

Keywords: indospicine; 2-aminopimelamic acid; 2-aminopimelic acid; in vivo; foregut metabolites; camel; food safety

Key Contribution: The indospicine metabolites 2-aminopimelamic acid and 2-aminopimelic acid are formed during foregut microbial metabolism and can be detected in tissues of camels fed *Indigofera spicata*. These metabolites are not as persistent as indospicine in animal tissues after cessation of feeding, and could be used as markers for recent *Indigofera* consumption.

1. Introduction

There are more than 60 *Indigofera* species distributed throughout the arid and semiarid regions of Australia [1–4]. *Indigofera* spp. are leguminous shrubs and herbs which are high in protein, as well as highly palatable for animals. These plants are considered a nutritious animal fodder, however, some species contain indospicine, a non-proteinogenic arginine analogue which causes hepatotoxicosis in sheep, cows, rabbits, and dogs [5–9]. The introduced species, *Indigofera spicata*, has been found to contain high levels of this amino acid. Since there is no known mammalian enzyme which is capable of degrading indospicine [10], it can be toxic to simple-stomached animals. However, ruminants seem to be less susceptible to certain toxins due to the ability of rumen microflora to degrade some of those foodborne toxins. Feral camels in Australia seem to cope better in detoxification of plant toxins from their fodder, compared to domesticated animals [11].

Indospicine is a water-soluble free amino acid and rumen degradation processes compete with presumed rapid passage through the fermentation compartments of the digestive tract. This amino acid has been detected in meat, which indicates that at least a portion of indospicine can by-pass the fermentation processes of the foregut and be absorbed as is [12]. A number of dogs have died of secondary hepatotoxicosis after consuming indospicine-contaminated horse meat [7] and, more recently, camel meat [5], and this has raised food safety concerns. In a recent study, we have shown that indospicine accumulated in camel meat during a feeding trial in which six camels were fed a diet containing *Indigofera spicata* for 32 days, and that indospicine can be detected in plasma as long as three months after removing *Indigofera* from the diet [13].

We also reported previously that microflora of both the bovine rumen and camel foregut fluids have the ability to degrade indospicine in vitro within an incubation period of 48 h [14]. However, the in vitro degradability of indospicine is indicative of the potential degradability, and not the actual degradability, that occurs in the animal system. Factors including the microbial community, residence time of the solid fraction of digesta, and outflow rate of the fluid phase all play an important role. Camels are known to retain low quality fibre diets longer in the foregut compared with ruminant animals. Retention time is always shorter when the diet is of higher quality, which should be the case with lush early season pasture containing *Indigofera spicata* at the start of the wet season. Shift to such diet increases the outflow rate and allows more indospicine to enter the intestines where it then gets absorbed. Indospicine has been shown to be chemically stable and resistant to both acidic and base conditions [15,16]. Since the camel foregut fluid is only mildly acidic, it is most likely that rumen bacteria are responsible for the observed metabolism of indospicine (**1**) into its degradation product 2-aminopimelamic acid (**2**) and, further, to 2-aminopimelic acid (**3**) (Figure 1) [14,17].

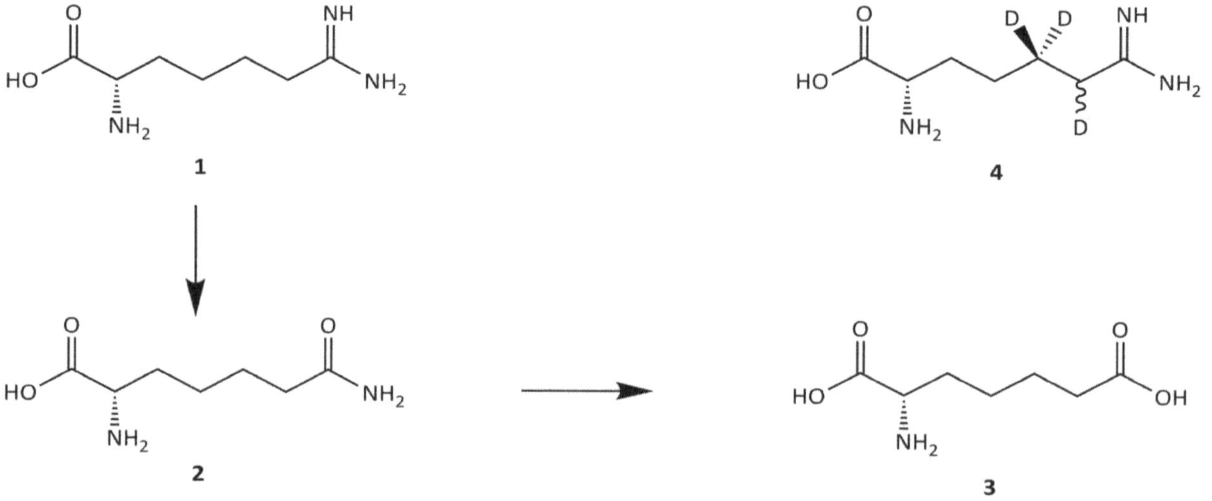

Figure 1. Chemical structures of indospicine (**1**) and its metabolites 2-aminopimelamic acid (**2**) and 2-aminopimelic acid (**3**), together with D_3-L-indospicine (**4**) which is used as an internal standard in LC-MS/MS analysis.

Although we could show previously that indospicine accumulated as a free amino acid in various animal tissues in vivo [13], it has also been demonstrated that indospicine can be metabolized in vitro by foregut microbiota [14]. These two processes of removal (outflow and absorption) and metabolism could be considered to operate in competition, and there is nothing known about the extent of in vivo metabolism of indospicine and whether the metabolites are also transported and accumulated in tissues. Hence, in the present study, we investigated the bioaccumulation and distribution, as well as the excretion, of the indospicine foregut metabolites, 2-aminopimelamic acid and 2-aminopimelic acid, in camels fed *Indigofera* plant material for 32 days.

2. Results and Discussion

2.1. Indospicine and Foregut Metabolites in Tissue Samples

It has previously been established that indospicine accumulates in muscle and other tissues of cattle [18] and camels [13] fed *Indigofera* plant material, however nothing is known about the fate of the indospicine metabolites, 2-aminopimelamic acid and 2-aminopimelic acid. In this study we have measured both indospicine and the two metabolite concentrations in tissues acquired during the previous camel feeding trial, where six young camels (camels 1–6) were fed *Indigofera spicata* for 32 days until indospicine levels in plasma plateaued. At this point, three animals (camels 1–3) were euthanized and the remaining camels (camels 4–6) fed an *Indigofera*-free diet for a further 100 days whilst monitoring decline in indospicine plasma levels [13]. The inclusion rate of *Indigofera spicata* was designed to provide 337 μg of indospicine per kg BW per day.

In accord with the previous analysis, the highest concentration of indospicine in necropsied camels (camels 1–3) was found, in this study, in the pancreas (5.06 ± 0.79 mg/kg FW), followed by the liver (3.57 ± 1.17 mg/kg FW), heart (2.32 ± 0.48 mg/kg FW), kidney (1.48 ± 0.11 mg/kg FW), muscle (1.24 ± 0.26 mg/kg FW), and spleen (0.96 ± 0.34 mg/kg FW) (Figure 2). However, if we look at the deamino metabolites, the highest concentration of the intermediate metabolite 2-aminopimelamic acid was found in the kidney (0.96 ± 0.12 mg/kg FW), followed by the pancreas (0.36 ± 0.11 mg/kg FW), liver (0.27 ± 0.06 mg/kg FW), spleen (0.20 ± 0.08 mg/kg FW), and heart (0.15 ± 0.02 mg/kg FW). Neither 2-aminopimelamic acid nor 2-aminopimelic acid could be detected in the muscle tissue. Only low concentrations of the second metabolite 2-aminopimelic acid could be found in kidney (0.34 ± 0.10 mg/kg FW), pancreas (0.11 ± 0.03 mg/kg FW), and liver (0.01 ± 0.02 mg/kg FW).

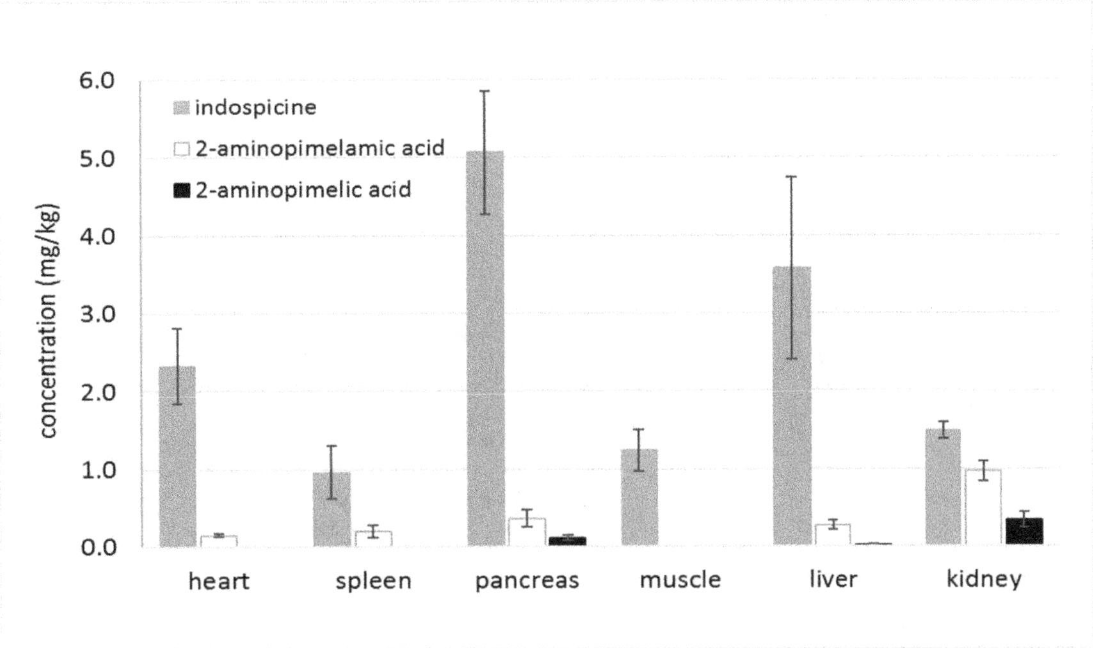

Figure 2. Mean concentrations (mg/kg FW) of indospicine, 2-aminopimelamic acid, and 2-aminopimelic acid for camels 1–3 necropsy tissues (heart, spleen, pancreas, muscle, liver and kidney) after 32-day feeding period.

The kidney, spleen, heart, liver, and pancreas are all organs where arginine metabolism plays an important role in metabolic pathways. Indospicine has, for example, been shown to be a competitive inhibitor of hepatic arginase [19], but arginase does not occur only in the liver. Arginase has two isoforms: arginase I, a cytoplasmic enzyme, which is highly expressed in the liver, and arginase II, a mitochondrial enzyme more widely distributed in extrahepatic tissues, which is expressed primarily in the kidney [20]. Arginase is a key enzyme of the urea cycle and is also present in other tissues such as the spleen, heart, kidney, and pancreas [20–22]. Arginase activity is reportedly blocked through the competitive binding of indospicine [19], which is consistent with the observed elevated presence of indospicine in these same tissues (Figure 2). It is worth noting that 2-aminopimelamic acid is an analogue of citrulline, in the same manner that indospicine is an analogue of arginine. We can hypothesize then that the high concentration of 2-aminopimelamic acid found in the kidney, spleen, heart, liver, and pancreas could likewise be due to interference in citrulline metabolism within the urea cycle, potentially blocking the enzyme argininosuccinate synthetase which utilises citrulline in the synthesis of argininosuccinic acid [22]. The kidney is noted to have a significant capacity for citrulline metabolism [22], and the highest level of accumulation of the citrulline analogue 2-aminopimelamic acid may reflect interference in this metabolism. Such interference could then be consistent with the dramatically elevated citrulline levels reported in indospicine treated rats [23].

2.2. *Indospicine and Foregut Metabolites in Plasma*

The concentration of indospicine [13], as well as that of 2-aminopimelamic acid, was observed to rapidly increase in plasma of camels 1–6 in the first 13–20 days after commencement of the feeding trial (Figure 3). The indospicine concentration reached a plateau phase before it decreased slowly after ceasing the *Indigofera* uptake. The concentration of the 2-aminopimelamic acid similarly rose in the first 13 days, plateauing for about two weeks, before it slightly decreased and rose to a second maximum at 32 days, the point at which the *Indigofera* intake was stopped. In the depletion phase, when the remaining camels (camels 4–6) were fed an *Indigofera*-free diet, the 2-aminopimelamic acid concentration followed the indospicine concentration in a slow decrease until no 2-aminopimelamic acid was detectable at 62 days (30 days after cessation of *Indigofera* consumption).

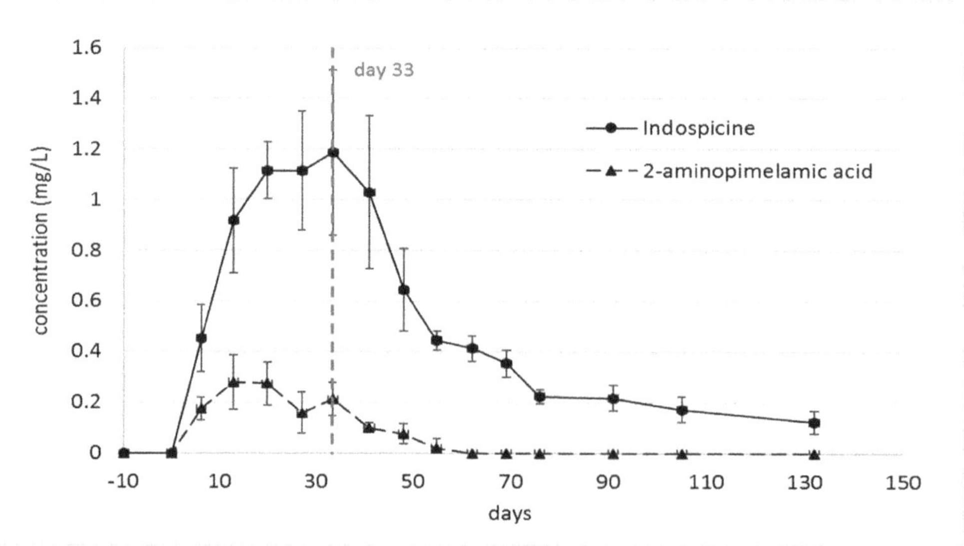

Figure 3. Comparison of indospicine and 2-aminopimelic acid concentrations in plasma (mean ± SD) during the first 32 days (n = 6) of the treatment and after cessation of *Indigofera spicata* feeding (n = 3). Camels 1–3 were autopsied at day 33.

The plasma concentrations of the 2-aminopimelamic acid were much lower than those of indospicine. As already described, there was a considerable variation in the indospicine plasma levels between the individual camels [13], but the individual indospicine plasma curves all followed a similar pattern. By comparison, the pattern of the 2-aminopimelamic acid plasma content proved to be quite variable between the camels (Figure 4). All camels showed a rapid increase within the first 13 or 20 days, to levels between 0.19–0.44 mg/L, with considerable variation between individuals. During the depletion phase, the plasma levels of the three camels (camels 4–6) followed a steady decrease until day 55 or 60, with 2-aminopimelamic acid being eliminated from the system much quicker than indospicine, which is still detected in plasma at 130 days—100 days after cessation of *Indigofera* intake (Figure 3).

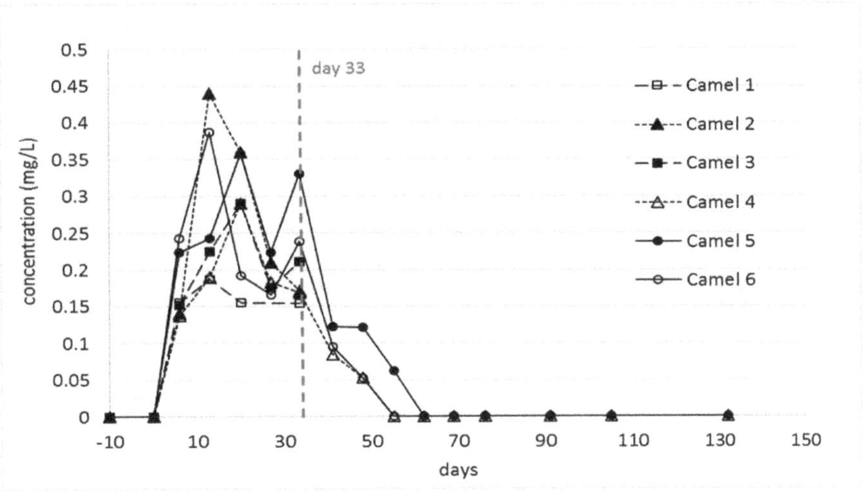

Figure 4. Individual 2-aminopimelamic acid concentrations of plasma during the first 32 days (n = 6) of the treatment and after cessation of *Indigofera spicata* feeding (n = 3). Camels 1–3 were autopsied at day 33.

The rapid increase in plasma levels of the 2-aminopimelamic acid during the feeding phase followed the rapid increase of indospicine, which can be metabolized in the foregut to 2-aminopimelamic acid and, further, to 2-aminopimelic acid. This second metabolite 2-aminopimelic acid was not detectable in any of the plasma samples. Since no 'fresh' indospicine was uptaken after cessation of the *Indigofera* feeding, it is apparent that the metabolites were eliminated much faster from the camel's system, compared to the indospicine. The lengthier persistence of indospicine residues is attributed to the slow release of this arginine analogue from organ tissue, such as the pancreas, where it was accumulated.

3. Conclusions

These results demonstrate that foregut metabolism of indospicine does occur in vivo, and that the two metabolites, 2-aminopimelamic acid and 2-aminopimelic acid, are both absorbed and bioaccumulate in a range of tissues. After consumption of *Indigofera* plants, the intermediate metabolite 2-aminopimelamic acid was present in all camel tissues except muscle, with highest levels measured in the kidney, pancreas, and liver. Lower levels of 2-aminopimelic acid were also found in these same three tissues. Cytotoxicity studies conducted by Sultan et al. [24] demonstrated that the second metabolite, 2-aminopimelic acid, is less toxic than indospicine. However, nothing to date is known about the toxicity of its precursor, 2-aminopimelamic acid, so there is a possibility that this intermediate metabolite, which is an analogue of citrulline, also contributes to the effects of *Indigofera* poisoning. The clearance of this metabolite from plasma occurs faster compared to indospicine (after cessation of *Indigofera* intake), therefore the presence of 2-aminopimelamic acid in plasma could potentially be used as an indicator of whether the indospicine contamination is recent, within the past month, or longer. The actual route of excretion of indospicine (or its metabolites) has not been investigated, and it is recommended that further trials be conducted to determine levels of indospicine and its metabolites in both milk and urine of animals consuming *Indigofera*.

4. Materials and Methods

4.1. Standards and Reagents

Indospicine (**1**) and 2-aminopimelamic acid (**2**) (>99% pure), both external standards, as well as D_3-L-indospicine (**4**) (>99% pure) as internal standard, were synthesized and provided by Prof. James De Voss and Dr. Robert Lang, The University of Queensland [16,25]. Another external standard, 2-aminopimelic acid (**3**) (>99% pure), and heptafluorobutyric acid (HFBA), ion chromatography grade, were purchased from Sigma Aldrich (Castle Hill, NSW, Australia). External (0.005–2 mg/L) and internal (1 mg/L) standard solutions were prepared in 0.1% HFBA in Milli-Q water.

4.2. Tissue Collection

Tissue samples of a previous animal trial [13] (which had been stored frozen at −20 °C) were re-analyzed to focus on the accumulation, distribution, and persistence of foregut deamino metabolites of indospicine during and after cessation of the feeding period.

Briefly, in this study six camels (2–4 years) with a weight between 270–387 kg were fed with dried and chaffed *Indigofera spicata* to deliver 337 μg indospicine/kg BW/day for 32 days. The camels were sourced from the feral population in central Australia and purchased from a commercial supplier. Animal protocols for this study were approved by the Animal Ethics Committee of University of Queensland, Queensland, Australia (AEC Approval Number: SAFS/047/14/SLAI). The *Indigofera* plant material was fed daily in two equal meals at 9:00 am and 2:00 pm. On day 33, three of the animals (camels 1–3) were euthanized and tissues from 6 organs (muscle, heart, spleen, pancreas, liver, and kidney) were collected during necropsy and frozen (−20 °C) until needed.

To exclude previous exposure to indospicine (or metabolites), the plasma collection started 10 days before the experimental feeding phase. Venous blood samples were collected from the jugular vein from all six animals (camels 1–6) in weekly intervals during the treatment phase. This was continued with the remaining three animals (camels 4–6) in weekly intervals until day 76, and then in fortnightly intervals until the end of the experiment (day 132), 100 days after cessation of the *Indigofera* feeding. Blood was collected in lithium heparin containers and spun for 10 min at 4400 rpm at 19 °C (Sigma 4K10, Osterode am Harz, Germany), after which the plasma was collected and frozen at −30 °C until analysis.

4.3. Indospicine Extraction

The tissue samples were extracted as previously described [26]. Camel organ tissues were thawed; 0.5 g was mixed with 25 mL 0.1% HFBA and homogenized for 15 s using a Polytron T25 homogenizer (Labtek, Brendale, Australia). The homogenates were cooled for 20 min at 4 °C before centrifugation (4500 rpm, 10 min, 18 °C). One mL of the supernatant was spiked with the internal standard, and an aliquot of 450 μL was transferred into a pre-rinsed Amicon Ultra, 0.5 mL, 3 K centrifugal filter (Merck, Millipore, Kilsyth, VIC, Australia), which was centrifuged at 10,000 rpm for 20 min. The filtrate was transferred into an autosampler vial for UHPLC-MS/MS analysis.

Plasma samples were thawed and diluted 25 times with 0.1% HFBA. An aliquot of the diluted plasma sample was spiked with the internal standard, and 450 μL were transferred to a pre-rinsed Amicon Ultra centrifugal filter. The filtrate was transferred into an autosampler vial for UHPLC-MS/MS analysis.

4.4. UHPLC-MS/MS Analysis

Quantification of the compounds was done according to a previously validated UHPLC-MS/MS method, with modifications [26]. Separation and quantification were carried out on a Shimadzu Nexera X2 UHPLC system (Shimadzu, Rydalmere, NSW, Australia), combined with a Shimadzu LCMS-8050 triple quadrupole mass spectrometer, equipped with an electrospray ionization (ESI) source operated in positive mode. Indospicine, 2-aminopimelamic acid, and 2-aminopimelic acid were separated at

35 °C on an Acquity BEH C18 column (Waters, Rydalmere, NSW, Australia) (100 mm × 2.1 mm id, 1.7 μm) with 0.1% HFBA and 100% acetonitrile as mobile phases A and B, respectively. The flow rate was set to 0.3 mL/min and the injection volume was 2 μL. The following gradient was applied: 1–70% B (3 min), 70% B (isocratic for 1 min), 70–100% B (0.5 min), the total run time was 6 min. The interface temperature was set to 300 °C, and the heating block to 400 °C. Nitrogen was used as nebulizing gas (2.0 L/min) and drying gas (5.0 L/min), compressed air was used as heating gas (10.0 L/min), and argon was used as the CID gas maintained at 270 kPa.

Indospicine was quantified using stable isotope dilution assay with two MS/MS transitions for each compound, *m/z* 174.2 → 84.0 and *m/z* 174.2 → 111.1 for indospicine, and *m/z* 177.2 → 113.0 and *m/z* 177.2 → 114.1 for the internal standard D_3-L-indospicine.

External calibration curves were used for the quantification of 2-aminopimelamic acid and 2-aminopimelic acid. Specific SRMs were used for the identification: transition of *m/z* 175.00 → 112.05 (verified by transition of *m/z* 175.00 → 67.05) for 2-aminopimelamic acid, and transition of *m/z* 175.90 → 112.20 (verified by transition of *m/z* 175.90 → 69.15) for 2-aminopimelic acid. Unit resolution was used for both precursor and product *m/z* values. The collision energies (shown in Table 1) were optimized for each transition for maximum sensitivity.

Table 1. Collision energy for quantifier and verifying single reaction monitoring (SRM) transitions for compounds **1–4**.

Compound	Collison Energy (eV)	
	Quantifier SRM	Verifier SRM
Indospicine (**1**)	−23.0	−17.0
D_3-L-indospicine (**4**)	−17.0	−16.5
2-aminopimelamic acid (**2**)	−15.0	−28.0
2-aminopimelic acid (**3**)	−15.0	−20.0

Author Contributions: Study design, R.A.J., E.T.T.T. and M.T.F.; investigation and formal analysis, G.N., M.Y., C.G. and K.W.Y.; writing—original draft preparation, G.N.; writing—review and editing, M.T.F., R.A.J., C.G., and E.T.T.T.; project administration, M.T.F.; funding acquisition, M.T.F.

References

1. Tan, E.T.T.; Materne, C.M.; Silcock, R.G.; D'Arcy, B.R.; Al Jassim, R.; Fletcher, M.T. Seasonal and species variation of the hepatotoxin indospicine in Australian *Indigofera* legumes as measured by UPLC-MS/MS. *J. Agric. Food. Chem.* **2016**, *64*, 6613–6621. [CrossRef]
2. Wilson, P.G.; Rowe, R. A revision of the Indigofereae (Fabaceae) in Australia. 1. *Indigastrum* and the simple or unifoliolate species of *Indigofera*. *Telopea* **2004**, *10*, 651–682.
3. Wilson, P.G.; Rowe, R. A revision of the *Indigofereae* (Fabaceae) in Australia. 2. *Indigofera* species with trifoliolate and alternately pinnate leaves. *Telopea* **2008**, *12*, 293–307. [CrossRef]
4. Wilson, P.G.; Rowe, R. Three new species of *Indigofera* (Fabaceae: Faboideae) from Cape York Peninsula. *Telopea* **2008**, *12*, 285–292. [CrossRef]
5. FitzGerald, L.M.; Fletcher, M.T.; Paul, A.E.H.; Mansfield, C.S.; O'Hara, A.J. Hepatotoxicosis in dogs consuming a diet of camel meat contaminated with indospicine. *Aust. Vet. J.* **2011**, *89*, 95–100. [CrossRef]
6. Fletcher, M.T.; Al Jassim, R.A.M.; Cawtdell-Smith, A.J. The occurrence and toxicity of indospicine to grazing animals. *Agriculture* **2015**, *5*, 427–440. [CrossRef]
7. Hegarty, M.P.; Kelly, W.R.; McEwan, D.; Williams, O.J.; Cameron, R. Hepatotoxicity to dogs of horse meat contaminated with indospicine. *Aust. Vet. J.* **1988**, *65*, 337–340. [CrossRef] [PubMed]
8. Hutton, E.M.; Windrum, G.M.; Kratzing, C.C. Studies on the toxicity of *Indigofera endecaphylla*: I. Toxicity for rabbits. *J. Nutr.* **1958**, *64*, 321–337. [CrossRef] [PubMed]
9. Nordfeldt, S.; Henke, L.A.; Morita, K.; Matsumoto, H.; Takahash, M.; Younge, O.R.; Willers, E.H.; Cross, R.F. Feeding tests with *Indigofera endecaphylla* Jacq. (Creeping indigo) and some observations on its poisonous effects on domestic animals. *Hawaii Agric. Exp. Station Coll. Agric. Univ. Hawaii Tech. Bull.* **1952**, *15*, 5–23.
10. Hegarty, M.P. Toxic amino acids in foods of animals and man. *Proc. Nutr. Soc. Australia* **1986**, *11*, 73–81.

11. Fowler, M.E. Plant poisoning in free-living wild animals: A review. *J. Wildl. Dis.* **1983**, *19*, 34–43. [CrossRef]
12. Tan, E.T.T.; Al Jassim, R.; D'Arcy, B.R.; Fletcher, M.T. Level of natural hepatotoxin (Indospicine) contamination in Australian camel meat. *Food Addit. Contam. Part A* **2016**, *33*, 1587–1595. [CrossRef] [PubMed]
13. Tan, E.T.T.; Al Jassim, R.; Cawdell-Smith, A.J.; Ossedryver, S.M.; D'Arcy, B.R.; Fletcher, M.T. Accumulation, persistence, and effects of indospicine residues in camels fed *Indigofera* plant. *J. Agric. Food. Chem.* **2016**, *64*, 6622–6629. [CrossRef] [PubMed]
14. Tan, E.T.T.; Al Jassim, R.; D'Arcy, B.R.; Fletcher, M.T. In vitro biodegradation of hepatotoxic indospicine in *Indigofera spicata* and its degradation derivatives by camel foregut and cattle rumen fluids. *J. Agric. Food. Chem.* **2017**, *65*, 7528–7534. [CrossRef] [PubMed]
15. Sultan, S.; Giles, C.; Netzel, G.; Osborne, S.A.; Netzel, M.E.; Fletcher, M.T. Release of indospicine from contaminated camel meat following cooking and simulated gastrointestinal digestion: Implications for human consumption. *Toxins* **2018**, *10*, 356. [CrossRef] [PubMed]
16. Tan, E.T.T.; Yong, K.W.L.; Wong, S.H.; D'Arcy, B.R.; Al Jassim, R.; De Voss, J.J.; Fletcher, M.T. Thermo-alkaline treatment as a practical degradation strategy to reduce indospicine contamination in camel meat. *J. Agric. Food. Chem.* **2016**, *64*, 8447–8453. [CrossRef]
17. Hegarty, M.P.; Pound, A.W. Indospicine, a hepatotoxic amino acid from *Indigofera spicata*: Isolation, structure, and biological studies. *Aust. J. Biol. Sci* **1970**, *23*, 831–842. [CrossRef]
18. Fletcher, M.T.; Reichmann, K.G.; Ossedryver, S.M.; McKenzie, R.A.; Carter, P.D.; Blaney, B.J. Accumulation and depletion of indospicine in calves (*Bos taurus*) fed creeping indigo (*Indigofera spicata*). *Anim. Prod. Sci.* **2018**, *58*, 568–576. [CrossRef]
19. Madsen, N.P.; Hegarty, M.P. Inhibition of rat liver homogenate arginase activity *in vitro* by the hepatotoxic amino acid indospicine. *Biochem. Pharmacol.* **1970**, *19*, 2391–2393. [CrossRef]
20. Biczó, G.; Hegyi, P.; Berczi, S.; Dósa, S.; Hracskó, Z.; Varga, I.S.; Iványi, B.; Venglovecz, V.; Wittmann, T.; Takács, T.; et al. Inhibition of arginase activity ameliorates L-arginine-induced acute pancreatitis in rats. *Pancreas* **2010**, *39*, 868–874.
21. Emmanuel, B. Urea cycle enzymes in tissues (liver, rumen epithelium, heart, kidney, lung and spleen) of sheep (*Ovis aries*). *Comp. Biochem. Physiol. B Biochem. Mol. Biol.* **1980**, *65*, 693–697. [CrossRef]
22. Morris, S.M. Regulation of enzymes of urea and arginine synthesis. *Annu. Rev. Nutr.* **1992**, *12*, 81–101. [CrossRef]
23. Hegarty, M.P.; Court, R.D. Indigofera spicata. In *Tropical Crops and Pastures Division of CSIRO 1975-1976 Annual Report*; Commonwealth Scientific and Industrial Research Organisation: Brisbane, QLD, Australia, 1976; p. 70.
24. Sultan, S.; Osborne, S.A.; Addepalli, R.; Netzel, G.; Netzel, M.E.; Fletcher, M.T. Indospicine cytotoxicity and transport in human cell lines. *Food Chem.* **2018**, *267*, 119–123. [CrossRef] [PubMed]
25. Lang, C.S.; Wong, S.H.; Chow, S.; Challinor, V.L.; Yong, K.W.L.; Fletcher, M.; Arthur, D.M.; Ng, J.C.; De Voss, J.J. Synthesis of L-indospicine, [5,5,6-$^2H_3$]-L-indospicine and L-norindospicine. *Org. Biomol. Chem.* **2016**, *14*, 6826–6832. [CrossRef]
26. Tan, E.T.; Fletcher, M.T.; Yong, K.W.; D'Arcy, B.R.; Al Jassim, R. Determination of hepatotoxic indospicine in Australian camel meat by ultra-performance liquid chromatography-tandem mass spectrometry. *J. Agric. Food. Chem.* **2014**, *62*, 1974–1979. [CrossRef] [PubMed]

14

Perfluorooctane Sulfonate (PFOS), Perfluorooctanoic Acid (PFOA), Brominated Dioxins (PBDDs) and Furans (PBDFs) in Wild and Farmed Organisms at Different Trophic Levels

Elena Fattore [1], Renzo Bagnati [1], Andrea Colombo [1], Roberto Fanelli [1], Roberto Miniero [2], Gianfranco Brambilla [2], Alessandro Di Domenico [2], Alessandra Roncarati [3] and Enrico Davoli [1,*]

[1] Department of Environmental Health Sciences, Istituto di Ricerche Farmacologiche Mario Negri IRCCS, 20156 Milano, Italy; elena.fattore@marionegri.it (E.F.); renzo.bagnati@marionegri.it (R.B.); andrea.colombo@marionegri.it (A.C.); roberto.fanelli@marionegri.it (R.F.)

[2] Toxicological Chemistry Unit, Environment Department, Italian National Institute of Health, 00161 Rome, Italy; roberto.miniero@iss.it (R.M.); gianfranco.brambilla@iss.it (G.B.); alessandro.didomenico@iss.it (A.D.D.)

[3] School of Biosciences and Veterinary Medicine, University of Camerino, I-62024 Matelica, Italy; alessandra.roncarati@unicam.it

* Correspondence: enrico.davoli@marionegri.it

Abstract: The present study shows the results of perfuorooctane sulfonate (PFOS), perfluorooctanoic acid (PFOA), brominated dioxins (PBDDs) and furans (PBDFs) measured in several marine fish and seafood of commercial interest at different trophic levels of the food chain. The aims were to investigate the level of the contamination in Mediterranean aquatic wildlife, and in farmed fish, to assess human exposure associated to fishery products consumption. Samples of wild fish were collected during three different sampling campaigns in different Food and Agriculture Organization (FAO) 37 areas of the Mediterranean Sea. In addition, farmed fish (gilthead sea bream and European sea bass) from off-shore cages from different marine aquaculture plants. Results showed contamination values of PFOS and PFOA were lower than those detected in sea basins other than the Mediterranean Sea. Concentration values of PFOS were generally higher than those of PFOA; moreover, levels in farmed fish were lower than in wild samples from the Mediterranean Sea. Intake of PFOS and PFOA through fishery products consumption was estimated to be 2.12 and 0.24 ng/kg·BW·day, respectively, for high consumers (95th percentile). Results of 2,3,7,8-substituted congeners of PBDDs and PBDFs were almost all below the limit of detection (LOD), making it difficult to establish the contribution of these compounds to the total contamination of dioxin-like compounds in fish and fishery products.

Keywords: perfluorooctane sulfonate; perfluorooctane acid; PFOS; PFOA; mediteranean fish; toxicological risk

---

1. Introduction

Perfluorooctane sulfonate (PFOS) and perfluorooctanoic acid (PFOA) are two chemicals included in the large group of the perfluorinated compounds (PFC) which have been widely produced for industrial purposes since 1950 [1]. They are characterized by a fully fluorinated hydrophobic chain and a hydrophilic head and these properties, in combination with their high chemical stability, make these compounds unique for their ability to repel both water and oils. They have been used in many applications, such as surface treatments for coatings, clothes, carpets, packaging products, cookware, food contact papers, and as additives in the fire-fighting foam. They are considered emerging pollutants, since they have been detected in human tissues and wildlife with increasing trends [2–4] and seem to meet the criteria of persistence, biomagnification, and long-distance transport, to be included in the persistent organic pollutants (POPs) under the Stockholm convention. Liver is the target organ of toxicity of these chemicals. Toxicity of PFOS and PFOA includes developmental effects, interference with transport and metabolism of fatty acids, immune-suppression, and interference with thyroid hormones. PFOA shows the typical effects of the peroxisome proliferator-activated receptor alpha (PPAR-α) agonists, which include increase of β-oxidation and cytochrome P450 mediated reactions [5]. For both compounds, carcinogenicity has been shown in animal study mediated by a non-genotoxic mechanism.

PBDDs and PBDFs are another group of POPs formed as byproducts of other brominated compounds, such as brominated flame retardants (BFRs) or brominated biphenyls (PBBs), or ex novo in the combustion processes starting from brominated precursors. In addition, for PBDDs, a biogenic origin in the marine environment has been hypothesized [6]. These compounds are of concern because they seem to have the same mechanism of toxicity of the highly toxic 2,3,7,8-substituted congeners of the polychlorinated dibenzo-p-dioxins (PCDDs) and furans (PCDFs) through the binding to the aryl hydrocarbon receptor (AhR) [7,8], which is the protein mediating the dioxin-like toxicity [9]. Indeed, the classical fingerprint of the dioxin-like biological effects, such as wasting syndrome, thymic atrophy, chloracne, teratogenesis, reproductive effects, and immunotoxicity have been observed [10].

One of the main research gaps related to these pollutants is to what extent exposure for humans and other living organisms occurs [11], since few data on environmental occurrence are available, especially for PBDDs and PBDFs. Fish and fishery products are a known source of dietary intake of POPs for general population, since seas and oceans represent the final accumulation tank of such compounds and their tendency to bioaccumulate.

Within a more extensive research project on the welfare of wildlife and farmed fauna in the Mediterranean Sea [12], this paper reports detailed results species and location specific for PFOS, PFOA, PBBD and PBDF analysis in fish and other aquatic organisms collected in different areas of the Mediterranean Sea. The aims were to investigate the level of the contamination in farmed and wild fish at different trophic levels of the food chain and to assess human exposure associated to fish and fishery products consumption.

2. Materials and Methods

2.1. Sampling

Wild aquatic organisms, at different levels of the food chain, were collected during different sampling campaigns in May, November and January, in three areas in the Mediterranean Sea. The different sampling areas were selected based on the anthropic level of the coasts and were located close to: Monopoli, in the Adriatic sea, south Bari; Porto Palo, in the Ionian sea, in front of the city of Pachino; and Bagnara Calabra, in the Tyrrhenian Sea nearby the Eolie Islands. During the first two sampling campaigns, farmed fish (gilthead sea bream and European sea bass) from off-shore cages have been also collected from three different aquaculture plants. In total. 61 samples of aquatic organisms were analyzed for PFOS, PFOA, PBDDs and PBDFs. The species analyzed with the corresponding sampling areas are shown in the Table 1.

For fish of larger sizes (>100 g), the fillet was isolated and analyzed, whereas for species of smaller size, where it was difficult to separate the fillet, the analysis was performed on the whole fish without head, tail and entrails. For shrimps, the analyzed samples consisted of the body without exoskeleton.

Table 1. Aquatic species, with the corresponding sampling areas analyzed in the present study.

No of Samples	Species	Sampling Area
5	Gilthead sea bream (*Sparus aurata* L.)	Farmed fish
5	European sea bass (*Dicentrarchus labrax* L.)	Farmed fish
9	Red Mullet (*Mullus surmuletus* L.)	PP, BC, MO
9	Anchovy (*Engraulis encrasicholus* L.)	PP, BC, MO
3	Pilchard (*Sardina pilchardus* Walb.)	PP
3	Pink shrimp (*Parapenaeus longirostris* Lucas)	PP
4	Bonito (*Sarda sarda* L.)	PP, BC, MO
2	Mackerel (*Scomber scombrus* L.)	PP
9	Hake (*Merluccius merluccius* L.)	PP, BC, MO
3	Horse mackerel (*Trachurus trachurus* L.)	MO, BC
2	Norway lobster (*Nephrops norvegicus* L.)	MO
3	Bullet tuna (*Auxis rochei* Risso)	BC, PP
2	Swordfish (*Xiphias gladius* L.)	PP, BC
2	Bluefin Tuna (*Thunnus thynnus* L.)	PP, BC

Legend: PP, Porto Palo; BC, Bagnara Calabra; MO, Monopoli.

2.2. PFOS and PFOA Analytical Procedure

Fresh samples (0.5 g), after spiking of the internal standard 13C12 PFOS and 13C12 PFOA (Wellington Laboratories, Guelph, ON, Canada) in methanol, were extracted by ultrasounds for 40 min, and centrifugated at 2800 rpm for 10 min. The supernatant (0.5 mL) was transferred to glass vials and added to 0.5 mL Milli-Q water. Instrumental analysis was performed by high pressure liquid chromatography-tandem mass spectrometry (HPLC-MS-MS) Perkin-Elmer Series 200 (Waltham, MA, USA), Applied Biosystem API 3000 (Concord, ON, Canada) with electrospray ionization (ESI). The HPLC conditions were the following: chromatographic column XTerra MS C18 2.1 × 100 mm, 3.5 μm. The mobile phase A was 5 mM ammonium acetate and the mobile phase B was acetonitrile and the flow rate 200 μL/min. Spectrometric conditions have been optimized in multiple reaction monitoring (MRM) mode using a continuous direct infusion of a solution of the analytes. Detailed analytical methodology for PFOS and PFOA quantification will be published elsewhere.

2.3. PBDD and PBDF Analytical Procedure

Homogenized samples (20–60 g) were lyophilized (Thermo MicroModulyo Freeze Dryer, Fisher Scientific, Hampton, NH, USA) and spiked with a mixture of the following labeled internal standard: 2,3,7,8-tetrabromodibenzofuran (2,3,7,8-TBDF)-13C12; 1,2,3,7,8-pentabromodibenzo-p-dioxin (1,2,3,7,8-PBDD)-13C12, 1,2,3,7,8-pentabromodibenzofuran (1,2,3,7,8-PBDF)-13C12, 2,3,4,7,8-pentabromodibenzofuran (2,3,4,7,8-PBDF)-13C12, and 1,2,3,4,7,8-hexbromodibenzofuran (1,2,3,4,7,8-HBDF)-13C12. Labeled and native analytical standards (congeners 2,3,7,8-substituted from tetra to hexa for dioxins and furan and 1,2,3,4,6,7,8-heptabromodibenzofuran) were purchased from Cambridge Isotope Laboratories Inc. (Tewksbury, MA, USA). Lyophilized samples were extracted using an accelerated solvent extractor ASE300 (Dionex, Sunnyvale, CA, USA) by a mixture of n-hexane:acetone (9:1) and three extraction cycles using a 60% vessel flush at 80 °C and 1500 psi. The extracts were completely evaporated until dryness by rotary evaporator and the fat content was determined gravimetrically. Cleanup was carried out overnight adding sulfuric acid on an Extrelut column and subsequently by alumina column, adapting the clean-up procedure for chlorinated dioxins.

Quantification has been performed by high resolution gas chromatography–high resolution mass spectrometry HRGC-HRMS (Thermo Fisher, Waltham, MA, USA) using a thermo Finnigan MAT95 XP mass spectrometry with GC PAL, CTC Analytics auto sampler, in EI^+ and SIM modes, electron energy 38 eV, ion source temperature 280 °C, resolution power 8000–10,000. The selected ions used for quantification and confirmation were M+2 and M+4 for TBDF and TBDD; M+4 and M+6 for PBDF,

PBDD, HeBDF and HeBDD; and M+6 and M+8 for HpBDF. The chromatographic conditions were: capillary column J&W DB-5MS, 30 m × 0.25 mm, film thickness 0.1 μm. Temperature program: 80 °C, 25 °C/min until 180 °C; 3 °C/min until 280 °C; 6 °C/min until 310 °C for 7 min. For the limit of detection (LOD), a signal-to-noise ratio of 3:1 was chosen.

3. Results and Discussion

Sampling details and descriptive statistics for complete sample dataset has been reported elsewhere [12]. Concentration values for PFOA were below the LOD (0.05 ng/g fresh weight) in 37 samples and for PFOS in 11 samples, out of 65, corresponding to a total below LOD of 57% for PFOA and 17% for PFOS of negative results, respectively. Levels ranged from <0.05 to 1.89, and from <0.05 to 5.96 ng/g fresh weight (fw) for PFOA and PFOS, respectively. Levels ranged from < 0.05 to 1.89, and from <0.05 to 5.96 ng/g fresh weight (fw) for PFOA and PFOS, respectively. Mean ± standard deviation concentrations for PFOA and PFOS in the wildlife of the three sampling areas of the Mediterranean Sea were, respectively, 0.09 ± 0.11 and 1.19 ± 0.91 in Porto Palo, 0.06 ± 0.08 and 1.27 ± 1.36 in Bagnara Calabra, and 0.20 ± 0.47 and 1.25 ± 1.06 in Monopoli. As already reported in other investigations on levels of these chemicals in aquatic organisms, concentrations of PFOS resulted higher than those of PFOA [13]. In addition, concentration levels measured in this study are markedly lower than those reported in other geographical areas but comparable to those in muscle fish and other aquatic organisms sampled in the Mediterranean Sea [1,14,15].

Figure 1 shows the PFOA and PFOS concentrations in the samples analyzed. From that figure, it is evident that levels of both chemicals in the farmed sea bass (*Dicentrarchus labrax* L.) and sea bream (*Sparus aurata* L.) are at least one order of magnitude lower than those measured in the wildlife of the Mediterranean Sea. These results seem to indicate that contamination from PFOS and PFOA arises from food through biomagnification rather than directly from water (bioconcentration) since the farmed fish analyzed have been sampled in offshore cages of the aquaculture plants. A larger set of samples should be performed in order to provide better statistics, especially if there is an interest in Species-Specific risk assessment.

Among the wildlife, the highest concentrations of PFOS were found in anchovy (*Engraulis encrasicholus* L.), 2.7 ± 1.5 ng/g fw, followed from horse mackerel (*Trachurus trachurus* L.), 2.4 ± 0.4 ng/g fw, whereas the lowest in hake (*Merluccius merluccius* L.), 0.46 ± 0.29 ng/g fw, and Atlantic mackerel (*Scomber scombrus* L.), 0.29 ± 0.13 ng/g fw (Figure 2). Thus, from these data, it is not evident that organisms at higher position in the food chain have higher contamination levels of PFOS and PFOA. That could be due to the different behavior of these chemicals in the bioaccumulation process, since PFOS and PFOA do not accumulate into the fat tissues, as the typical POPs, but they bind serum and liver proteins. However, further studies with a greater number of samples should be performed to confirm this hypothesis.

Analysis of the edible part allows estimating the human exposure to these pollutants through fish and other sea food consumption. Combining food consumption data for the Italian general population [16] and average PFOS and PFOA concentration, as measured in the present study, the human intake associated to average fishery products consumption were 0.82 and 0.09 ng/kg BW·day, respectively, whereas the intake associated to high consumers (95th percentile) were 2.1 and 0.24 ng/kg·BW·day, respectively. These latest figures represent 0.01% and 1.3% of the corresponding tolerable daily intake (TDI) for PFOS and PFOA, respectively, established by the European Food Safety Authority (EFSA) in 2008 [17]. However, in 2016, US-EPA issued a RfD of 20 ng/kg·BW·day valid both for PFOS and PFOA [18], while EFSA is re-evaluating its former 2008 opinion on the basis of epidemiological evidences in PFOS and PFOA exposed groups [19]. Under such scenarios, the seafood consumption could cover approximately the 10% of the PFOS tolerable intakes [20].

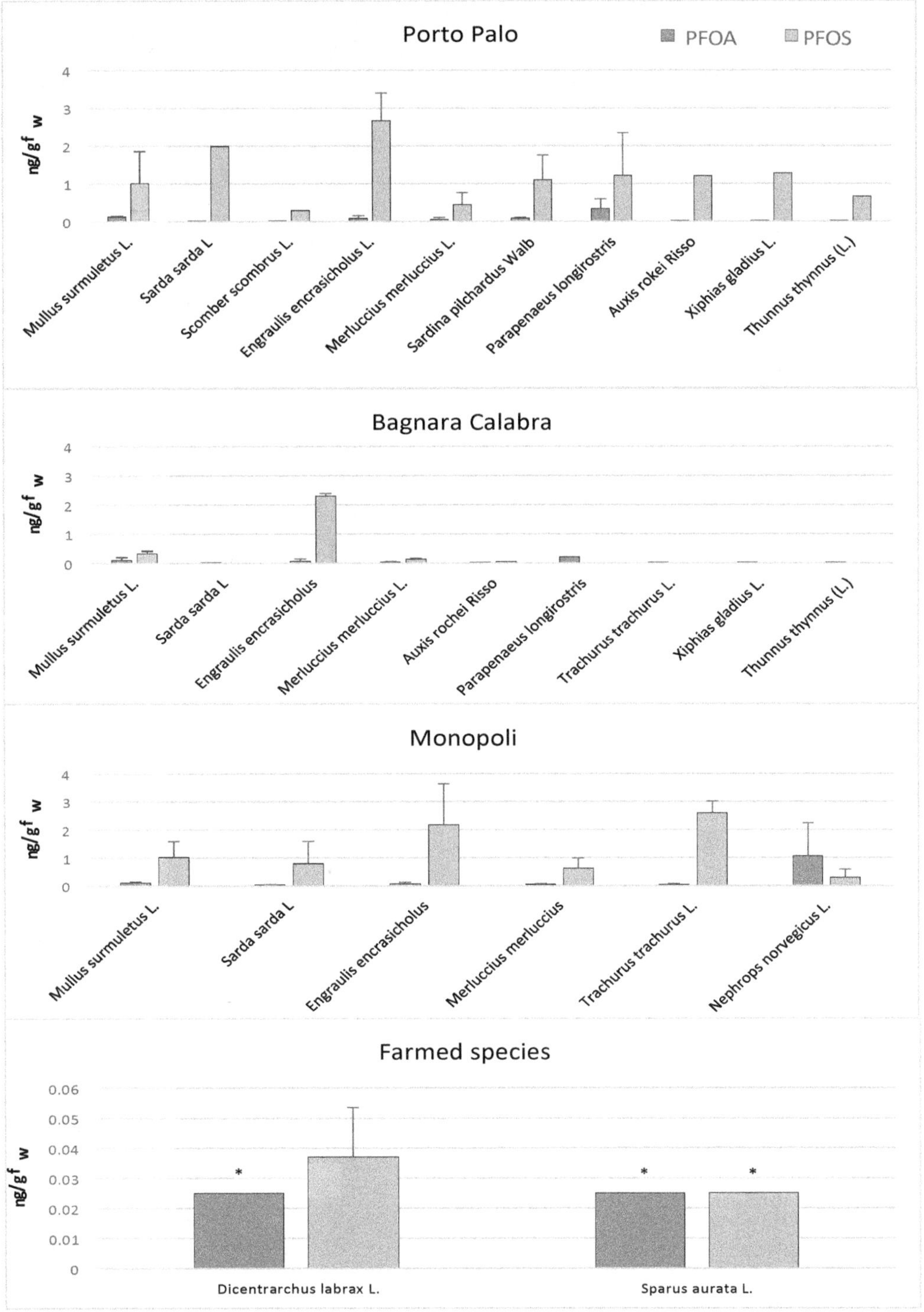

Figure 1. PFOS (perfuorooctane sulfonate) and PFOA (perfluorooctanoic acid) mean concentrations ± standard deviation expressed as ng/g of fresh weight (fw) in wild aquatic organisms from Porto Palo, Bagnara Calabra and Monopoli in the Mediterranean Sea, in farmed fish. * 1/2 Limit of Detection (LOD).

Results of PBDD and PBDF analysis showed concentration values above the LOD for at least a congener in only in 5 samples out of 65. The congeners detected above the LOD values were the 2,3,7,8-TBDF, 2,3,7,8-TBDD, 1,2,3,7,8-PBDF, 2,3,4,7,8-PBDF, 1,2,3,7,8-PBDD, and 1,2,3,4,6,7,8-HpBDF. Concentration values of these congeners ranged from 0.01 to 0.89 pg/g fw and the highest values were found in a wild species, (*Sarda sarda* L.), collected in the sampling area of Monopoli.

Overall, these results indicate that contamination level due to the 2,3,7,8-sostituted congeners of PBDDs and PBDFs, if occurs, is in the range of low ppt; moreover, because of the limited number of samples above the LOD, and the relatively high LOD values (range 0.003–4.5 pg/g fw), it is not possible to make a reliable estimation of the contribution of these compounds to the total contamination of dioxin-like compounds in fish and other aquatic organisms.

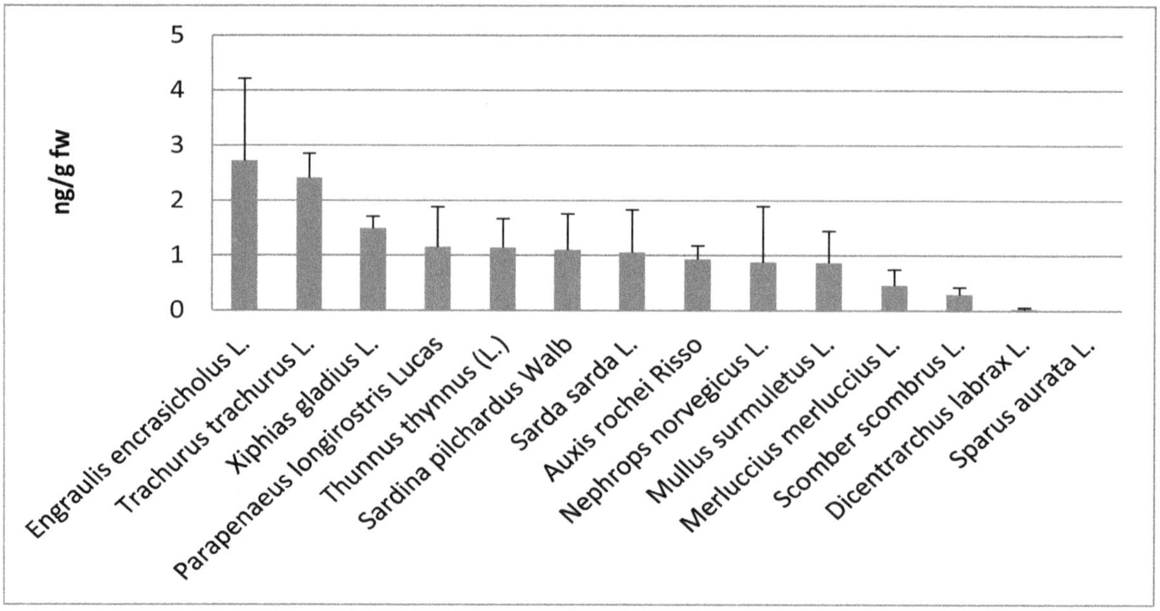

Figure 2. PFOS and PFOA mean concentrations ± standard deviation expressed as ng/g of fresh weight (fw) in different aquatic species from the Mediterranean Sea.

Author Contributions: A.D.D. and R.M. conceived and designed the project. Experiments were conceived, performed and executed by R.F., E.F., R.B., A.C., R.M., G.B., A.R. and E.D., E.F. and R.M. analyzed the data. E.F. and E.D. wrote the manuscript that has been discussed with all authors.

Acknowledgments: This study was carried out within the framework of the project "Added value to national seafood, from the geo-referenced assessment of the environmental exposure to micro-contaminants, from the nutritional fatty acid profiling of fillets, and from the welfare evaluation in farmed species". We are greatly thankful to Alessandro di Domenico from the Italian National Institute for Health for the high scientific quality of his coordinating work of the project.

References

1. Fromme, H.; Tittlemier, S.A.; Völkel, W.; Wilhelm, M.; Twardella, D. Perfluorinated compounds–Exposure assessment for the general population in western countries. *Int. J. Hyg. Environ. Health* **2009**, *212*, 239–270. [CrossRef] [PubMed]
2. Houde, M.; Martin, J.W.; Letcher, R.J.; Solomon, K.R.; Muir, D.C.G. Biological monitoring of polyfluoroalkyl substances: A review. *Environ. Sci. Technol.* **2006**, *40*, 3463–3473. [CrossRef] [PubMed]
3. Kannan, K.; Corsolini, S.; Falandysz, J.; Fillmann, G.; Kumar, K.S.; Loganathan, B.G.; Mohd, M.A.; Olivero, J.; Van Wouwe, N.; Yang, J.H.; et al. Perfluorooctanesulfonate and related fluorochemicals in human blood from several countries. *Environ. Sci. Technol.* **2004**, *38*, 4489–4495. [CrossRef] [PubMed]
4. Schiavone, A.; Corsolini, S.; Kannan, K.; Tao, L.; Trivelpiece, W.; Torres, D.; Focardi, S. Perfluorinated contaminants in fur seal pups and penguin eggs from South Shetland, Antarctica. *Sci. Total Environ.* **2009**, *407*, 3899–3904. [CrossRef] [PubMed]
5. Olsen, G.W.; Huang, H.Y.; Helzlsouer, K.J.; Hansen, K.J.; Butenhoff, J.L.; Mandel, J.H. Historical comparison of perfluorooctanesulfonate, perfluorooctanoate, and other fluorochemicals in human blood. *Environ. Health Perspect.* **2005**, *113*, 539–545. [CrossRef] [PubMed]
6. Malmvärn, A.; Zebühr, Y.; Jensen, S.; Kautsky, L.; Greyerz, E.; Nakano, T.; Asplund, L. Identification of polybrominated dibenzo-p-dioxins in blue mussels (Mytilus edulis) from the Baltic Sea. *Environ. Sci. Technol.*

2005, *39*, 8235–8242. [CrossRef] [PubMed]

7. Mason, G.; Zacharewski, T.; Denomme, M.A.; Safe, L.; Safe, S. Polybrominated dibenzo-p-dioxins and related compounds: Quantitative in vivo and in vitro structure-activity relationships. *Toxicology* **1987**, *44*, 245–255. [CrossRef]
8. Samara, F.; Gullett, B.K.; Harrison, R.O.; Chu, A.; Clark, G.C. Determination of relative assay response factors for toxic chlorinated and brominated dioxins/furans using an enzyme immunoassay (EIA) and a chemically-activated luciferase gene expression cell bioassay (CALUX). *Environ. Int.* **2009**, *35*, 588–593. [CrossRef] [PubMed]
9. Birnbaum, L.S. Evidence for the role of the Ah receptor in response to dioxin. *Prog. Clin. Biol. Res.* **1994**, *387*, 139–154. [PubMed]
10. WHO. Environmental Health Criteria 205. 1998. Available online: http://www.inchem.org/documents/ehc/ehc/ehc205.htm (accessed on 19 August 2018).
11. Birnbaum, L.S.; Staskal, D.F.; Diliberto, J.J. Health effects of polybrominated dibenzo-p-dioxins (PBDDs) and dibenzofurans (PBDFs). *Environ. Int.* **2003**, *29*, 855–860. [CrossRef]
12. Miniero, R.; Abate, V.; Brambilla, G.; Davoli, E.; De Felip, E.; De Filippis, S.P.; Dellatte, E.; De Luca, S.; Fanelli, R.; Fattore, E.; et al. Persistent toxic substances in Mediterranean aquatic species. *Sci. Total Environ.* **2014**, *494–495*, 18–27. [CrossRef] [PubMed]
13. Yeung, L.W.Y.; Yamashita, N.; Taniyasu, S.; Lam, P.K.S.; Sinha, R.K.; Borole, D.V.; Kannan, K. A survey of perfluorinate compounds in surface water and biota including dolphins from the Ganges River and in other waterbodies in India. *Chemosphere* **2009**, *76*, 55–62. [CrossRef] [PubMed]
14. Corsolini, S.; Guerranti, C.; Perra, G.; Focardi, S. Polybrominated diphenyl ethers, perfluorinated compounds and chlorinated pesticides in swordfish (*Xiphias gladius*) from the Mediterranean Sea. *Environ. Sci. Technol.* **2008**, *42*, 4344–4349. [CrossRef] [PubMed]
15. Nania, V.; Pellegrini, G.E.; Fabrizi, L.; Sesta, G.; Sanctis, P.D.; Lucchetti, D.; Pasquale, M.D.; Coni, E. Monitoring of perfluorinated compounds in edible fish from the Mediterranean Sea. *Food Chem.* **2009**, *115*, 951–957. [CrossRef]
16. Turrini, A.; Saba, A.; Perrone, D.; Cialfa, E.; D'Amicis, A. Food consumption patterns in Italy: The INN-CA Study 1994-1996. *Eur. J. Clin. Nutr.* **2001**, *55*, 571–588. [CrossRef] [PubMed]
17. European Food Safety Authority (EFSA). Opinion of the Scientific Panel on Contaminants in the Food chain on Perfluorooctane sulfonate (PFOS), perfluorooctanoic acid (PFOA) and their salts. *EFSA J.* **2008**, *653*, 1–131. [CrossRef]
18. Drinking Water Health Advisories for PFOA and PFOS. Available online: https://www.epa.gov/ground-water-and-drinking-water/drinking-water-health-advisories-pfoa-and-pfos (accessed on 19 August 2018).
19. Brambilla, G.; Italian National Institute of Health, Rome, Italy. Personal communication, 2018.
20. European Food Safety Authority (EFSA). Minutes of the Working Group on PFOS and PFOA. 2018. Available online: www.efsa.europa.eu/sites/default/files/contamwgpfasfood.pdf (accessed on 19 August 2018).

15

Application of Near-Infrared Hyperspectral Imaging with Machine Learning Methods to Identify Geographical Origins of Dry Narrow-Leaved Oleaster (*Elaeagnus angustifolia*) Fruits

Pan Gao [1,2,†], Wei Xu [3,4,†], Tianying Yan [1,2], Chu Zhang [5,6], Xin Lv [2,3] and Yong He [5,6,*]

[1] College of Information Science and Technology, Shihezi University, Shihezi 832000, China; gp_inf@shzu.edu.cn (P.G.); yantianying@163.com (T.Y.)
[2] Key Laboratory of Oasis Ecology Agriculture, Shihezi University, Shihezi 832003, China; lxshz@126.com
[3] College of Agriculture, Shihezi University, Shihezi 832003, China; xu_wei082@163.com
[4] Xinjiang Production and Construction Corps Key Laboratory of Special Fruits and Vegetables Cultivation Physiology and Germplasm Resources Utilization, Shihezi 832003, China
[5] College of Biosystems Engineering and Food Science, Zhejiang University, Hangzhou 310058, China; chuzh@zju.edu.cn
[6] Key Laboratory of Spectroscopy Sensing, Ministry of Agriculture and Rural Affairs, Hangzhou 310058, China
* Correspondence: yhe@zju.edu.cn
† These two authors contributed equally to this manuscript.

Abstract: Narrow-leaved oleaster (*Elaeagnus angustifolia*) fruit is a kind of natural product used as food and traditional medicine. Narrow-leaved oleaster fruits from different geographical origins vary in chemical and physical properties and differ in their nutritional and commercial values. In this study, near-infrared hyperspectral imaging covering the spectral range of 874–1734 nm was used to identify the geographical origins of dry narrow-leaved oleaster fruits with machine learning methods. Average spectra of each single narrow-leaved oleaster fruit were extracted. Second derivative spectra were used to identify effective wavelengths. Partial least squares discriminant analysis (PLS-DA) and support vector machine (SVM) were used to build discriminant models for geographical origin identification using full spectra and effective wavelengths. In addition, deep convolutional neural network (CNN) models were built using full spectra and effective wavelengths. Good classification performances were obtained by these three models using full spectra and effective wavelengths, with classification accuracy of the calibration, validation, and prediction set all over 90%. Models using effective wavelengths obtained close results to models using full spectra. The performances of the PLS-DA, SVM, and CNN models were close. The overall results illustrated that near-infrared hyperspectral imaging coupled with machine learning could be used to trace geographical origins of dry narrow-leaved oleaster fruits.

Keywords: narrow-leaved oleaster fruits; near-infrared hyperspectral imaging; geographical origin; convolutional neural network; effective wavelengths

---

1. Introduction

Narrow-leaved oleaster (*Elaeagnus angustifolia*) is a shrub-like plant of Elaeagnus, which is widely distributed from the Mediterranean region to the northern hemisphere, including in northern Russia and northwestern China. Narrow-leaved oleaster fruits contain a variety of functional health components; in particular, they contain polysaccharides, phenolic acids, and flavonoids. Therefore, narrow-leaved oleaster fruits, as a traditional medicine, are used to treat many diseases in nations and countries from

Central Asia to West Asia. As a medicine and food, the fruit of narrow-leaved oleaster fruits is not only a raw material for food industry processing but also a raw material for functional food and new drugs [1–11]. It has good prospects for development and utilization in arid and semi-arid regions of Northwest China. Its unique habitat environment and long history of planting have produced unique qualities of narrow-leaved oleaster fruits in different producing areas. The qualities of narrow-leaved oleaster fruits are different depending on their place of origin, so it is urgent to establish effective methods for identification of the place of origin of narrow-leaved oleaster fruits.

At present, different scholars have isolated the bioactive components of narrow-leaved oleaster fruits [12], studied the physical and chemical properties and antioxidant properties of narrow-leaved oleaster fruits [13], used Gas Chromatography-Mass Spectrometer (GC-MS) to analyze the components of narrow-leaved oleaster fruit oil [14], and studied the diseases of narrow-leaved oleaster fruits [15]. However, there have been few studies on differentiation of the origins of narrow-leaved oleaster fruits. It is feasible to differentiate narrow-leaved oleaster fruits from different producing areas by synthesizing external morphological and microscopic characteristics and physicochemical identification of fruit powder. Manual sorting has many drawbacks, such as involving monotonous work and strong subjectivity, and being time-consuming and difficult to quantify. Physical and chemical index testing is destructive, and requires complicated sample pretreatment, a long detection cycle, and so on. It also has higher professional requirements for testers. These methods are time-consuming and laborious and cannot achieve the goal of fast and non-destructive classification. In view of the drawbacks of traditional detection methods, many applications use hyperspectral imaging for non-destructive detection due to its advantages of non-destructive, rapid, and accurate measurement, which has broad prospects.

Near-infrared hyperspectral imaging is a chemical analysis tool that can detect different absorption frequencies of specific molecules in substances. Near-infrared hyperspectral imaging can acquire spectral and image information of samples simultaneously. It can obtain comprehensive spectral information of samples. It has the characteristics of fastness and high accuracy. Near-infrared hyperspectral imaging has been widely used in geographical origins and variety identification of food [16]. C. Ru et al. used the hyperspectral imaging method of spectral image fusion in the range of visible and near-infrared (VNIR) and shortwave infrared (SWIR) to classify the geographical origin of Rhizoma Atractylodis Macrocephalae [17]. A. Noviyanto et al. used hyperspectral imaging and machine learning to distinguish honey botanical origins [18]. S. Minaei et al. used visible-near-infrared (VIS-NIR) hyperspectral imaging combined with a machine learning algorithm to predict honey floral origins [19]. M. Puneet et al. used near-infrared hyperspectral imaging to identify six different tea products [20]. Our research team has used near-infrared hyperspectral imaging for varietal and geographical origin identification of agricultural and food materials. C. Zhang et al. used near-infrared hyperspectral imaging to identify coffee bean varieties from different locations [21]. W. Yin et al. used near-infrared hyperspectral imaging to identify geographical origins of Chinese wolfberries [22]. S. Zhu et al. used near-infrared hyperspectral imaging to identify cotton seed varieties [23]. These researchers obtained good performances and illustrated the feasibility of using near-infrared hyperspectral imaging to identify the varietal and geographical origin of agricultural and food materials.

In this study, a near-infrared hyperspectral imaging system covering the spectral range of 874–1734 nm was used. This spectral range is related to various chemical compounds. Researchers have used hyperspectral imaging at this spectral range to obtain good performances for determining contents of protein [24], oil [25], water [26], total iron-reactive phenolics, anthocyanins and tannins [27], and flavanol [28], etc. Previous studies have shown that near-infrared hyperspectral imaging can achieve target classification, but there is no relevant research on the place of origin classification of dry narrow-leaved oleaster fruits. The main purpose of this study was to detect the geographical origin of dry narrow-leaved oleaster fruits based on near-infrared hyperspectral imaging technology, combined with characteristic wavelength selection and machine learning algorithms, including deep learning, providing theoretical methods and a basis for distinguishing the different producing areas of narrow-leaved oleaster fruits.

2. Materials and Methods

2.1. Sample Preparation

Dry narrow-leaved oleaster fruits from three different geographical origins, including Miqin County, Gansu province (Gansu), China (103°4′48″ E, 38°37′12″ N); Zhongwei City, Ningxia Hui Autonomous Region (Ningxia), China (105°10′48″ E, 37°30′36″ N); and Aksu City, Xinjiang Uygur Autonomous Region (Xinjiang), China (80°17′24″ E, 41°9′00″ N), were collected. For each geographical origin, fully matured fruits were harvested in October 2018 and air-dried for consumption and trade. For each geographical origin, intact, clean, and dry narrow-leaved oleaster fruits were collected for hyperspectral image acquisition. In total, 1105, 1205, and 962 intact fruits were obtained from Gansu, Ningxia, and Xinjiang, respectively. The convolutional neural network (CNN) was trained with an independent validation set. To build discriminant models, the samples were randomly split into calibration, validation, and prediction sets. There were 539, 602, and 481 samples from Gansu, Ningxia, and Xinjiang in the calibration set, 291, 303, and 241 samples from Gansu, Ningxia, and Xinjiang in the validation set, and 275, 300, and 240 samples from Gansu, Ningxia, and Xinjiang in the prediction set, respectively. Samples of each geographical origin for hyperspectral imaging acquisition are placed and presented in Figure 1.

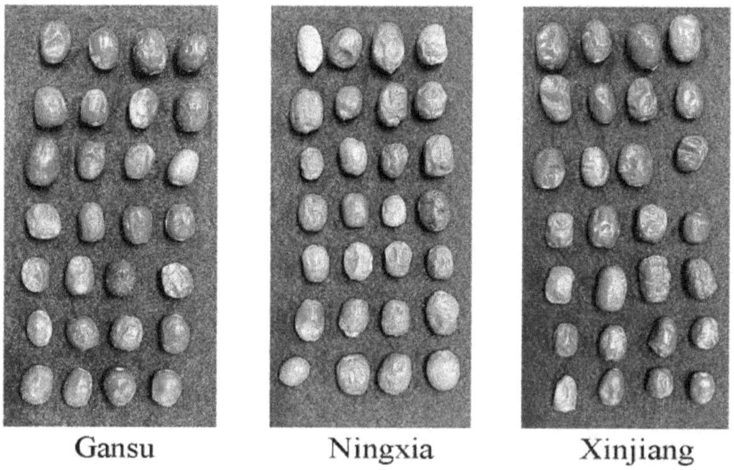

Figure 1. Samples of each geographical origin for hyperspectral imaging acquisition.

2.2. Hyperspectral Image Acquisition and Correction

A near-infrared hyperspectral imaging system was used to acquire hyperspectral images of single narrow-leaved oleaster fruits. This hyperspectral imaging system consisted of four major modules, including an imaging module, an illumination module, a sample motion module, and a software module. The imaging module consisted of an imaging spectrograph (ImSpector N17E, Spectral Imaging Ltd., Oulu, Finland) coupled with an InGaAs camera (Xeva 992, Xenics Infrared Solutions, Leuven, Belgium). The spectral range of the hyperspectral imaging system was 874–1734 nm, the spectral resolution 5 nm, and the number of wavebands 256. The lens for the camera was OLES22 (Spectral Imaging Ltd., Oulu, Finland). The illumination module had a 3900 light source (Illumination Technologies Inc., New York, NY, USA). The sample motion module was formed by an IRCP0076 electric displacement table (Isuzu Optics Corp., Taiwan, China) and samples were placed in the motion platform for line-scan. The software module was used to control the image acquisition and motion platform. The structure of the acquired hyperspectral image was able to be expressed as 320 pixels × L pixels × 256 (wavebands), where 320 pixels was the width of the image, the number 256 was the number of wavebands, and L pixels was the length of the image. L was manually determined during the image acquisition to ensure all samples in one plate were covered in one image.

The image quality, which was determined by the distance between the sample and the lens, the moving speed of the motion platform, and the camera exposure time, was determined by setting these parameters as 12.6 cm, 11 mm/s, and 3000 μs, respectively. In this study, intact narrow-leaved oleaster fruits were placed separately on a black plate for image acquisition. For each image, a random number of fruits was placed there (as shown in Figure 1), and there were at least twenty fruits in an image. During image acquisition, the imaging conditions and system parameters always remained. After image acquisition, the raw hyperspectral images were corrected into reflectance images according to the equation

$$I_c = \frac{I_r - I_d}{I_w - I_d}, \quad (1)$$

where I_c is the corrected image, I_r is the raw original image, I_d is the dark reference image and I_w is the white reference image.

2.3. Spectral Data Extraction

After image correction, spectral data were extracted from each narrow-leaved oleaster fruit. The hyperspectral imaging system collected reflectance spectra of the samples, and reflectance spectra were used for analysis in this study. Each single narrow-leaved oleaster fruit was defined as a region of interest (ROI). A binary image was formed of each hyperspectral image by binarizing the gray-scale image at 1119 nm, in which the narrow-leaved oleaster fruits region was '1' and the background region was '0'. The binary image was then applied to the gray-scale images at each gray-scale image to remove background information. Considering that obvious noises existed at the beginning and end of the spectra, only spectra in the range 975–1646 nm (waveband numbers 31 to 230) were studied, resulting in 200 wavelength variables in the spectral range. Pixel-wise spectra were preprocessed by wavelet transform (wavelet function Daubechies 6 with decomposition level 3) to reduce random noise and area normalization to reduce the influence of sample shape. Pixel-wise spectra within one narrow-leaved oleaster fruit were averaged to represent the sample.

2.4. Data Analysis Methods

2.4.1. Principal Component Analysis

Principal component analysis (PCA) is a widely used qualitative analysis and feature extraction method for spectral data analysis. PCA projects the original spectral data to some new principal component variables (PCs) through linear transformation. Each principal component is linearly combined with the original data. The PCs are ranked by the explained variance. The first PC (PC1) explains the largest of the total variance, followed by PC2 and PC3 and so on. In general, the first few PCs could explain most of the total variance and these few principal components with the largest variance could reflect the data information. In general, the scores of scatter plots which are obtained by projecting scores of one PC onto another PC are used to explore clusters of samples from different classes. In this study, PCA was used to explore qualitative discrimination of narrow-leaved oleaster fruit samples from Gansu, Ningxia, and Xinjiang.

2.4.2. Partial Least Squares Discriminant Analysis

The partial least squares discriminant analysis (PLS-DA) algorithm is based on the PLS regression model to discriminate the target, where the variables in the X block (spectral data) are related to the category values corresponding to the classes contained in the Y vector [29–35]. The integer values are assigned to each class. The category values can be assigned as real integer numbers or they can be formed by dummy variables (0 and 1). PLS regression is firstly conducted on X and Y and the decimal prediction results are transformed into category values according to certain rules.

2.4.3. Support Vector Machine

The support vector machine (SVM) system has been widely applied in statistics, especially for classification. The main idea of SVM is to find the most distinguishable hyperplane by maximizing the margin between the closest points in each class [34–38]. By choosing and optimizing parameters such as penalty factor and kernel function, the discriminant model established by small data samples can still produce small errors for independent test sets. In this paper, the parameter penalty coefficient C of SVM model was searched, and the optimum range was 10^{-8} to 10^{8}. The kernel function was a radial basis function (RBF) and the searching range of the width of the kernel function (g) was 10^{-8} to 10^{8}.

2.4.4. Convolutional Neural Network

The convolutional neural network has been proved as a data processing method with high efficiency and high performance for hyperspectral data analysis due to its ability to aid automatic feature learning [39]. In this study, a simplified CNN architecture based on the model proposed in [40] was designed for narrow-leaved oleaster fruit discrimination.

Figure 2 shows the CNN architecture used in this research. It consisted of two main parts. The first part included two one-dimensional convolution layers (Conv1D, represented by a box with a green background), each of which having been followed by a ReLU activation (yellow box), a one-dimension MaxPooling layer (MaxPool1D, blue box) and a batch-normalization (white box) process. The other part included a fully connected network which was constructed by three Dense layers (light red box) and a SoftMax layer (gray box). The numbers of kernels in the convolution layers were 64 and 32, respectively, with a kernel size of 3 and stride of 1 without padding. MaxPooling layers were configured with a pool size of 2 and stride of 2. The numbers of neurons in the Dense layers were defined as 512, 128, and 3, in order. The first two Dense layers were activated by the ReLU function and followed by a batch-normalization process.

The training procedure was implemented by minimizing the SoftMax Cross Entropy Loss using a stochastic gradient descent (SGD) algorithm. The learning rate was optimized and set as 0.0005. The batch size was set as 400. The train epoch was defined as 400.

2.4.5. Optimal Wavelength Selection

Extracted spectra data contain redundant and collinear information, and some of the wavelengths are uninformative. These uninformative wavelengths may result in unstable calibrations. Moreover, a large number of wavelengths for calibration may result in a complex model structure. Selecting the most informative wavelengths is an important step for further multivariate analysis.

In this study, second derivative spectra were used to select the optimal wavelengths for narrow-leaved oleaster fruits. The second derivative is a widely used spectral preprocessing method which can highlight spectral peaks and suppress background information. In second derivative spectra, the background information is quite small and close to zero, and the positive and negative peaks with greater differences among different categories of samples are manually selected as optimal wavelengths [41].

2.5. Software and Model Evaluation

In this study, PCA, PLS-DA, and SVM were executed on a Matlab R2014b (The Math Works, Natick, MA, USA), the second derivate was conducted on Unscrambler 10.1 (CAMO AS, Oslo, Norway), and the CNN model was performed on Python 3 and MXNET framework (Amazon, Seattle, WA, USA). PCA and PLS-DA was computed using leave-one-out cross validation, SVM was computed using five-fold cross validation, and CNN was computed using an independent validation set. Model performances were evaluated by their classification accuracy, which was calculated as the ratio of the number of correctly classified samples to the total number of samples.

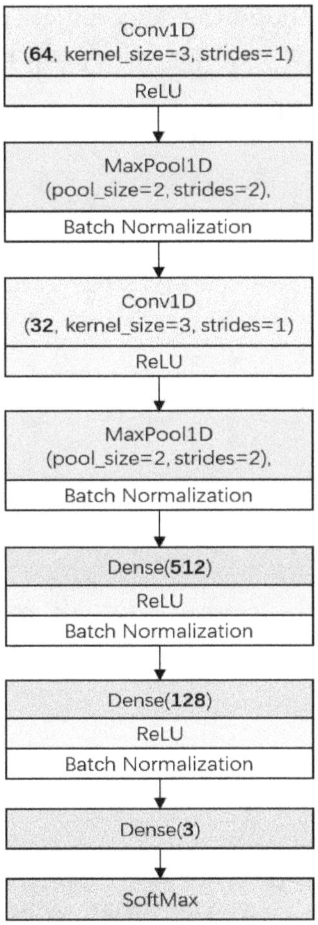

Figure 2. The proposed convolutional neural network (CNN) architecture for narrow-leaved oleaster fruit identification. Conv1D denotes 1-dimension convolution layer, ReLU (Rectified Linear Unit) is the activation function, MaxPool1D denotes 1-dimension max pooling layer, Dense denotes densely-connected neural network layer. The parameter of Conv1D which is defined as 'Channels' is the number of the kernels or filters. The parameter of Dense which is defined as 'units' is the number of the neurons.

3. Results

3.1. Spectral Profiles and Effective Wavelength Identification

Figure 3 shows the average spectra with standard deviation of each wavelength of narrow-leaved oleaster fruits from Gansu, Ningxia, and Xinjiang. Slight differences in reflectance values exist in the average spectra. The differences exist across the whole spectral ranges. However, the overlaps can be observed according to the standard deviation in Figure 3. With these overlaps, the samples from different geographical origins cannot simply be identified by observing their spectral differences. Figure 4 shows the second derivative spectra of the average spectra of narrow-leaved oleaster fruit samples from Gansu, Ningxia and Xinjiang. There are wavelengths with differences. Wavelengths corresponding to the peaks and valleys with greater differences were manually identified. As shown in Figure 4, a total of 22 wavelengths can be identified: 995, 1022, 1032, 1042, 1056, 1072, 1089, 1136, 1190, 1244, 1274, 1284, 1315, 1352, 1365, 1375, 1402, 1433, 1456, 1487, 1500, and 1632 nm. These wavelengths were selected as the effective wavelengths for geographical identification. In this study, the full spectra were used to conduct PCA for qualitative analysis of the sample cluster within one geographical origin and sample separability among different geographical origins. The full spectra were also used to build machine learning models to quantitatively assess the sample separability among different geographical origins. To reduce redundant and collinear information which are informative in full spectra, simplify the models and improve model robustness, the selected effective wavelengths were used to build machine learning models for comparison with the full-spectra-based models.

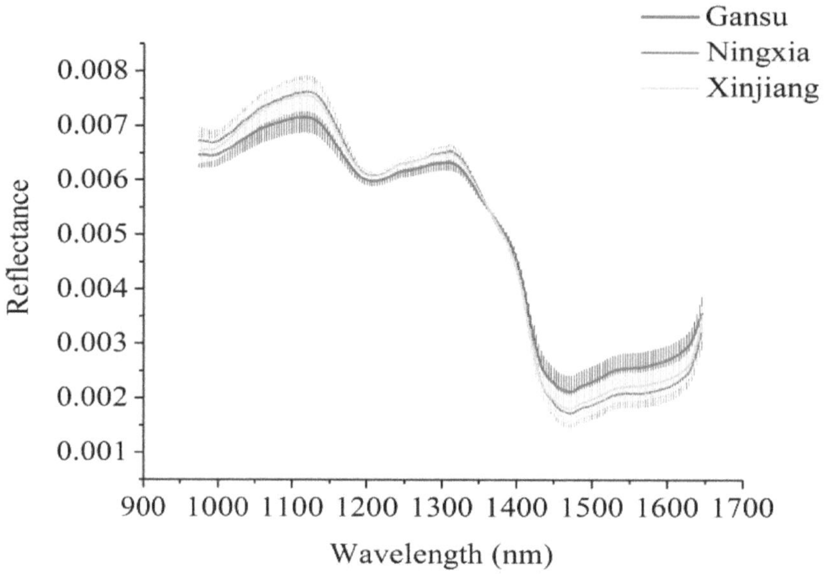

Figure 3. Average spectra with standard deviation of each wavelength of narrow-leaved oleaster fruits from Gansu, Ningxia, and Xinjiang.

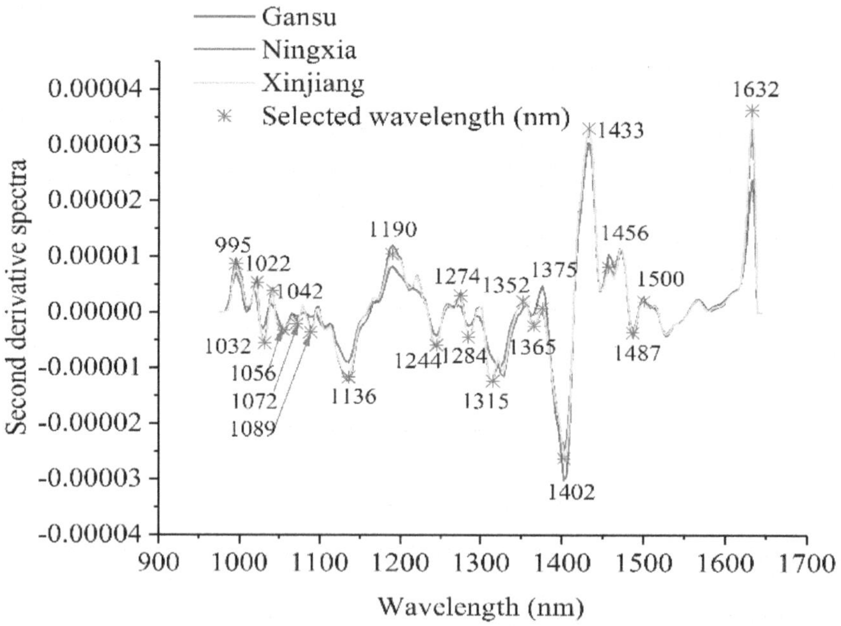

Figure 4. Effective wavelength selection using the second derivative spectra of average spectra of the samples from Gansu, Ningxia, and Xinjiang.

3.2. Principal Component Analysis

PCA was conducted to qualitatively cluster the samples in the scoring spaces. PCA was conducted on the full spectra of the calibration set, and the spectral data were centered for PCA analysis. The first three PCs explain most of the total variance, which was over 99% (PC1: 97.34%, PC2: 1.24%, PC3: 0.63%). Score scatter plots of two different PCs are shown in Figure 5. Samples from the same geographical origins are marked with the same color, as well as the confidence ellipse (confidence level at 0.95). As shown in the score scatter plot of PC1 versus PC2, samples from each geographical origin are able to cluster well. Overlaps exist among the samples from Gansu, Ningxia, and Xinjiang.

In the score scatter plot of PC1 versus PC3, samples from each geographical origin are able to cluster well. Samples from Gansu show greater overlaps with samples from the other geographical origins, and samples from Ningxia and Xinjiang are able to separate well. In the score scatter plot of PC2 versus PC3, samples from each geographical origin are able to cluster well. Samples from Gansu show greater overlaps with samples from the other geographical origins, and samples from Ningxia and Xinjiang are able to separate well. The score scatter plots in Figure 5 showed that the samples from different geographical origins are able to be well clustered and that they have great potential to be correctly identified.

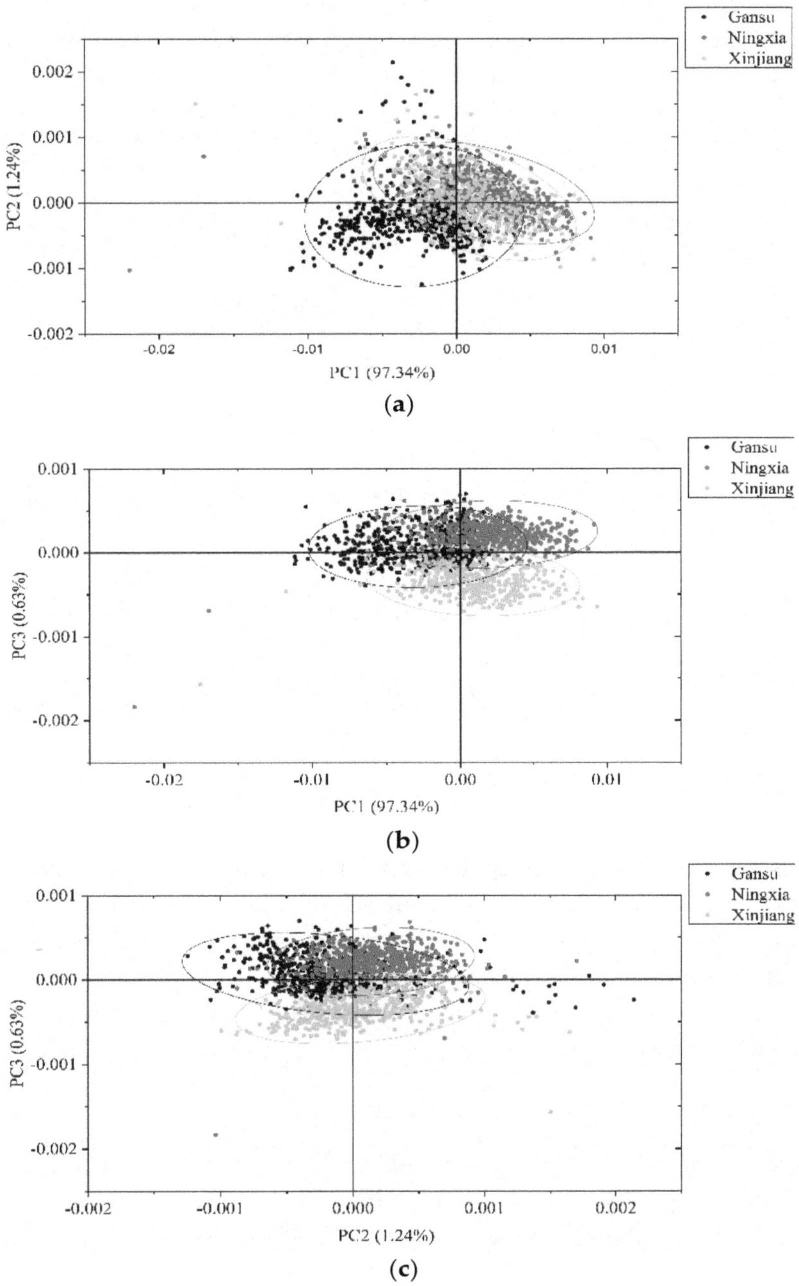

Figure 5. Principal component analysis (PCA) score scatter plots of (**a**) PC1 versus PC2; (**b**) PC1 versus PC3; and (**c**) PC2 versus PC3. The ellipse is the confidence ellipse (confidence level at 0.95).

3.3. *Classification Models Using Full Spectra*

PLS-DA, SVM, and CNN models were built using the full spectra. For the PLS-DA models, the category values of the samples from Gansu, Ningxia, and Xinjiang were labelled 001, 010, and 100.

For the SVM and CNN models, the category values of the samples from Gansu, Ningxia, and Xinjiang were labelled 0, 1, and 2.

The classification results of the three different models are shown in Table 1. All discriminant models obtained good performances, with the classification accuracy of the calibration, validation, and prediction sets all over 90%. For the PLS-DA model, the optimal number of latent variables (LVs) was 12, and good classification performance was obtained. Classification accuracies of the calibration, validation, and prediction sets were all over 99%. For the SVM model, the model parameters (C, g) were optimized as (100, 10,000). The classification accuracy of the calibration set was 100%, while the classification accuracy of the validation and prediction sets was found to be lower. For the CNN model, the classification accuracy of the calibration, validation, and prediction sets were determined to be all over 97%. With regard to all three models, the PLS-DA model performed the best, the CNN model obtained results quite close to and slightly worse than those for PLS-DA, and the SVM model performed the worst.

Table 1. Confusion matrix of the partial least squares discriminant analysis (PLS-DA), support vector machine (SVM) and convolutional neural network (CNN) models using full spectra.

Model	Category Values	Calibration				Validation				Prediction			
		0	1	2	Total (%)	0	1	2	Total (%)	0	1	2	Total (%)
PLS	0 *	539	0	0		291	0	0		268	0	7	
	1	0	601	1		0	303	0		0	299	1	
	2	0	0	481		0	0	241		0	0	240	
	Total (%)				99.94				100				99.02
SVM	0	539	0	0		289	0	2		224	0	51	
	1	0	602	0		0	303	0		0	300	0	
	2	0	0	481		0	0	241		0	0	240	
	Total (%)				100				99.76				93.74
CNN	0	539	0	0		289	0	2		253	0	22	
	1	1	601	0		0	303	0		0	300	0	
	2	6	0	475		4	0	237		0	0	240	
	Total (%)				99.57				99.28				97.30

* 0, 1, and 2 are the assigned category values of the samples from Gansu, Ningxia, and Xinjiang, respectively.

When using the PLS-DA model, samples from Ningxia were misclassified as samples from Xinjiang and samples from Gansu were misclassified as samples from Xinjiang; when using the SVM model, samples from Gansu were misclassified as samples from Xinjiang; and when using the CNN model, samples from Gansu and Xinjiang were misclassified as each other. The overall classification results indicated good separability among the samples from the three geographical origins. Samples from Gansu and Xinjiang were more likely to be misclassified, due to the results of the three discriminant models.

3.4. Classification Models Using Optimal Wavelengths

After effective wavelength selection, the PLS-DA, SVM, and CNN models were built using the selected effective wavelengths. The results of the three discriminant models are shown in Table 2. Good performances were obtained by the three models, with the classification accuracy of the calibration, validation, and prediction sets all over 95%. For the PLS-DA model, the optimal number of LVs was found to be 17. The classification accuracies of the calibration, validation, and prediction sets were all over 99%. For the SVM model, the model parameters (C, g) were optimized as (100, 108). The classification accuracies of the calibration, validation, and prediction sets were all over 95%. For the CNN model, the classification accuracies of the calibration, validation, and prediction sets were all over 97%.

Table 2. Confusion matrices of the PLS-DA, SVM, and CNN models using effective wavelengths.

Model	Category Values	Calibration				Validation				Prediction			
		0	1	2	Total (%)	0	1	2	Total (%)	0	1	2	Total (%)
PLS	0 *	538	0	1		291	0	0		272	0	3	
	1	1	601	0		0	303	0		0	300	0	
	2	1	0	480		0	0	241		0	0	240	
	Total (%)				99.92				100				99.63
SVM	0	539	0	0		271	0	20		238	0	37	
	1	0	602	0		0	303	0		0	300	0	
	2	2	0	479		1	0	240		1	0	239	
	Total (%)				99.88				97.49				95.34
CNN	0	539	0	0		287	0	4		263	0	12	
	1	0	602	0		0	303	0		0	299	1	
	2	4	0	477		8	0	233		5	0	235	
	Total (%)				99.75				98.56				97.79

* 0, 1, and 2 are the assigned category values of the samples from Gansu, Ningxia, and Xinjiang, respectively.

When using the PLS-DA model, samples from Gansu and Xinjiang were misclassified as each other, and one sample from Ningxia was misclassified as a sample from Gansu. When using the SVM model, it was observed that samples from Gansu and Xinjiang were misclassified as each other. When using the CNN model, samples from Gansu and Xinjiang were misclassified as each other, and one sample from Ningxia was misclassified as a sample from Xinjiang. The confusion matrices of the three models illustrate that samples from Gansu and Xinjiang were more likely to be misclassified.

The PLS-DA, SVM, and CNN models using effective wavelengths obtained similar results to those using effective wavelengths, illustrating the effectiveness of effective wavelength selection. The overall classification accuracy of all models indicates that there are great differences existing in narrow-leaved oleaster fruits from the three different geographical origins considered. As shown in Tables 1 and 2, the PLS-DA models performed slightly better than the CNN models, and the CNN models performed slightly better than the SVM models. Although differences existed in these model performances, the differences were quite small. The results illustrate that CNN models could be used for narrow-leaved oleaster fruit geographical origin identification. Moreover, the results of the discriminant models using full spectra and effective wavelengths all showed that samples from Gansu and Xinjiang were more likely to be misclassified.

4. Conclusions

In this work, near-infrared hyperspectral imaging was successfully used to identify the geographical origins of narrow-leaved oleaster fruits from Gansu, Ningxia, and Xinjiang. PCA score scatter plots showed the separability of the samples from the three geographical origins. PLS-DA, SVM, and CNN models were established using full spectra and effective wavelengths selected by second derivative spectra. The high classification accuracy, which was over 90% for models using full spectra and effective wavelengths, illustrates that the proposed method can effectively distinguish narrow-leaved oleaster fruits from different geographical origins. The performances of the models using effective wavelengths were similar to those using full spectra. Moreover, deep CNN models obtained close results to the PLS-DA and SVM models, showing good performances of deep learning for narrow-leaved oleaster fruit geographical origin detection. According to the discriminant models, samples from Gansu and Xinjiang were more likely to be misclassified. These results indicate that it would be possible to develop online systems for narrow-leaved oleaster fruit origin detection using near-infrared hyperspectral imaging and machine learning methods.

Author Contributions: Conceptualization, P.G.; data curation, P.G. and C.Z.; formal analysis, W.X.; funding acquisition, W.X.; investigation, C.Z.; methodology, C.Z., X.L., and Y.H.; project administration, P.G.; resources,

T.Y.; software, T.Y.; supervision, Y.H.; validation, X.L.; visualization, T.Y.; writing—original draft, P.G., W.X., and C.Z.; writing—review and editing, X.L. and Y.H.

Acknowledgments: The authors want to thank L.Z., a Ph.D candidate in College of Biosystems Engineering and Food Science, Zhejiang University, China, for providing help on data analysis.

References

1. Zhang, L.; Alvarez, L.V.; Bonthond, G.; Tian, C.; Fan, X. Cytospora elaeagnicola sp. nov. Associated with Narrow-leaved oleaster fruits Canker Disease in China. *Mycobiology* **2019**, *47*, 1–10. [CrossRef] [PubMed]
2. Zhanga, X.; Lia, G.; Sheng, D. Simulating the potential distribution of *Elaeagnus angustifolia* L. based on climatic constraints in China. *Ecol. Eng.* **2018**, *113*, 27–34. [CrossRef]
3. Lin, J.; Li, J.P.; Yuan, F.; Yang, Z.; Wang, B.S.; Chen, M. Transcriptome profiling of genes involved in photosynthesis in *Elaeagnus angustifolia* L. under salt stress. *Photosynthetica* **2018**, *56*, 1–12. [CrossRef]
4. Chen, X.; Yushuang, L.; Guangying, C.; Chi, G.; Shengge, L.; Huiming, H.; Tao, Y. Angustifolinoid A, a macrocyclic flavonoid glycoside from *Elaeagnus angustifolia* flowers. *Tetrahedron Lett.* **2018**, *59*, 2610–2613. [CrossRef]
5. Du, H.; Chen, J.; Tian, S.; Gu, H.; Li, N.; Sun, Y.; Ru, J.; Wang, J. Extraction optimization, preliminary characterization and immunological activities in vitro of polysaccharides from *Elaeagnus angustifolia* L. pulp. *Carbohydr. Polym.* **2016**, *151*, 348–357. [CrossRef]
6. Mcshane, R.; Auerbach, D.; Friedman, J.M.; Auble, G.T.; Shafroth, P.B.; Merigliano, M.; Scott, M.; Poff, N. Distribution of invasive and native riparian woody plants across the western USA in relation to climate, river flow, floodplain geometry and patterns of introduction. *Ecography* **2016**, *38*, 1254–1265. [CrossRef]
7. Collette, L.K.D.; Pither, J. Insect assemblages associated with the exotic riparian shrub Russian olive (Elaeagnaceae), and co-occurring native shrubs in British Columbia, Canada. *Can. Entomol.* **2016**, *148*, 316–328. [CrossRef]
8. Tredick, C.A.; Kelly, M.J.; Vaughan, M.R. Impacts of large-scale restoration efforts on black bear habitat use in Canyon de Chelly National Monument, Arizona, United States. *J. Mammal.* **2016**, *97*, gyw060. [CrossRef]
9. Khamzina, A.; Lamers, J.P.A.; Martius, C. Above- and belowground litter stocks and decay at a multi-species afforestation site on arid, saline soil. *Nutr. Cycl. Agroecosyst.* **2016**, *104*, 187–199. [CrossRef]
10. Singh, A.; Singh, N.B.; Hussain, I.; Singh, H.; Yadav, V.; Singh, S.C. Green synthesis of nano zinc oxide and evaluation of its impact on germination and metabolic activity of *Solanum lycopersicum*. *J. Biol.* **2016**, *233*, 84–94. [CrossRef]
11. Singh, A.; Singh, N.B.; Afzal, S.; Singh, T.; Hussain, I. Zinc oxide nanoparticles: A review of their biological synthesis, antimicrobial activity, uptake, translocation and biotransformation in plants. *J. Mater. Sci.* **2017**, *53*, 185–201. [CrossRef]
12. Hassanzadeh, Z.; Hassanpour, H. Evaluation of physicochemical characteristics and antioxidant properties of *Elaeagnus angustifolia* L. *Sci. Hortic.* **2018**, *238*, 83–90. [CrossRef]
13. Waili, A.; Yili, A.; Maksimov, V.V.; Mijiti, Y.; Atamuratov, F.N.; Ziyavitdinov, Z.F.; Mamadrakhimov, A.; Asia, H.A.; Salikhov, S.I. Erratum to: Isolation of Biologically Active Constituents from Fruit of *Elaeagnus angustifolia*. *Chem. Nat. Compd.* **2016**, *52*, 776. [CrossRef]
14. Wei, Q.; Wei, Y.; Wu, H.; Yang, X.; Zhang, H. Chemical Composition, Anti-oxidant, and Antimicrobial Activities of Four Saline-Tolerant Plant Seed Oils Extracted by SFC. *J. Am. Oil Chem. Soc.* **2016**, *93*, 1–10. [CrossRef]
15. Morehart, A.L. Phomopsis canker and dieback of *Elaeagnus angustifolia*. *Plant Dis.* **2015**, *64*, 66. [CrossRef]
16. Marena, M. Near-infrared spectroscopy and hyperspectral imaging: Non-destructive analysis of biological materials. *Chem. Soc. Rev.* **2014**, *43*, 8200–8214.
17. Ru, C.; Li, Z.; Tang, R. A Hyperspectral Imaging Approach for Classifying Geographical Origins of Rhizoma Atractylodis Macrocephalae Using the Fusion of Spectrum-Image in VNIR and SWIR Ranges (VNIR-SWIR-FuSI). *Sensors* **2019**, *19*, 2045. [CrossRef]
18. Noviyanto, A.; Abdulla, W.H. Honey botanical origin classification using hyperspectral imaging and machine learning. *J. Food Eng.* **2019**, *265*, 109684. [CrossRef]
19. Minaei, S.; Shafiee, S.; Polder, G.; Moghadam-Charkari, N.; Van Ruth, S.; Barzegar, M.; Zahiri, J.; Alewijn, M.; Kuś, P.M. VIS/NIR imaging application for honey floral origin determination. *Infrared Phys. Technol.* **2017**, *86*, 218–225. [CrossRef]

20. Puneet, M.; Alison, N.; Julius, T.; Guoping, L.; Sally, R.; Stephen, M. Near-infrared hyperspectral imaging for non-destructive classification of commercial tea products. *J. Food Eng.* **2018**, *238*, 70–77. [CrossRef]
21. Zhang, C.; Liu, F.; He, Y. Identification of coffee bean varieties using hyperspectral imaging: Influence of preprocessing methods and pixel-wise spectra analysis. *Sci. Rep.* **2018**, *8*, 2166. [CrossRef] [PubMed]
22. Yin, W.; Zhang, C.; Zhu, H.; Zhao, Y.; He, Y. Application of near-infrared hyperspectral imaging to discriminate different geographical origins of chinese wolfberries. *PLoS ONE* **2017**, *12*, e0180534. [CrossRef] [PubMed]
23. Zhu, S.; Zhou, L.; Gao, P.; Bao, Y.; He, Y.; Feng, L. Near-Infrared Hyperspectral Imaging Combined with Deep Learning to Identify Cotton Seed Varieties. *Molecules* **2019**, *24*, 3268. [CrossRef] [PubMed]
24. Mahesh, S.; Jayas, D.S.; Paliwal, J.; White, N.D.G. Comparison of partial least squares regression (plsr) and principal components regression (pcr) methods for protein and hardness predictions using the near-infrared (nir) hyperspectral images of bulk samples of Canadian wheat. *Food Bioprocess Technol.* **2015**, *8*, 31–40. [CrossRef]
25. Weinstock, B.A.; Janni, J.; Hagen, L.; Wright, S. Prediction of oil and oleic acid concentrations in individual corn (*Zea mays* L.) kernels using near-infrared reflectance hyperspectral imaging and multivariate analysis. *Appl. Spectrosc.* **2006**, *60*, 9. [CrossRef]
26. Sun, J.; Lu, X.; Mao, H.; Wu, X.; Gao, H. Quantitative determination of rice moisture based on hyperspectral imaging technology and bcc-ls-svr algorithm. *J. Food Process Eng.* **2016**, *40*. [CrossRef]
27. Zhang, N.; Liu, X.; Jin, X.; Li, C.; Wu, X.; Yang, S.; Ning, J.; Yanne, P. Determination of total iron-reactive phenolics, anthocyanins and tannins in wine grapes of skins and seeds based on near-infrared hyperspectral imaging. *Food Chem.* **2017**, *237*, 811–817. [CrossRef]
28. Rodríguez-Pulido, F.J.; Hernández-Hierro, J.M.; Nogales-Bueno, J.; Gordillo, B.; González-Miret, M.L.; Heredia, F.J. A novel method for evaluating flavanols in grape seeds by near infrared hyperspectral imaging. *Talanta* **2014**, *122*, 145–150.
29. Mazivila, S. Discrimination of the type of biodiesel/diesel blend (B5) using mid-infrared spectroscopy and PLS-DA. *Fuel* **2015**, *142*, 222–246. [CrossRef]
30. Botelho, B.G.; Reis, N.; Oliveira, L.S.; Sena, M.M. Development and analytical validation of a screening method for simultaneous detection of five adulterants in raw milk using mid-infrared spectroscopy and PLS-DA. *Food Chem.* **2015**, *181*, 31–37. [CrossRef]
31. Balage, J.M.; Amigo, J.M.; Antonelo, D.S.; Mazon, M.R.; e Silva, S.D. Shear force analysis by core location in Longissimus steaks from Nellore cattle using hyperspectral images—A feasibility study. *Meat Sci.* **2018**, *143*, 30–38. [CrossRef] [PubMed]
32. Melucci, D.; Bendini, A.; Tesini, F.; Barbieri, S.; Zappi, A.; Vichi, S.; Conte, L.; Gallina, T.T. Rapid direct analysis to discriminate geographic origin of extra virgin olive oils by flash gas chromatography electronic nose and chemometrics. *Food Chem.* **2016**, *204*, 263–273. [CrossRef] [PubMed]
33. Da Costa, G.B.; Fernandes, D.D.S.; Gomes, A.A.; De Almeida, V.E.; Veras, G. Using near infrared spectroscopy to classify soybean oil according to expiration date. *Food Chem.* **2016**, *196*, 539–543. [CrossRef]
34. Du, L.; Lu, W.; Cai, Z.J.; Bao, L.; Hartmann, C.; Gao, B.; Yu, L.L. Rapid detection of milk adulteration using intact protein flow injection mass spectrometric fingerprints combined with chemometrics. *Food Chem.* **2017**, *240*, 573–578. [CrossRef]
35. Schmutzler, M.; Beganovic, A.; Böhler, G.; Huck, C.W. Methods for detection of pork adulteration in veal product based on FT-NIR spectroscopy for laboratory, industrial and on-site analysis. *Food Control* **2015**, *57*, 258–267. [CrossRef]
36. Yang, H.X.; Fu, H.B.; Wang, H.D.; Jia, J.W.; Sigrist, M.W.; Dong, F.Z. Laser-induced breakdown spectroscopy applied to the characterization of rock by support vector machine combined with principal component analysis. *Chin. Phys. B* **2016**, *25*, 065201. [CrossRef]
37. Li, J.L.; Sun, D.W.; Pu, H.; Jayas, D.S. Determination of trace thiophanate-methyl and its metabolite carbendazim with teratogenic risk in red bell pepper (*Capsicum annuum* L.) by surface-enhanced Raman imaging technique. *Food Chem.* **2017**, *218*, 543–552. [CrossRef]
38. Ropodi, A.I.; Panagou, E.Z.; Nychas, G.J.E. Multispectral imaging (MSI): A promising method for the detection of minced beef adulteration with horsemeat. *Food Control* **2017**, *73*, 57–63. [CrossRef]
39. Wu, N.; Zhang, C.; Bai, X.; Du, X.; He, Y. Discrimination of Chrysanthemum Varieties Using Hyperspectral Imaging Combined with a Deep Convolutional Neural Network. *Moleclues* **2018** *23*, 2831. [CrossRef]

40. Qiu, Z.; Chen, J.; Zhao, Y.; Zhu, S.; He, Y.; Zhang, C. Variety Identification of Single Rice Seed Using Hyperspectral Imaging Combined with Convolutional Neural Network. *Appl. Sci.* **2018**, *8*, 212. [CrossRef]
41. Zhang, C.; Feng, X.; Wang, J.; Liu, F.; He, Y.; Zhou, W. Mid-infrared spectroscopy combined with chemometrics to detect Sclerotinia stem rot on oilseed rape (*Brassica napus* L.) leaves. *Plant Methods* **2017**, *13*, 39. [CrossRef] [PubMed]

Detection of Shiga Toxin 2 Produced by *Escherichia coli* in Foods using a Novel AlphaLISA

Cheryl M. Armstrong [1], Leah E. Ruth [2], Joseph A. Capobianco [1], Terence P. Strobaugh Jr. [1], Fernando M. Rubio [2] and Andrew G. Gehring [1,*]

1 Molecular Characterization of Foodborne Pathogens Research Unit, United States Department of Agriculture, Eastern Regional Research Center, Wyndmoor, PA 19038, USA; Cheryl.Armstrong@ars.usda.gov (C.M.A.); Joseph.Capobianco@ars.usda.gov (J.A.C.); Terence.Strobaugh@ars.usda.gov (T.P.S.J.)
2 Abraxis, Inc., Warminster, PA 18974, USA; lruth@abraxiskits.com (L.E.R.); frubio@abraxiskits.com (F.M.R.)
* Correspondence: Andrew.Gehring@ars.usda.gov

Abstract: Amplified luminescent proximity homogenous assay-linked immunosorbent assay (AlphaLISA) is comprised of a bead-based immunoassay that is used for small molecule detection. In this study, a novel AlphaLISA was developed and optimized for the detection of Shiga-toxin 2 (Stx2). Efficacy and sensitivity trials showed the AlphaLISA could detect ≥0.5 ng/mL of purified Stx2, which was comparable to the industry-standard enzyme-linked immunosorbent assay (ELISA) tests for Stx2 detection. In addition, evaluation of Shiga toxin-producing *Escherichia coli* (STEC)-inoculated Romaine lettuce and ground beef samples demonstrated that both the AlphaLISA and the ELISA were able to discern uninoculated samples from 1× and 10× diluted samples containing ~10 CFU/mL of STEC enriched in modified tryptic soy broth with mitomycin C for 16 h. Overall, the increased signal-to-noise ratios indicated a more robust signal was produced by the AlphaLISA compared to the ELISA and the delineation of higher toxin concentrations without the need for sample dilution implied a greater dynamic range for the AlphaLISA. Implementation of the newly developed AlphaLISA will allow for more rapid analysis for Stx2 with less manual manipulation, thus improving assay throughput and the ability to automate sample screening while maintaining detection limits of 0.5 ng/mL.

Keywords: amplified luminescent proximity homogenous assay-linked immunosorbent assay (AlphaLISA); detection; enzyme-linked immunosorbent assay (ELISA); *E. coli*; Shiga toxin; STEC; Stx2

Key Contribution: There are many benefits to the newly developed AlphaLISA including a shorter and simpler homogeneous protocol that requires less manual manipulation and no wash steps compared to the ELISA. Overall, this AlphaLISA allows for the rapid detection of bacterial-generated Stx2 in food matrices using a highly sensitive, robust, and automatable process that is amendable to high-throughput screening.

---

1. Introduction

Food poisoning due to Shiga toxin-producing *Escherichia coli* (STEC) is a consistent cause for concern in the United States because of its association with hemorrhagic colitis. Clinically, hemorrhagic colitis is characterized by the onset of a variety of symptoms including nausea, vomiting, abdominal pain, diarrhea, and bloody stools with approximately 5% of cases progressing to a more severe form of clinical disease known as hemolytic uremic syndrome (HUS) [1]. Expenses related to STEC illnesses are difficult to quantify; however, estimates from summary health measures such as the annual number of illnesses, hospitalizations, and deaths as estimated by the Centers for Disease Control and Prevention

(CDC) [2] can be used to approximate the costs to be around $300 M and a loss of 1700 quality-adjusted life years annually [3].

Rapid and accurate identification of STEC is imperative for the protection of several facets of human health. For example, stopping STEC contamination in food production is critical for curtailing the number of infections and preventing full-scale outbreaks from occurring, while obtaining an accurate diagnosis upon infection is crucial for ensuring proper care for patients. Because treatment of STEC infections with antibiotics greatly increases the risk of serious complications resulting from the development of the condition known as hemolytic uremic syndrome (HUS) [4,5] and Shiga toxin is known to be essential for the development of HUS from STEC infections [6]; the production and subsequent release of Shiga toxin has been a topic of intense investigation. Production of Shiga toxin is both a distinguishing feature and an important virulence factor for STEC strains. It can be expressed by strains carrying the *stx* gene, which is encoded for by a lambdoid bacteriophage [7,8]. Current research separates Shiga toxin into two main antigenically distinct groups, Shiga toxin 1 (Stx1) and Shiga toxin 2 (Stx2) [9]. Both toxins consist of a single A-subunit of ~32 kDa and five B-subunits of ~7.7 kDa each [10]. However, marked differences in the toxicity level exist between the two groups. For example, Stx2 types have an LD50 approximately 400 times lower than Stx1 [11] despite having ~60% identity at the amino acid level between the two groups [10]. Each of the two main groups can be further subdivided into the following variants: Stx1a, Stx1c, Stx1d, and Stx2a, Stx2b, Stx2c, Stx2d, Stx2e, Stx2f, and Stx2g [12]. The Food Safety and Inspection Service, an agency of the United States Department of Agriculture, has undertaken routine screening of meat samples for the presence of *stx* in order to identify the presence/absence of STEC in the food supply chain [13,14].

Because culture-based STEC detection can be long and laborious, alternative methods for identifying strains that produce Shiga toxin have emerged. More rapid antibody-based detection tests, specifically enzyme-linked immunosorbent assays (ELISAs) [15,16], antibody-based lateral flow assays (LATs) [17], and immunomagnetic separation assays (IMS) [18,19] have been employed as common testing platforms for the detection of STECs and Shiga toxin in lieu of traditional microbiological detection methods such as colony plating on selective media. A new platform known as the AlphaLISA, or amplified luminescent proximity homogenous assay-linked immunosorbent assay, utilizes both bead- and antibody-based technologies for pathogen or toxin detection. In this system, two different antibody-coated beads (aka. "donor" and "acceptor" beads) are used to bind either the same antigenic region on a target or two distinct antigenic regions of a target that are within the vicinity of one another. The donor beads contain a photosensitizing agent, which upon laser excitation at 680 nm, causes the donor bead to emit ~60,000 singlet oxygen molecules per second [20]. These singlet oxygen molecules then react with thioxene derivatives within the acceptor beads, resulting in the production of a chemiluminescent signal at 340–350 nm [21]. This signal activates fluorophores, also contained within the acceptor beads, resulting in a detectable signal consisting of emitted fluorescence with a narrow bandwidth centered around 615 nm. Because the half-life of the singlet oxygen is 4 microseconds, its diffusion distance is limited to approximately 200 nm in aqueous solutions. Thus, if the beads are not in close enough proximity, the singlet oxygen molecules from the donor bead decay to their ground state prior to reaching the acceptor beads and no signal is emitted.

The AlphaLISA has many advantages over the ELISA; namely its high sensitivity, relatively quick testing time, reduced hands-on workflow resulting from the ability to sequentially overlay the reagents, and it is easily adaptable to automation and high-throughput screening [22]. In this study, the first AlphaLISA for the detection of a bacterial toxin in foods was developed using the presence of Shiga toxin 2 (Stx2) as a marker for STEC contamination in foods. Comparisons were made between the newly developed AlphaLISA and an industry-standard ELISA in two different food matrices, Romaine lettuce and ground beef, to further evaluate the assay's utility as a rapid detection method.

2. Results

2.1. Antibody Configuration, Titration, and Gain Settings for Amplified Luminescent Proximity Homogenous Assay-Linked Immunosorbant Assay (AlphaLISA)

The AlphaLISA design presented here (Figure 1) utilizes two different monoclonal antibodies with specific affinity to Stx2 for incorporation onto the donor and acceptor beads. Antibody M-1005 binds the Shiga toxin 2 A-subunit while M-1003 binds the Shiga toxin 2 B-subunits. These antibodies were chosen so that the two beads would bind to different areas of the toxin, as antibodies with the same target site may compete with each other rather than trap the toxin between them. During assay development, antibody M-1003 was first used as the biotinylated antibody for association with the streptavidin-coated donor bead while the M-1005 antibody was conjugated to the acceptor bead (Figure 1A). Then, antibody M-1003 was conjugated to the acceptor bead while the M-1005 antibody was used as the biotinylated antibody for association with the streptavidin-coated donor bead (Figure 1B). In addition to testing both antibody configurations, the final concentration of the biotinylated antibody was also varied (0, 0.3, 1.0, and 3.0 nM) while keeping the acceptor-conjugated antibody beads constant to ensure optimal performance would be achieved for the detection of Stx2 in the AlphaLISA format. Results of this assay (Figure 2) showed that using 0.3 nM of biotinylated M-1005 antibody for association with the donor, while conjugating antibody M-1003 to the acceptor produced the largest signal intensities. Furthermore, this condition demonstrated the ability to differentiate between all levels of Stx2 as the *p*-values of independent Student's *t*-tests were less than 0.003. Since the response for this condition displayed the largest signal amplitudes and the ability to differentiate between each level of Stx2 tested, this configuration was subsequently used in the remainder of the assays presented. The final parameter examined in an effort to maximize the sensitivity and dynamic range of the AlphaLISA was the gain setting for the photomultiplier tube on the plate reader. A multifactorial test was performed using 0.3 nM of biotinylated M-1005 as the donor and conjugating M-1003 to the acceptor bead in an assay evaluating four levels of Stx2 (0, 3, 30, and 300 ng/mL) at three different gain settings (gain = 100, 125, and 150). While we observed larger signal intensities with the higher gains, we also noted that for some conditions, the assay was unable to differentiate between levels of Stx2 using a Student's *t*-test with an alpha value of 0.05. From this, it was determined that a gain setting of 100 was optimal because it demonstrated the greatest precision and was the best at resolving the lower levels of toxin (data not shown).

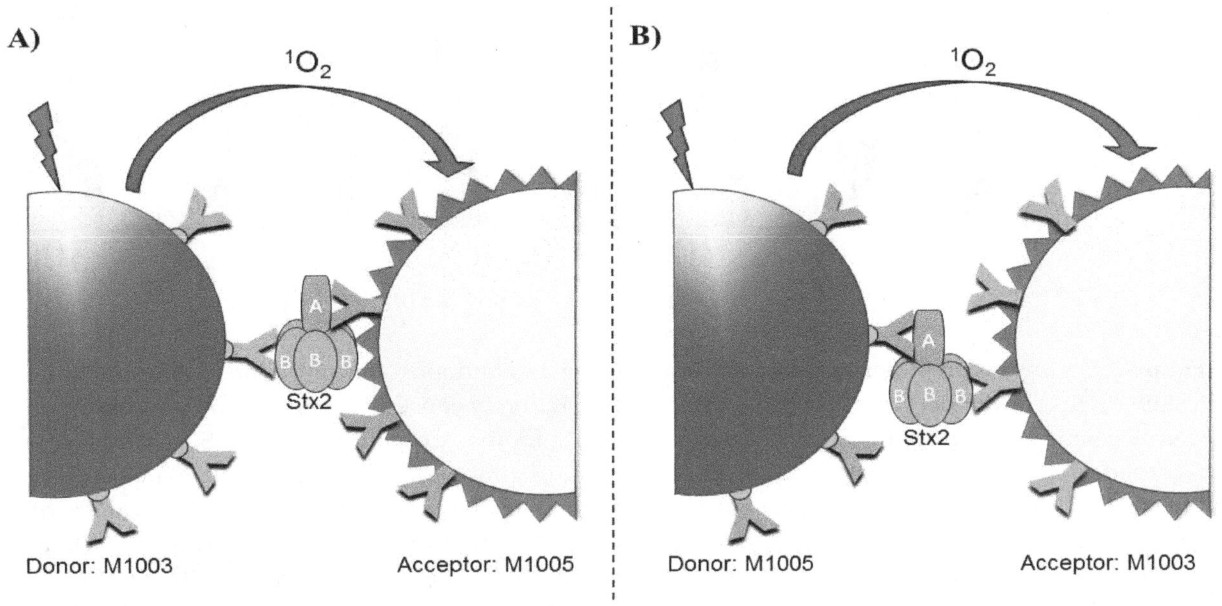

Figure 1. Schematic representation of the sandwich amplified luminescent proximity homogenous

assay-linked immunosorbent assay (AlphaLISA) for Shiga toxin detection. Two sets of monoclonal antibodies were employed in this assay (orange and gray Y's): one specific for the (**A**) subunit (M-1005) and other for the (**B**) subunit (M-1003) of Stx2. One set of antibodies was biotinylated (green hemisphere) to allow association with streptavidin-coated donor beads (blue sphere), while the other antibody set was directly conjugated to the acceptor beads (yellow sphere). When Stx2 is present, the donor and acceptor beads are colocalized through binding of the antibodies to the different subunits of the common antigen. Upon excitation by an Alpha laser at 680 nm (lightning bolt), the donor beads emit singlet oxygen molecules that react with the acceptor beads. Ultimately, energy transfer results in the emission of light by the acceptor beads, thus creating a detectable fluorescence signal at 615 nm. The following two formats were analyzed during assay development: (**A**) Antibody M-1003 was biotinylated and thus associated with the donor beads while M-1005 was conjugated to the acceptor beads and (**B**) antibody M-1005 was biotinylated and thus associated with the donor beads while M-1003 was conjugated to the acceptor beads.

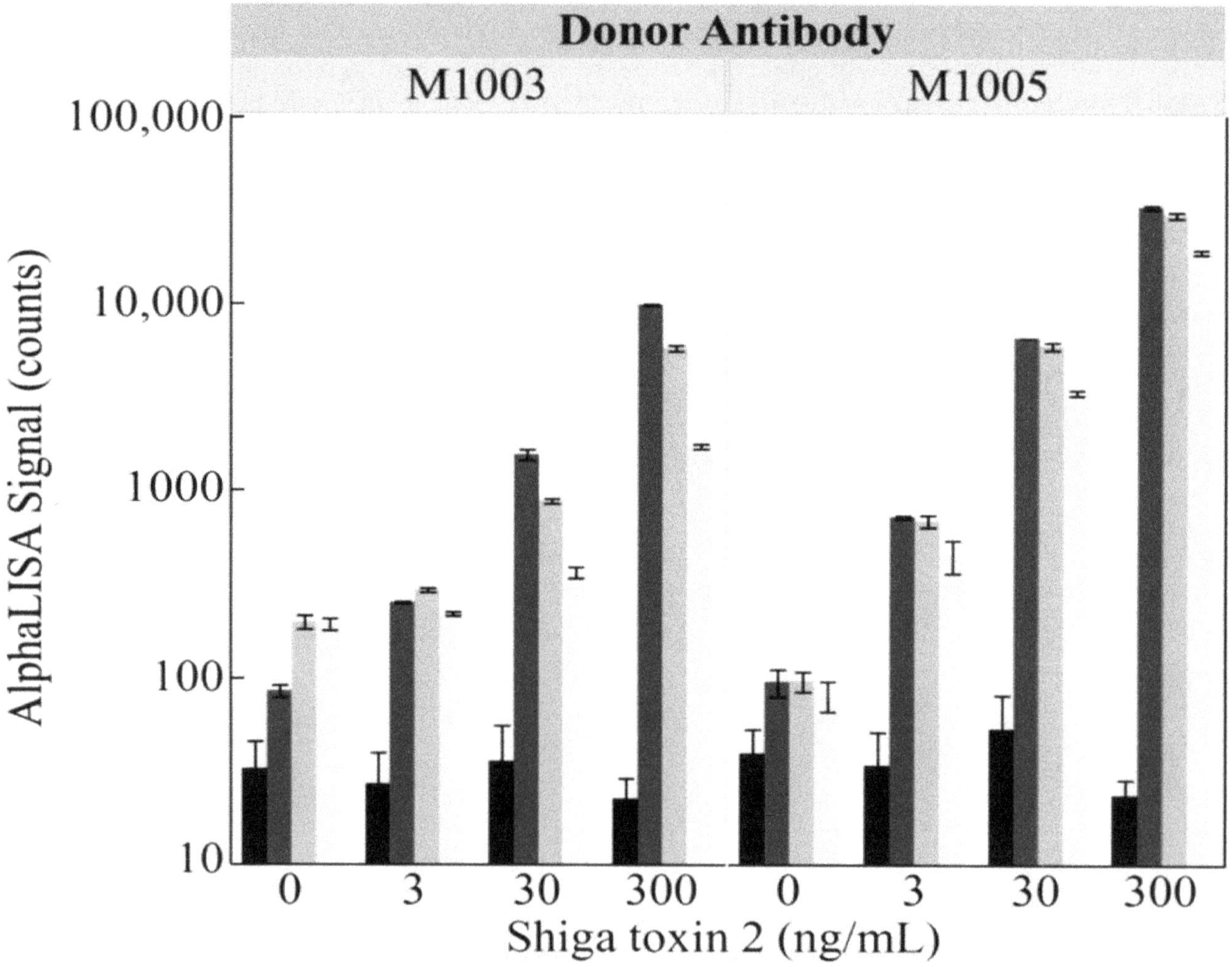

Figure 2. Optimization of donor/acceptor bead pairing and antibody concentration. A multifactorial test involving the donor/acceptor bead pairing, donor antibody concentration, and Stx2 concentration was performed to identify the optimal assay parameters for the AlphaLISA. AlphaLISA signals were recorded for the following factors: (a) bead pairings consisting of donor M-1003 with 25 μg acceptor M-1005 (left panel) versus donor M-1005 with 25 μg acceptor M-1003 (right panel); (b) donor antibody concentrations of 0 (black bars), 0.3 (dark gray bars), 1.0 (medium gray bars), and 3.0 nM (light gray bars); and (c) Stx2 concentrations of 0, 3 ng/mL, 30 ng/mL, and 300 ng/mL. Emitted light was measured on a BioTek Cytation 5 in Alpha mode at a gain setting of 100. Bars represent the mean values (n = 4) with error bars denoting the standard deviation.

2.2. Sensitivity of the AlphaLISA Compared to the Enzyme-Linked Immunosorbent Assay (ELISA)

Using the optimized parameters, purified Stx2 was then used to determine the sensitivity for the newly developed AlphaLISA. Varying concentrations of Stx2 (0, 0.5, 1, 3, and 10 ng/mL) were assayed using the AlphaLISA and the results recorded (Figure 3 black line). To benchmark the newly developed AlphaLISA against similar technologies, the same concentrations of Stx2 were also assayed using a commercially available ELISA (Figure 3 gray line). Six independent trials for each assay were conducted and the results were analyzed via independent Student's *t*-tests with alpha <0.05. The analysis demonstrated the ability of both the AlphaLISA and the ELISA to differentiate samples that do not contain Stx2 from samples containing 0.5, 1, 3, and 10 ng/mL of Stx2. In addition, independent *t*-tests (alpha < 0.05) demonstrated the ability of the assay to differentiate between all levels of Stx2 concentrations. Although the signal for both the AlphaLISA and the ELISA increased as the concentration of Stx2 increased, it is worth noting that the signal for the ELISA appeared to reach the maximum absorbance at 3 ng/mL of Stx2. (Note the maximum absorbance signal measurable by the instrument is 4.0.) The AlphaLISA, however, continued to show a marked increase in signal from 3 to 10 ng/mL.

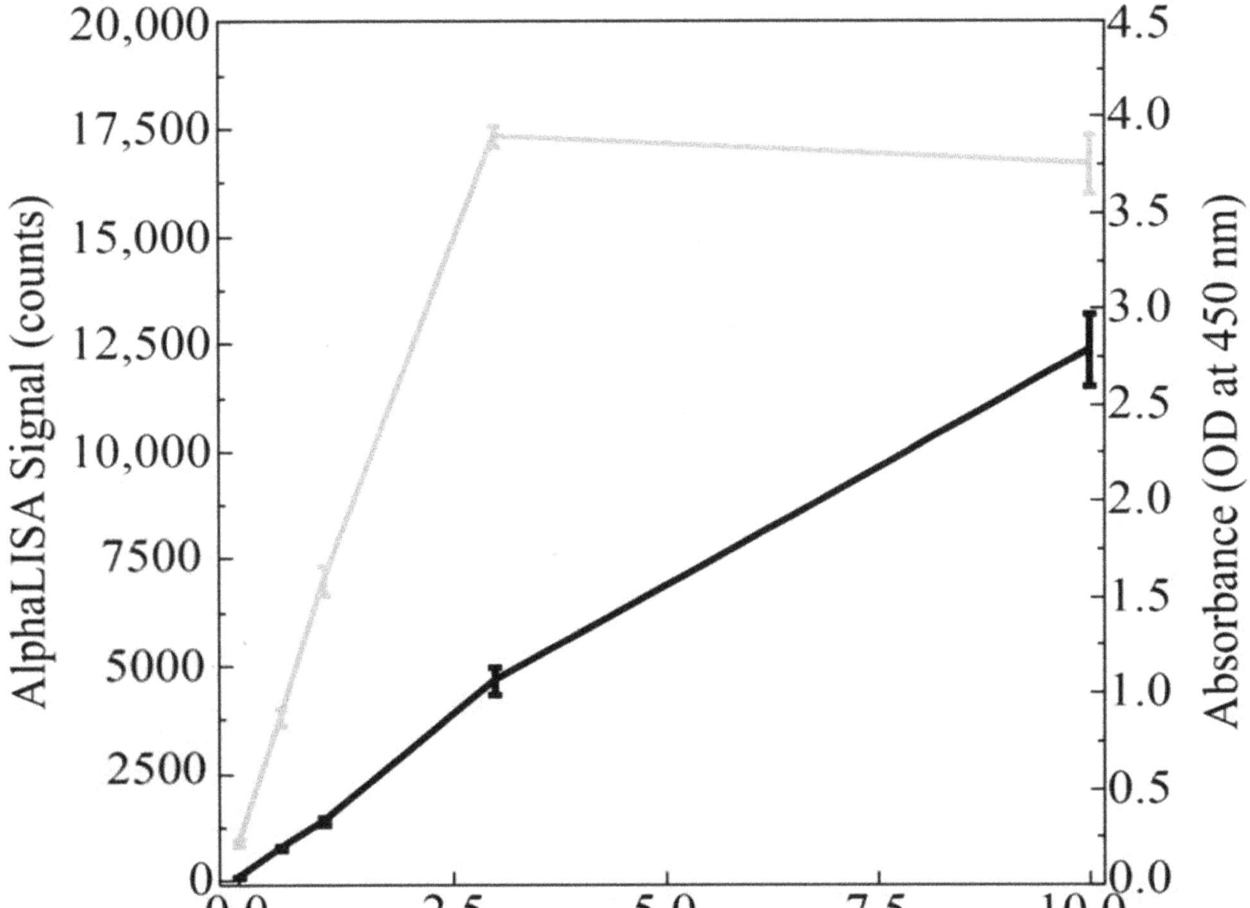

Figure 3. Detection of purified Stx2 using the AlphaLISA and the enzyme-linked immunosorbent assay (ELISA). Varying concentrations of purified Stx2 ranging from 0–10 ng/mL were analyzed using both the AlphaLISA (black line) and the ELISA (gray line). The response from both assays were quantified on a BioTek Cytation 5 with the emitted light from the AlphaLISA measured via the Alpha mode at a gain setting of 100 (left *y*-axis) and the absorbance values of the ELISA measured at 450 nm (right *y*-axis). Mean values from 6 trials were plotted with error bars denoting the standard deviation.

2.3. Detection of Stx2 in Food Matrices Using the AlphaLISA

In addition to the testing performed on purified toxin, the ability of the AlphaLISA to detect the presence of Stx2 in inoculated lettuce and ground beef samples was also examined to ascertain the functionality of the AlphaLISA in food matrices. These two food matrices were chosen because of their previous implication in STEC outbreaks [23–26]. For comparison purposes, lettuce inoculated with 9 CFU/mL of a Shiga toxin 2-producing *E. coli* O145 strain was tested by both AlphaLISA and ELISA for the presence of Stx2. During an overnight incubation, expression of Stx2 was induced in samples through the addition of mitomycin C. Samples consisting of 1× (undiluted) material, a 10× dilution, or a 100× dilution were then assayed (Figure 4A). Lettuce samples that had not been inoculated with STEC but were incubated in the same broth and under the same conditions were used as controls for this study. Because the output of the ELISA and the AlphaLISA differ, with the ELISA being a colorimetric assay measuring absorbance while the AlphaLISA yields emitted fluorescence, the numeric values are not directly comparable. However, both assays demonstrate a dose-dependent response. Both the AlphaLISA and ELISA were repeated 3 times in duplicate, and the data was analyzed using Student's *t*-tests, alpha = 0.05. In both the AlphaLISA and the ELISA, the 1× lysate dilution, the 10× lysate dilution and the control, were significantly different ($p < 0.0001$). However, the 100× lysate dilution could not be differentiated from the uninoculated control in either the AlphaLISA or the ELISA as $p = 0.73$ and 0.34, respectively. This implies that the AlphaLISA and the ELISA displayed similar Stx2 detection capabilities in lettuce.

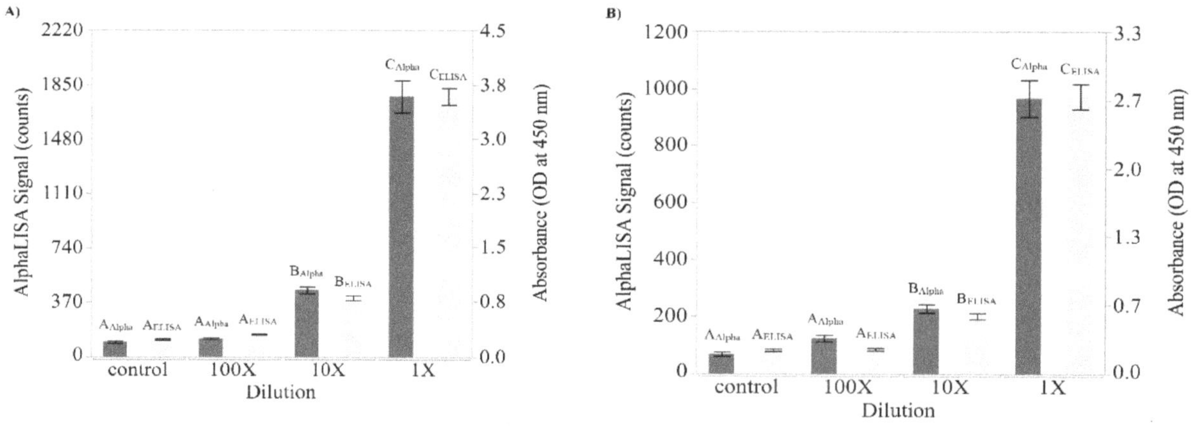

Figure 4. Detection of Stx2 in Shiga toxin-producing *Escherichia coli* (STEC)-inoculated foods using the AlphaLISA and ELISA. Stx2 production was induced using mitomycin C in STEC-inoculated (**A**) Romaine lettuce and (**B**) ground beef. Undiluted (1×) or diluted (10× and 100×) samples were subsequently assayed for the presence of Stx2 using both the AlphaLISA (dark gray bars) and the ELISA (light gray bars). Control samples contained uninoculated lettuce and ground beef samples, respectively. Responses were quantified on a BioTek Cytation 5 with the emitted light from the AlphaLISA measured via Alpha mode at a gain setting of 100 (left *y*-axis) and the absorbance values of the ELISA measured at 450 nm (right *y*-axis). Mean values from 3 independent trials containing 2 replicates were plotted with error bars denoting the standard deviation. Significance between AlphaLISA signal or absorbance values are denoted by dissimilar letters as determined by a Student's *t*-test at a 95% confidence level.

Testing of the inoculated ground beef samples was performed in a similar fashion to the lettuce samples using both the AlphaLISA and the ELISA to validate the presence of Stx2 (Figure 4B). The resulting data was also similar with both the AlphaLISA and the ELISA demonstrating a significant difference between the 1× lysate dilution, the 10× lysate dilution and the control ($p < 0.0001$). However, the 100× lysate dilution could not be differentiated from the uninoculated control in either the AlphaLISA or the ELISA as $p = 0.18$ and 0.91, respectively. Taken together, the data from both the lettuce and the ground beef illustrate that the AlphaLISA is comparable to the ELISA in terms of sensitivity and the ability to detect the presence of Stx2 in inoculated food matrices.

2.4. AlphaLISA Sensitivity Using Increased Reagent Volumes

In an attempt to enhance the sensitivity of the AlphaLISA, a series of experiments were conducted using increased reagent volumes for the AlphaLISA (Figure 5). The premise behind these experiments was that the small sample volume of the AlphaLISA (5 μL) could be the limiting factor with regards to the sensitivity of the overall assay. Therefore, by simply increasing all assay components, the limit of detection may also increase accordingly. The ability of the AlphaLISA to detect 0, 0.5, 1.0, 3.0, or 10.0 ng/mL of purified Stx2 in the presence of a food matrix (lettuce or ground beef) using either the original assay components (1×), doubling the assay components (2×), or tripling the assay components (3×) was determined. From this, it was shown that when the assay components were increased from 1× to 2× that a corresponding increase in AlphaLISA signal was also detected in both lettuce and ground beef. However, the results were not consistent when assay components were increased to 3×. When the components of the assay were tripled, the AlphaLISA signal was higher than that seen for the 1× but not the 2× amounts in the lettuce; whereas this level resulted in the lowest signal in ground beef when compared to both the 1× and 2× assays.

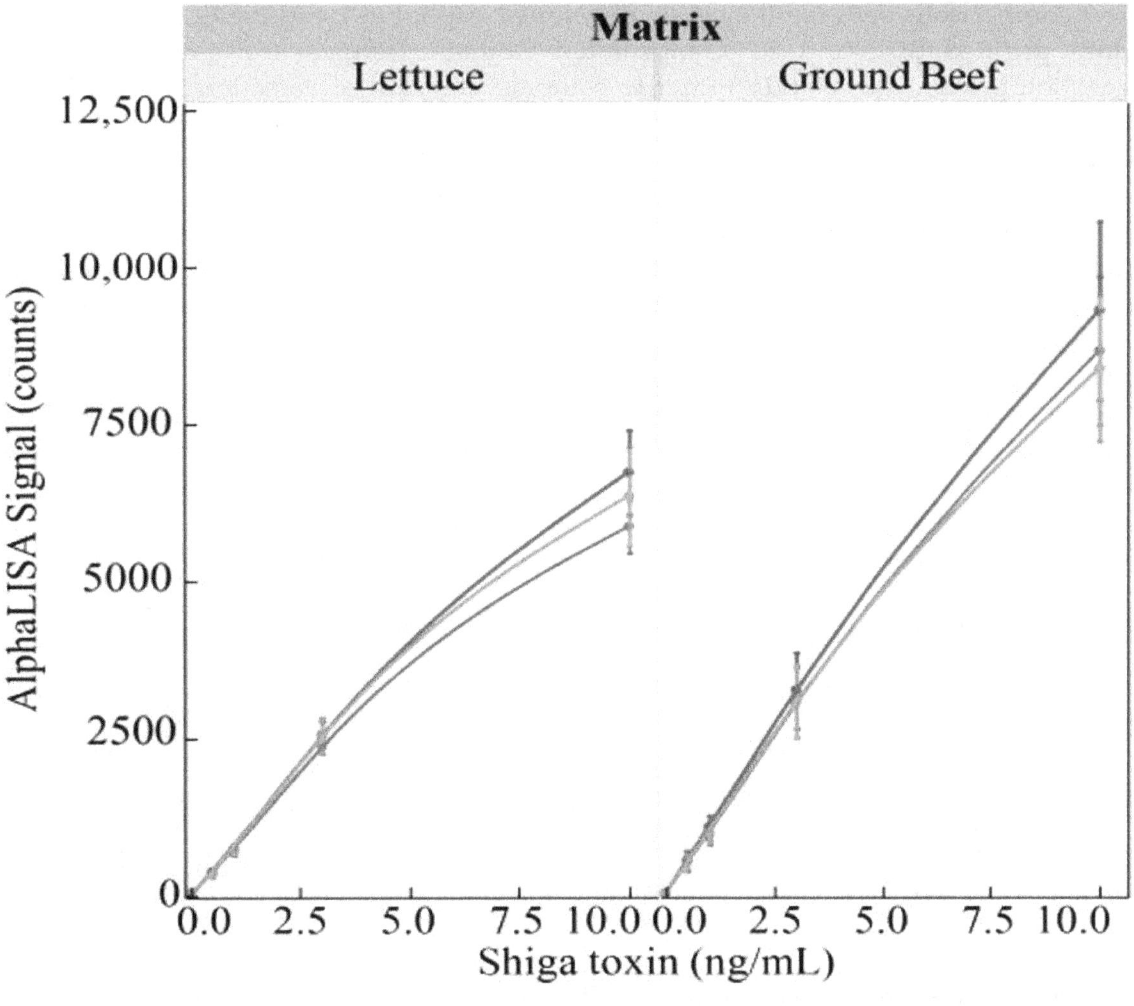

Figure 5. The effect of increasing reaction components on the sensitivity of the AlphaLISA for Stx2 in food matrices. Purified Stx2 (0, 0.5, 1.0, 3.0 or 5.0 ng/mL) was added to uninoculated lettuce (**left**) panel or ground beef (**right**) panel and the effect of increasing all of the reaction components (1×-blue line, 2×-red line, or 3×-green line) in the AlphaLISA signal was determined. Responses were quantified on a BioTek Cytation 5 with the emitted light from the AlphaLISA measured via Alpha mode at a gain setting of 100. Mean values from 2 independent trials, each containing 2 replicates are plotted with error bars denoting the standard deviation.

3. Discussion

AlphaLISA is a versatile technology that employs oxygen-channeling chemistry [27,28] and has been used for the detection of a wide variety of analytes from proteins to peptides to other small molecules. Currently, hundreds of different AlphaLISA biomarker detection kits are commercially available ranging in application from medicine (for tracing unwanted host cell proteins during industrial scale production of biotherapeutics), to agriculture (for the detection of aflatoxins in foods and animal feed), to basic research (as an alternative to the electrophoretic mobility shift assay for the detection of DNA-protein interactions). The AlphaLISA is considered a "no wash" alternative to the ELISA because it does not require any wash or separation steps [22]. The reagents are simply overlaid in succession, which greatly simplifies the protocol and makes the AlphaLISA highly amendable to high-throughput automated screening. Because of this, the AlphaLISA would be the superior assay when a considerable number of samples are to be analyzed under a standard operating procedure, such as is the case with the assessment of food quality and safety around the globe. Here, an AlphaLISA was developed to identify the presence of foodborne pathogens through the detection of Shiga toxin 2 in food samples.

Direct comparisons between the newly developed AlphaLISA and the ELISA were performed using both purified Stx2 and STEC-inoculated Romaine lettuce and ground beef samples. The suggestion provided by Perkin Elmer to estimate the limit of detection is to add 3 standard deviations to the mean of the "zero analyte" condition. Using this suggestion and the data presented in Figure 2, Table 1 was generated. Although the limits of detection were only slightly better for the AlphaLISA compared to the ELISA in terms of toxin concentration (0.10 ng/mL versus 0.14 ng/mL, respectively), when one considers that the ELISA employs a 100 μL sample while the AlphaLISA employs a 5 μL sample, quantitation of the amount of toxin on a per weight basis demonstrated that the AlphaLISA is the superior assay (0.5 pg for the AlphaLISA versus 14.3 pg for the ELISA). Interestingly, an attempt was made to increase the sensitivity of the AlphaLISA by doubling and tripling the components in the reaction including the amount of sample per assay (Figure 5). Although a corresponding increase in signal was detected when the reaction components were doubled (1× to 2× comparison), a tripling of the reaction components showed a decrease in signal intensity (2× to 3× comparison). The reason behind this deleterious effect with regard to signal production upon tripling of the reaction components is unknown and was not further pursued since the mean signal produced by the 1×, 2×, and 3× component mixtures were not statistically different.

Table 1. Comparisons of the limit of detection and signal-to-noise ratios of the AlphaLISA versus the ELISA.

Assay	Average Zero Signal	3× SD	LOD (ng/mL)	Signal-to-Noise (S:N) [a]			
				0.5 ng/mL	1 ng/mL	3 ng/mL	10 ng/mL
AlphaLISA	75.92	63.285	0.100	10	18	61	163
ELISA	0.21767	0.1665	0.143	4	7	18	17

SD = standard deviation, LOD = limit of detection. [a] Signal-to-noise ratios were calculated by dividing the signal mean of the stated assay upon the addition of either 0.5, 1, 3, or 10 ng/mL purified Stx2 to phosphate buffered saline (PBS) by the signal mean in samples with no Stx added to PBS.

The AlphaLISA is also superior because ELISAs typically have a narrower dynamic range, thus requiring the testing of multiple sample dilutions to accurately measure antigen concentration. As can be seen from Figure 3, samples within the range of 0.5 ng/mL to 3 ng/mL were detectable by the ELISA while samples containing 10 ng/mL of Stx2 fell outside of the absorbance range of the Cytation 5 microtiter plate reader, and thus could not be accurately measured unless diluted. Conversely, the AlphaLISA was able to measure concentrations within the 10 ng/mL range and possibly higher without dilution given the fact that AlphaLISA counts of ~1,600,000 have been recorded at higher gains using the Cytation 5 by Biotek (data not shown). Given that the saturation point was reached for ELISA but not for the AlphaLISA suggests that the dynamic range of the AlphaLISA is greater than that of the ELISA. While the observation of a broader dynamic range is consistent with what has been previously reported [22], larger scale studies would need to be conducted to precisely define the dynamic range of the assays. Regardless, the signal-to-noise ratio for the AlphaLISA was much larger than that of the ELISA (Table 1), indicating a better specification since there is more desired signal compared to background noise.

Antibody selection is crucial for the development of a functional AlphaLISA, including the fact that the specificity of the antibodies ultimately determines the accuracy of selection for the assay. The design presented here is specific for Stx2 (the more toxic of the two Stx types [11]) because it utilizes two different monoclonal antibodies with the donor antibody M-1005 specifically recognizing the Stx2 A-subunit and the acceptor M-1003 specifically recognizing the Stx2 B-subunit. Although the determination of which antibody was selected as the donor compared to the acceptor was done empirically, it has been hypothesized that this configuration was the most robust based on the following. First, the selection of two different antibodies ensures that the binding reaction can proceed without interference from the antibodies competing for binding sites, and allows the binding reaction to be more efficient compared to assays which utilize either the same antibody or antibodies that recognize different peptides within the same protein. Second, the AlphaLISA signaling cascade allows a single donor to activate multiple acceptors, and thus the creation of a more intense output signal compared to that of only a single acceptor in the presence of multiple donors. Because Stx2 consists of an A subunit noncovalently joined to a pentamer of identical B subunits, there are theoretically five times the number of acceptor particles surrounding every donor particle in the assay. Taken together, the configuration shown in Figure 1B would likely emit the most intense signal because the single A subunit donor could emit to multiple B subunit acceptors, which corresponded to the signal intensity observed experimentally. This result was analogous to what was seen for the detection of *Bacillus anthracis* spores via an AlphaLISA [29].

Ultimately, we developed a novel AlphaLISA for the detection of Stx2 and demonstrated its ability to identify the presence of Stx2 not only in phosphate buffered saline (PBS) but in two food matrices as well. Because the AlphaLISA can be performed in ~1.5 h (compared to the ~2.5 h needed for the ELISA), does not require the plate manipulation needed to wash the sample wells like the ELISA, and appears to have a larger dynamic range; the AlphaLISA is the superior method for the detection of Stx2 when large numbers of samples are to be queried. Other potential benefits may also be realized such as the fact that the AlphaLISA requires the toxin to be intact since the signal is dependent upon the A and B subunits being within a set distance to one another. This AlphaLISA might also be applicable for quantifying the amount of Stx2 in other matrices, indicating a use not only for detecting the presence of STECs in food as presented here but for use in clinical samples consisting of serum/stool [30,31] or for research investigating variables related to *stx* induction as well [32].

4. Materials and Methods

4.1. Preparation of Antibodies

Antibodies known to detect a majority of the Stx2 subtypes were selected from purified Abraxis (Warminster, PA, USA) in-house stocks stored in PBS (Abraxis LLC). The monoclonal antibody M-1003 binds the Shiga toxin B-subunits, while the monoclonal antibody M-1005 binds the Shiga toxin A-subunit. These two antibody stocks were also biotinylated using the Thermo Fisher Scientific (Waltham, MA, USA) EZ-Link Sulfo-NHS-LC-LC-Biotin (prod. #21338) according to manufacturer instructions, with a one-hour incubation at room temperature followed by removal of excess biotin with a Zeba Spin Desalting Column (Thermo Fisher Scientific prod. #89893). Biotinylated antibodies were stored in PBS at −20 °C. Antibody concentrations for both the biotinylated and stock antibodies were measured at 280 nm using a DeNovix microspectrophotometer (DeNovix Inc., Wilmington, DE, USA).

4.2. Preparation of AlphaLISA Donor and Acceptor Beads

Uncoupled AlphaLISA donor and acceptor beads were purchased from PerkinElmer, Inc. (Waltham, MA, USA; acceptor prod. #6772001 and donor prod. #6760002S). The streptavidin-coated donor beads did not need additional modification; however, the acceptor beads required antibody coupling prior to the start of the AlphaLISA assay. Two sets of acceptor beads were prepared (M-1003 and M-1005) using a modified version of the procedures found in the PerkinElmer "ELISA to Alpha Immunoassay Conversion Guide," section "Protocol for Direct Conjugation of an Antibody to an AlphaLISA Acceptor Bead" [33]. Briefly, two sets of 25 μL of 20 mg/mL acceptor beads (equaling 0.5 mg each) were washed with 25 μL of PBS in 1.5 mL microfuge tubes, centrifuged at 16,000× *g* for 15 min (Eppendorf centrifuge), and the supernatants discarded. To each set of 0.5 mg beads, 0.05 mg of either M-1003 or M-1005 antibody (non-biotinylated), 0.625 μL of 10% Tween-20 (Sigma-Aldrich, St. Louis, MO, USA; prod. #P7949), and 5 μL of 25 mg/mL $NaBH_3CN$ (Acros Organics, Thermo Fisher Scientific; prod #168550100) was added and brought to a final volume of 100 μL with 130 mM sodium phosphate (Sigma-Aldrich prod. #S0876/0). The bead pellet was resuspended in this mixture by pipetting, and incubated statically for 18 h at 37 °C. Blocking of unreacted sites was then performed by adding 5 μL of 65 mg/mL carboxy-methoxylamine (CMO) (Sigma-Aldrich prod. #C13408) in 0.8 M sodium hydroxide (Alfa Aesar, Haverhill, MA, USA; prod. #35631) to the previous reactions, and incubating for 1 h at 37 °C with an occasional gentle vortex. Finally, the beads underwent a series of wash steps as follows. The two conjugated, blocked bead solutions were centrifuged at 16,000× *g* for 15 min and the supernatants were discarded. The pellets were resuspended in 100 μL of 0.1 M Tris-HCl pH 8.0 (Sigma-Aldrich prod. #T3253) and centrifuged again at 16,000× *g* for 15 min. The previous step was repeated twice more, and the beads were resuspended by pipetting up and down in 100 μL of PBS with 0.05% ProClin-300 (Sigma-Aldrich prod. #48914) added as a preservative, for a final concentration of 5 mg/mL. The two separate antibody-conjugated acceptor bead solutions were vortexed and sonicated with twenty 1-s pulses in a water bath sonicator and stored at 4 °C in the dark.

4.3. Preparation of Shiga Toxin 2 Standards

Shiga toxin 2 was purchased from Toxin Technology (Sarasota, FL, USA, prod. STX-2), at 0.5 mg/mL, and diluted to 3, 30, and 300 ng/mL in AlphaLISA buffer (25 mM HEPES pH 7.4 (VWR International, Radnor, PA, USA; prod. #0511) containing the following from Sigma-Aldrich 0.5% Triton-X 100, 0.1% casein (prod. #C-5890), 1 mg/mL dextran (prod. #31390), and 0.05% ProClin-300prod. #48914).

4.4. AlphaLISA Antibody Pair and Concentration Optimization

Once the reagents (biotinylated antibodies, antibody-conjugated acceptor beads, and donor beads) and Shiga toxin 2 standards were prepared, AlphaLISA optimization began. Stx2 standards at 0, 3, 30, and 300 ng/mL in AlphaLISA buffer were evaluated to determine optimal antibody pairing and titer level. As per the recommended PerkinElmer procedures in the ELISA to Alpha Immunoassay Conversion Guide, 5 μL of Stx2 standards were added to 1/2-area plate (Perkin Elmer prod. #6005560) wells. Then, 10 μL of biotinylated M-1003 or M-1005 antibody (diluted to a final concentration of 0, 0.3, 1, or 3 nM in a 50 μL total reaction volume) was added, followed by 10 μL of M-1003 or M-1005 antibody-conjugated acceptor beads (diluted to a final concentration of 10 μg/mL in 50 μL total reaction volume). M-1003-conjugated acceptor beads were only added to wells that had biotinylated-M-1005 added, and M-1005-conjuated acceptor beads were only added to wells that had biotinylated-M-1003 added in order to have complementary antibody pairs instead of competing pairs. The plate was incubated at room temperature for 1 h. Next, 25 μL of streptavidin-coated donor beads were added (for a final concentration of 40 μg/mL in 50 μL total reaction volume) under low light conditions to prevent photobleaching of the beads. This results in a final 50 μL total reaction volume. The plate was incubated in the dark at room temperature for 30 min. Total in-assay workflow time was approximately 1 h and 45 min. Microtiter plates were then read using a BioTek Cytation 5 (BioTek Instruments, Inc., Winooski, VT, USA) in Alpha measurement mode with a gain setting of 100.

4.5. Analysis of Purified Shiga Toxin 2

Shiga toxin 2 standards were prepared as described above. Samples consisting of 5 μL of the 0, 0.5, 3, 30, and 300 ng/mL Stx2 standards were assayed via the AlphaLISA using only the 0.3 nM of biotinylated M-1005 antibody for association with the donor and the acceptor-conjugated M-1003 antibody as described above. ELISA testing was conducted simultaneously using the Abraxis Shiga Toxin 2 ELISA kit (Abraxis prod. #542010) using 100 μL of the 0, 0.5, 3, 30, and 300 ng/mL Stx2 standards following the manufacturer's instructions. Absorbencies for the ELISA were also measured on the BioTek Cytation 5 at 450 nm five min after the addition of stop solution.

4.6. Preparation of Lettuce and Ground Beef Samples

Lettuce and ground beef samples were prepared following the procedures described previously [15]. Briefly, Romaine lettuce was purchased from a local grocery store, the outermost leaves were removed, while the remaining leaves were chopped into ~1 cm^2 pieces on a 70% ethanol sterilized surface. Ground beef was also purchased from a local grocery store, separated into 25 g quantities, and kept frozen until used. STEC inoculum was prepared from a frozen *E. coli* O145:H28 bacterial stock by growing a small chip of frozen stock in Tryptic Soy Broth (TSB) (Becton Dickinson, East Rutherford, NJ, USA; prod. #211825) overnight at 37 °C shaking at 120 rpm in a 15 mL Corning Falcon tube. The following day, the overnight culture was serially diluted in glass dilution tubes with peptone water (Becton Dickinson prod. #218105), and stored at 4 °C. The number of colony forming units (CFU) present were determined by plating 1 mL of each dilution onto a Tryptic Soy Agar (TSA) plate (Becton Dickinson prod. #236950), which was incubated at 37 °C overnight. Inoculum levels were 9 and 10 CFU/mL for the lettuce and beef samples, respectively.

Samples were incubated in broth at a 1:4 ratio as follows. Lettuce or ground beef partitioned into 25 g samples was placed into filter Stomacher bags (Seward Laboratory Systems Inc., Islandia, NY, USA). One milliliter of ~10 CFU/mL STEC inoculum was added to the samples, or 1 mL of peptone water for the uninoculated negative control samples. Next, 75 mL of modified mTSB (TSB with 10 g/L Casamino Acids (Neogen Corp., Lansing, MI, USA; prod.# 7229A) and 66.7 ng/L mitomycin C (Sigma-Aldrich #M-0503) antibiotic toxin inducer), pre-warmed to 42 °C, was measured into all samples and hand-massaged to mix well and break up any clumps. Samples were subsequently incubated overnight (16 h) statically at 42 °C. The following day, 10 mL samples were removed from the outer portion of the Stomacher bags and stored at −80 °C in 15 mL Falcon tubes till use. Dilutions of lettuce and ground beef samples were made in AlphaLISA buffer in a similar fashion to that of the Shiga toxin 2 standards.

4.7. AlphaLISA and ELISA Comparison Food Matrices

Testing of the inoculated lettuce and ground beef samples described above was performed as follows. For the AlphaLISA, 5 μL of either inoculated or uninoculated control samples were added to Perkin Elmer 1/2-area plates, and tests were performed as detailed in the AlphaLISA optimization section above using the optimized conditions of 0.3 nM biotinylated-M-1005 antibodies with M-1003-conjugated acceptor beads. ELISA testing was conducted simultaneously using the Abraxis Shiga Toxin 2 ELISA kit using 100 μL of sample and following the manufacturer's instructions. Microtiter plates were read on a BioTek Cytation 5 for either the fluorescence signal using the Alpha mode (gain setting of 100) for the AlphaLISA or for absorbance at 450 nm five minutes after the addition of stop solution for the ELISA.

4.8. AlphaLISA Using 1×, 2× and 3× Components

Stx2 samples containing 0, 0.5, 1.0, 3.0, and 10.0 ng/mL of the Stx2 standards were prepared in uninoculated lettuce as described above. From there 5, 10 or 15 μL of the Stx2 containing lettuce was assayed using 10, 20, or 30 μL of prepared biotinylated antibody and 10, 20, or 30 μL of the prepared acceptor-conjugated beads for the 1×, 2×, and 3× experiments, respectively. (The AlphaLISA beads for all assays consisted of the 0.3 nM of biotinylated M-1005 antibody for association with the donor- and the acceptor-conjugated M-1003 antibody.) The plate was incubated at room temperature for 1 h before 25, 50, or 75 μL of streptavidin-coated donor beads were added under low light conditions resulting in total reaction volumes of 50 μL (1×), 100 μL (2×), or 150 μL (3×). Microtiter plates were incubated as before with the resulting signal quantified on the BioTek Cytation 5 as previously described. Identical assays were performed except uninoculated ground beef was used as the matrix instead of lettuce for these studies.

Author Contributions: A.G.G., C.M.A., F.M.R., J.A.C. and L.E.R. conceived and designed the experiments; A.G.G., L.E.R. and T.P.S.J. performed the experiments; A.G.G., C.M.A., J.A.C. and L.E.R. analyzed the data; F.M.R. contributed reagents/materials/analysis tools; C.M.A. and L.E.R. wrote the paper; and A.G.G., C.M.A., F.M.R., J.A.C., L.E.R. and T.P.S.J. edited the manuscript.

References

1. Griffin, P.M. *Escherichia coli* O157:H7 and other enterohemorrhagic *Escherichia coli*. In *Infections of the Gastrointestinal Tract*; Blaser, M.J., Smith, P.D., Ravdin, J.I., Greenberg, H.B., Guerrant, R.L., Eds.; Raven Press Ltd.: New York, NY, USA, 1995.
2. Scallan, E.; Hoekstra, R.M.; Angulo, F.J.; Tauxe, R.V.; Widdowson, M.A.; Roy, S.L.; Jones, J.L.; Griffin, P.M. Foodborne illness acquired in the United States-major pathogens. *Emerg. Infect. Dis.* **2011**, *17*, 7–15. [CrossRef] [PubMed]
3. Hoffmann, S.; Batz, M.B.; Morris, J.G. Annual cost of illness and quality-adjusted life year losses in the United States due to 14 foodborne pathogens. *J. Food Protect.* **2012**, *75*, 1292–1302. [CrossRef] [PubMed]
4. Smith, K.E.; Wilker, P.R.; Reiter, P.L.; Hedican, E.B.; Bender, J.B.; Hedberg, C.W. Antibiotic treatment of Escherichia coli O157 infection and the risk of hemolytic uremic syndrome, Minnesota. *Pediatr. Infect. Dis. J.* **2012**, *31*, 37–41. [CrossRef] [PubMed]
5. Wong, C.S.; Jelacic, S.; Habeeb, R.L.; Watkins, S.L.; Tarr, P.I. The risk of the hemolytic-uremic syndrome after antibiotic treatment of Escherichia coli O157:H7 infections. *New Engl. J. Med.* **2000**, *342*, 1930–1936. [CrossRef] [PubMed]
6. Tarr, P.I.; Gordon, C.A.; Chandler, W.L. Shiga-toxin-producing *Escherichia coli* and haemolytic uraemic syndrome. *Lancet* **2005**, *365*, 1073–1086. [CrossRef]
7. Allison, H.E. Stx-phages: Drivers and mediators of the evolution of STEC and STEC-like pathogens. *Future Microbiol.* **2007**, *2*, 165–174. [CrossRef] [PubMed]
8. Los, J.M.; Los, M.; Wegrzyn, G. Bacteriophages carrying Shiga toxin genes: Genomic variations, detection and potential treatment of pathogenic bacteria. *Future Microbiol.* **2011**, *6*, 909–924. [CrossRef] [PubMed]
9. Obrien, A.D.; Holmes, R.K. Shiga and Shiga-like toxins. *Microbiol. Rev.* **1987**, *51*, 206–220.
10. Law, D. Virulence factors of *Escherichia coli* O157 and other Shiga toxin-producing *E. coli*. *J. Appl. Microbiol.* **2000**, *88*, 729–745. [CrossRef] [PubMed]
11. Tesh, V.L.; Burris, J.A.; Owens, J.W.; Gordon, V.M.; Wadolkowski, E.A.; Obrien, A.D.; Samuel, J.E. Comparison of the relative toxicities of Shiga-like toxins type-I and type-II for mice. *Infect. Immun.* **1993**, *61*, 3392–3402. [PubMed]
12. Scheutz, F.; Teel, L.D.; Beutin, L.; Pierard, D.; Buvens, G.; Karch, H.; Mellmann, A.; Caprioli, A.; Tozzoli, R.; Morabito, S.; et al. Multicenter evaluation of a sequence-based protocol for subtyping Shiga toxins and standardizing Stx nomenclature. *J. Clin. Microbiol.* **2012**, *50*, 2951–2963. [CrossRef] [PubMed]
13. Anonymous. Detection and Isolation of Non-O157 Shiga Toxin-Producing *Escherichia coli* (STEC) from Meat Products and Carcass and Environmental Sponges. Available online: https://www.fsis.usda.gov/wps/wcm/connect/7ffc02b5-3d33-4a79-b50c-81f208893204/MLG-5B.pdf?MOD=AJPERES (accessed on 10 August 2018).
14. Anonymous. Detection, Isolation and Identification of *Escherichia coli* O157:H7 from Meat Products and Carcass and Environmental Sponges. Available online: https://www.fsis.usda.gov/wps/wcm/connect/

51507fdb-dded-47f7-862d-ad80c3ee1738/MLG-5.pdf?MOD=AJPERES (accessed on 10 August 2018).
15. Gehring, A.G.; Fratamico, P.M.; Lee, J.; Ruth, L.E.; He, X.H.; He, Y.P.; Paoli, G.C.; Stanker, L.H.; Rubio, F.M. Evaluation of ELISA tests specific for Shiga toxin 1 and 2 in food and water samples. *Food Control* **2017**, *77*, 145–149. [CrossRef]
16. Kong, Q.; Patfield, S.; Skinner, C.; Stanker, L.H.; Gehring, A.; Fratamico, P.; Rubio, F.; Qi, W.; He, X. Validation of two new immunoassays for sensitive detection of a broad range of Shiga Toxins. *Austin Immunol.* **2016**, *1*, 1007.
17. Jung, B.Y.; Jung, S.C.; Kweon, C.H. Development of a rapid immunochromatographic strip for detection of *Escherichia coli* O157. *J. Food Protect.* **2005**, *68*, 2140–2143. [CrossRef]
18. Clotilde, L.M.; Bernard, C.T.; Hartman, G.L.; Lau, D.K.; Carter, J.M. Microbead-based immunoassay for simultaneous detection of Shiga toxins and isolation of *Escherichia coli* O157 in foods. *J. Food Prot.* **2011**, *74*, 373–379. [CrossRef] [PubMed]
19. DeCory, T.R.; Durst, R.A.; Zimmerman, S.J.; Garringer, L.A.; Paluca, G.; DeCory, H.H.; Montagna, R.A. Development of an immunomagnetic bead-immunoliposome fluorescence assay for rapid detection of *Escherichia coli* O157:H7 in aqueous samples and comparison of the assay with a standard microbiological method. *Appl. Environ. Microbiol.* **2005**, *71*, 1856–1864. [CrossRef] [PubMed]
20. Peppard, J.; Glickman, F.; He, Y.; Hu, S.I.; Doughty, J.; Goldberg, R. Development of a high-throughput screening assay for inhibitors of aggrecan cleavage using luminescent oxygen channeling (AlphaScreen (TM)). *J. Biomol. Screen* **2003**, *8*, 149–156. [CrossRef] [PubMed]
21. Roby, P.; Bosse, R.; Arcand, M. Multiplex Assay Methods and Compositions. U.S. Patent 8,486,719, 16 July 2013.
22. Eglen, R.M.; Reisine, T.; Roby, P.; Rouleau, N.; Illy, C.; Bosse, R.; Bielefeld, M. The use of AlphaScreen technology in HTS: Current status. *Curr. Chem. Genom.* **2008**, *1*, 2–10. [CrossRef] [PubMed]
23. Luna-Gierke, R.E.; Griffin, P.M.; Gould, L.H.; Herman, K.; Bopp, C.A.; Strockbine, N.; Mody, R.K. Outbreaks of non-O157 Shiga toxin-producing Escherichia coli infection: USA. *Epidemiol. Infect.* **2014**, *142*, 2270–2280. [CrossRef] [PubMed]
24. Robbins, A.; Anand, M.; Nicholas, D.C.; Egan, J.S.; Musser, K.A.; Giguere, S.; Prince, H.; Beaufait, H.E.; Sears, S.D.; Borda, J.; et al. Ground beef recall associated with Non-O157 Shiga toxin-producing Escherichia coli, United States. *Emerg. Infect. Dis.* **2014**, *20*, 165–167. [CrossRef] [PubMed]
25. Slayton, R.B.; Turabelidze, G.; Bennett, S.D.; Schwensohn, C.A.; Yaffee, A.Q.; Khan, F.; Butler, C.; Trees, E.; Ayers, T.L.; Davis, M.L.; et al. Outbreak of Shiga toxin-producing Escherichia coli (STEC) O157:H7 associated with romaine lettuce consumption, 2011. *PLoS ONE* **2013**, *8*, e55300. [CrossRef] [PubMed]
26. White, A.; Cronquist, A.; Bedrick, E.J.; Scallan, E. Food source prediction of Shiga toxin-producing Escherichia coli outbreaks using demographic and outbreak characteristics, United States, 1998–2014. *Foodborne Pathog. Dis.* **2016**, *13*, 527–534. [CrossRef] [PubMed]
27. Ullman, E.F.; Kirakossian, H.; Singh, S.; Wu, Z.P.; Irvin, B.R.; Pease, J.S.; Switchenko, A.C.; Irvine, J.D.; Dafforn, A.; Skold, C.N.; et al. Luminescent oxygen channeling immunoassay—Measurement of particle binding-kinetics by chemiluminescence. *Proc. Natl. Acad. Sci. USA* **1994**, *91*, 5426–5430. [CrossRef] [PubMed]
28. Ullman, E.F.; Kirakossian, H.; Switchenko, A.C.; Ishkanian, J.; Ericson, M.; Wartchow, C.A.; Pirio, M.; Pease, J.; Irvin, B.R.; Singh, S.; et al. Luminescent oxygen channeling assay (LOCI(TM)): Sensitive, broadly applicable homogeneous immunoassay method. *Clin. Chem.* **1996**, *42*, 1518–1526. [PubMed]
29. Mechaly, A.; Cohen, N.; Weiss, S.; Zahavy, E. A novel homogeneous immunoassay for anthrax detection based on the AlphaLISA method: Detection of *B. anthracis* spores and protective antigen (PA) in complex samples. *Anal. Bioanal. Chem.* **2013**, *405*, 3965–3972. [CrossRef] [PubMed]
30. Gould, L.H.; Bopp, C.; Strockbine, N.; Atkinson, R.; Baselski, V.; Body, B.; Carey, R.; Crandall, C.; Hurd, S.; Kaplan, R.; et al. Recommendations for diagnosis of shiga toxin—Producing *Escherichia coli* infections by clinical laboratories. *MMWR Recomm. Rep.* **2009**, *58*, 1–14. [CrossRef] [PubMed]
31. He, X.; Ardissino, G.; Patfield, S.; Cheng, L.W.; Silva, C.J.; Brigotti, M. An improved method for the sensitive detection of Shiga Toxin 2 in human serum. *Toxins* **2018**, *10*, 59. [CrossRef] [PubMed]
32. Marques, L.R.; Moore, M.A.; Wells, J.G.; Wachsmuth, I.K.; O'Brien, A.D. Production of Shiga-like toxin by *Escherichia coli*. *J. Infect. Dis.* **1986**, *154*, 338–341. [CrossRef] [PubMed]
33. PerkinElmer. ELISA to Alpha Immunoassay Conversion Guide. Available online: www.perkinelmer.com/lab-solutions/resources/docs/GDE_ELISAtoAlphaLISA.pdf (accessed on 7 September 2018).

Multi-Analyte MS based Investigation in Relation to the Illicit Treatment of Fish Products with Hydrogen Peroxide

Federica Dal Bello [1,*], Riccardo Aigotti [1], Michael Zorzi [1], Valerio Giaccone [2] and Claudio Medana [1]

[1] Molecular Biotechnology and Health Sciences Dept. Università degli Studi di Torino, Via Pietro Giuria 5, 10125 Torino, Italy; riccardo.aigotti@unito.it (R.A.); michael.zorzi@unito.it (M.Z.); claudio.medana@unito.it (C.M.)

[2] Department of Animal Medicine, Productions and Health, Università degli Studi di Padova, Viale dell'Università 16, 35020 Legnaro, PD, Italy; valerio.giaccone@unipd.it

* Correspondence: federica.dalbello@unito.it

Abstract: Fishery products are perishable due to the action of many enzymes, both endogenous and exogenous. The latter are produced by bacteria that may contaminate the products. When fishes age, there is a massive bacteria growth that causes the appearance of *off-flavor*. In order to obtain "false" freshness of fishery products, an illicit treatment with hydrogen peroxide is reported to be used. Residues of hydrogen peroxide in food may be of toxicology concern. We developed two mass spectrometry based methodologies to identify and quantify molecules related to the treatment of fishes with hydrogen peroxide. With ultra-high-performance liquid chromatography–mass spectrometry (UHPLC-MS) we evaluated the concentration of trimethylamine-N-oxide (TMAO), trimethylamine (TMA), dimethylamine (DMA), and cadaverine (CAD) in fish products. After evaluating LOQ, we measured and validated the lower limits of quantification (LLOQs as first levels of calibration curves) values of 50 (TMAO), 70 (TMA), 45 (DMA), and 40 (CAD) ng/mL. A high ratio between TMAO and TMA species indicated the freshness of the food. With a GC-MS method we confirmed the illicit treatment measuring the levels of H_2O_2 after an analytical reaction with anisole to give 2-hydroxyanisole as a marker. This latter product was detected in the headspace of the homogenized sample with simplification of the work-up. A LLOQ of 50 ng/mL was checked and validated. When fish products were whitened and refreshed with hydrogen peroxide, the detected amount of the product 2-hydroxyanisole could be very important, (larger than 100 mg/kg). The developed analytical methods were suitable to detect the illicit management of fishery products with hydrogen peroxide; they resulted as sensitive, selective, and robust.

Keywords: mass spectrometry methods; fishery product; hydrogen peroxide; illicit treatment

1. Introduction

Fishery products are defined by Regulation (EC) N 853/2004 of the European Parliament and the Council as "all seawater or freshwater animals (except for live bivalve mollusks, live echinoderms, live tunicates and live marine gastropods, and all mammals, reptiles and frogs) whether wild or farmed and including all edible forms, parts and products of such animals" [1]. The description includes all fishes (Osteichthyes, bony fishes, and Chondrichthyes, cartilaginous fishes), shellfish, and clams.

The fishery products are classified in four commercial categories: Fresh (no manipulation), prepared (any operations that affect the anatomical wholeness of the animal), frozen, and processed (any operation that transform the product, such as smoking, marinating, salting). The freshness of a fish product is the most important commercial quality factor for the consumer, because the safety is an essential prerequisite without which food cannot be placed on the market or further transformed. The Regulation (EC) N 2406/96 of the European Parliament and the Council defines four categories for fresh fish products (extra, a, b, and c); fishery products grouped in the last class must be judged as not suitable for human consumption and must be removed from the market [2].

Fishery products are perishable due to the action of many enzymes, both endogenous, located in the fish muscles, and exogenous, produced by intrinsic bacteria that are present and can contaminate in the products [3]. Consequently, the exponential growth of bacteria that triggers oxidative chemical reactions causes fading and opacification of the product, with appearance of *off-flavor*, that is an unpleased flavor caused by chemical lipid oxidation or non-protein nitrogen (NPN) degradation [3,4]. NPN in fishery products are distinguished in two structural categories: Volatile and nonvolatile compounds. Nonvolatile NPN compounds are mostly represented by heterocyclic metabolites while volatile ones own low molecular weight and are represented in fishery products mainly by ammonia, trimethylamine (TMA, C_3H_9N), and dimethylamine (DMA, C_2H_7N). TMA is an endogenous compound abundant in fishery products: It is a post-mortem product, deriving from trimethylamine-N-oxide (TMAO, C_3H_9NO) by an enzymatic activity [5–8].

TMAO is an amine oxide, less volatile, and less basic with respect to TMA, due to its oxidation. Under enzymatic activity, TMAO could generate several chemical compounds: DMA and formaldehyde (FA) from endogenous (aquatic environment) muscle bacteria (*Pseudomonas* and *Alteromonas*) activity, and TMA from exogenous (which accumulates in fish products after capture and is typical of the terrestrial environment) bacteria (*Salmonella, Vibrio*) activity [9]. Only a limited population of bacteria can cause deterioration of fishery products: They are called specific spoiling micro-organism (SSO) and are mostly Gram-negative micro-organisms. After fishing, SSO can contaminate fishery products on the surface and grow even at low temperatures [10–13]. DMA, FA, and TMA are products of enzymatic degradation by intrinsic bacteria together with biogenic amines (histamine, tyramine, phenylethylamine, putrescine, and cadaverine), thiols and hydrogen sulfide (H_2S). Moreover, in mollusks such as squid, the SSO enzymes can produce hypoxanthine and acetic acids salts, that contribute to the appearance of *off-flavor* [14,15].

The organo-nitrogen compounds derived from degradation of NPN and protein are quantifiable as a total volatile basic nitrogen TVB-N; for fresh fish products the TVB-N amount should be minor of 10 mg/kg, for aged ones the quantity ranges normally between 300–350 mg/kg (Regulation (EC) N 853/2004) [16–18]. Due to the unhealthiness, the *off-flavor* appearance, and color changes, aged fishery products must be retired from the market. However, some illicit treatments on these products might simulate a "false" freshness and one of these treatments is the use of hydrogen peroxide (H_2O_2). The illicit treatment with 0.5%–0.8% hydrogen peroxide water solutions is known and was reported in the literature [19]. Residues of hydrogen peroxide in food may be of toxicology concern.

Hydrogen peroxide is both an oxidant in aqueous solution with acidic pH and a reductant in alkaline water solution. When it is used as an illicit treatment for fishery products, the oxidizing properties are exploited. H_2O_2 can indeed convert TMA, a degradation product, to TMAO, the amine oxide naturally present in living fishes. TMAO is odorless and it has oxidizing properties giving to the fishery substrate high redox potential. This great redox potential is typical of muscle tissue of fresh fishery products, and decreases rapidly when TMAO is reduced to TMA by enzymatic activity [20–22].

If the TMAO amount is increased by H_2O_2 treatment, the proteins are stabilized [23]. For example, the amount of mucins, the main glycoproteins of mucus, is reduced on the fish skin after H_2O_2 treatment because of chemical degradation. The decrease in mucins concentration reduces viscosity and slows down the appearance of *off-flavor* [24].

Finally, the illicit treatment with H_2O_2 can cause a whitening and "re-freshing" effect on fishery products due to its oxidative properties (peroxidation of double bonds present in chromophores) [25,26].

The aim of the present study was the development of mass spectrometry (MS) based methods to evaluate the concentration of different compounds related to the illicit treatment of fish food based on the use of hydrogen peroxide (whitening and "re-freshing"). For this purpose, both liquid chromatography and gas chromatography hyphenated mass spectrometers were used. Mass spectrometry is recognized as one of the powerful and sensitive analytical techniques to identify, characterize, and quantify small molecules, such as amines and ethers. The use of this kind of MS based techniques is worthwhile in food analysis requiring complementary approaches to the detection of chemicals with different physical–chemical properties such as in the present work [27–29].

Two analytical methods were developed: The first one was a direct LC-MS/MS method for the determination of various amines and trimethylamine-oxide (TMAO); the second one was an indirect SPME-GC-MS method for the determination of residues of H_2O_2 on different fishery products matrices, by the hydroxylation reaction of anisole to 2-hydroxyanisole (guaiacol).

Another aim of this work was the application of the developed methods to investigate about H_2O_2 fish products treatment. The consequent alteration of the concentration ratio of TMAO and trimethylamine (TMA) which is a known fish product freshness parameter was evaluated.

2. Materials and Methods

All solvents and analytical standards of dimethylamine (DMA), trimethylamine (TMA), trimethylamine-N-oxide (TMAO), cadaverine (CAD), hydrogen peroxide solution, anisole, guaiacol, and p-xylene-d10 were purchased from Sigma Merck (Merck, Milan, MI, Italy). High-performance liquid chromatography (HPLC)-grade water was obtained from a MilliQ Academic water purification system (Millipore, Milan, Italy). Before use, solvents were filtered through a 0.45 μm filter and degassed for 10 min in an ultrasonic bath.

Fishery products samples were: Atlantic bonito (*Sarda sarda*) mackerel-like fish of the family Scombridae; European squid (*Loligo vulgaris*) belonging to the family Loliginidae. Atlantic bonito and European squid samples were purchased in a local market or caught fresh in Tyrrhenian Sea and brought in the laboratory no later than 5 h after fishing, in ice.

2.1. UHPLC-Tandem Mass Analysis of NPN

A Nexera (Shimadzu, Milan, MI, Italy) UHPLC (Ultra-high-performance liquid chromatography) coupled through an ESI source with a QTRAP-5500 tandem mass analyzer (Sciex, Milan, MI, Italy) was used to quantify the NPN in fishery samples.

The chromatographic separation was achieved using a RP C18 column (Kinetex EVO, 5 μm, 150 × 2.1 mm, Phenomenex, Castel Maggiore, BO, Italy) and heptafluoro-butanoic acid 10 mM in water (eluent A) and in methanol (eluent B). The gradient run started with 1% B, increased to 35% B in 8 min, up to 100% in 3 min, followed by reconditioning time. Flow rate and injection volume were 200 $\mu L\ min^{-1}$ and 5 μL, respectively.

The triple quadrupole was used in a MRM positive ion mode with the following source parameters: Curtain gas (arbitrary unit, arb), 25; spray voltage (V), 5500; gas1 (arb), 35; gas2 (arb), 40; capillary temperature (°C), 400. Nitrogen was used as curtain gas, gas 1, and gas 2. The MRM transitions, potentials, and collision energies were listed in Table 1.

Table 1. MRM (multiple reaction monitoring) parameters and voltages for tandem mass analysis in a positive ion mode of dimethylamine (DMA), trimethylamine (TMA), trimethylamine-N-oxide (TMAO), and cadaverine (CAD). DP: Declustering potential; EP: Entrance potential; CE: Collision energy; CXP: Collision exit potential.

Compound	Q1 *m/z*	Q3 *m/z*	DP (V)	EP (V)	CE (V)	CXP (V)
DMA	46	30	96	10	37	13.5
TMA	60	44	90	10	24	11
TMAO	76	58	90	12	22	15
CAD	103	86	60	10	13	23

The developed analytical method was validated evaluating selectivity, linearity, accuracy and precision, and lower limit of quantitation following the FDA guidelines [30].

2.2. Gas Chromatography–Mass Spectrometry Analysis of H_2O_2 Residue

For the GC-MS analysis we used a Varian Saturn 3900 (Agilent, Milan, MI, Italy) system, equipped with a 1177 injector. The separation column was a Zebron ZB-624 30 m, i.d. 0.25 mm (Phenomenex, Castel Maggiore, BO, Italy) applying a temperature gradient from 40 to 240 °C in 16 min. Injector temperature was 240 °C; split mode was employed and helium gas flow was 1.2 mL min^{-1}.

The mass spectrometry was a Varian Saturn 2100 T ion trap analyzer (Agilent, Milan, MI, Italy) equipped with an EI (electron ionization) source. The full mass acquisition range was from 40 to 500 *m/z*.

The developed methodology was based on the paper of Tanaka et al. [31] calibrating the modified procedure in order to measure H_2O_2 concentration in the range 0.05 to 1.00 μg/mL.

2.3. Sample Preparation for UHPLC/GC-MS Analysis of NPN

Samples of fish products muscles were weighted, minced, and extracted with pH 2.5, 0.1 M phosphate buffer: 40 mL of buffer were used for 8 g of fish sample; the suspensions were then centrifuged (2300 *g* for 10 min) and filtered (0.45 μm).

Some samples were treated with hydrogen peroxide to simulate the illicit treatment: After a total immersion of 8 g of fish products in H_2O_2 solution (0.8%) for 2 min, the liquid was removed and samples were rinsed with fresh water. Then, treated fishery products samples were extracted as just described.

For UHPLC-MS, 1 mL of extracted solution was diluted using the starting eluent mixture, placed in a vial, and analyzed in a MRM mode using the triple quadrupole in a positive ion mode.

For GC-MS, 2 mL of extracted solution were placed in a vial for headspace solid-phase microextraction (HS SPME) and added with 100 μL of potassium ferricyanide ($K_8Fe(CN)_6$) as catalyzer and 2 μL of anisole. The solution was heated at 60 °C in an oil bath for 1 h; the fiber for head space analysis was exposed for 30 min. A Supelco 75 μm Carboxen™-PDMS (polydimethylsiloxane) (Merck, Milan, MI, Italy) fiber was used. The extraction recovery of anisole was checked to be 80% by the use of p-xylene-d10 as an internal standard for 30 min fiber exposition.

3. Results

With the developed MS based methodologies we were able to quantify the amount of the amines dimethylamine (DMA), trimethylamine (TMA), trimethylamine-N-oxide (TMAO), and cadaverine (CAD) with the UHPLC-tandem mass analysis. The concentration of H_2O_2 residues was measured with the use of a HS SPME-GC-MS method.

3.1. Results of UHPLC-Tandem Mass Analysis of NPN

The chromatographic separation of the analyzed amines is shown in Figure 1. The ion pairing effect of the heptafluoro-butanoic acid present in the mobile phase allowed a valuable retention to obtain a satisfying separation of the analytes.

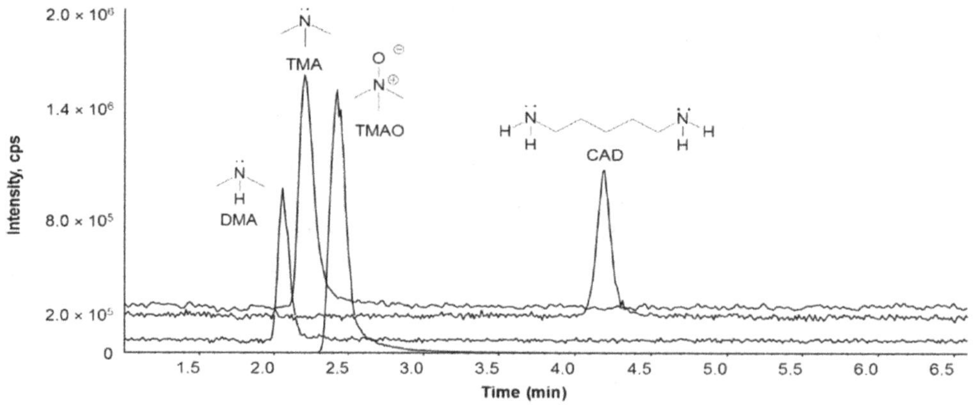

Figure 1. Chromatographic separation of dimethylamine (DMA, R_t 2.2 min), trimethylamine (TMA, R_t 2.25 min), trimethylamine-N-oxide (TMAO, R_t 2.5 min), and cadaverine (CAD R_t 4.3 min).

In order to quantify the amines in fresh fishery products samples or in fishery products subjected to an illicit treatment with H_2O_2, three calibration curves were prepared: (I) In pure pH 2.5, 0.1 M phosphate buffer, (II) In the extraction buffer fresh samples of Atlantic bonito, and (III) In the extraction buffer of fresh samples of European squid.

Each matrix material was weighted and extracted as previously described; since the matrix has a basal amount of amines, it was mandatory to prepare a matrix-blank without the addition of analyte (DMA, TMA, TMAO, and CAD) standards. Once obtained, the matrix-blank and the calibration curves were obtained by adding increasing amounts of amines as follows: 50, 100, 200, 400, 600, 800 ng/mL. A standard addition curve of TMAO in a real sample of European squid is shown in Figure 2. The curve had a positive intercept value because of the basal amounts of the analyte in fish.

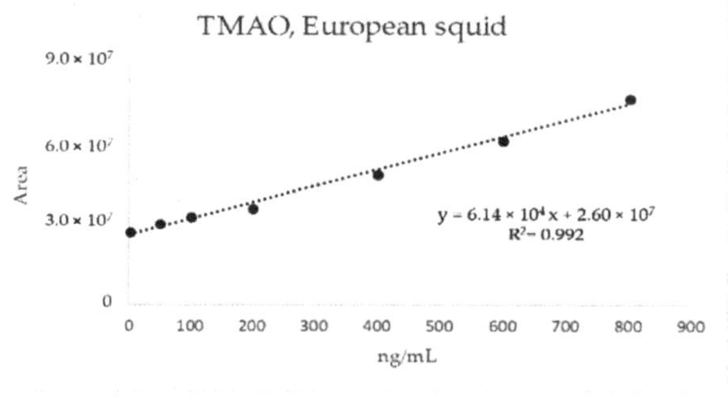

Figure 2. A standard addition curve of TMAO in a real sample of fresh European squid. The basal concentration of TMAO was 1150 ng/mL.

A full validation of the UHPLC-MS method for amines determination was performed. We followed the Food and Drug Administration (FDA) guidelines to evaluate the protocol [30] and used our previous work as an example for validation parameter definition [32]. Validation parameters were listed in Table 2 for all the UHPLC-MS analytes. LOD and LOQ were evaluated on the standard calibration curve by the signal to noise values of 3 and 10, respectively. LLOQ is the lower limit of quantification determined on the basis of a simple LOQ. To obtain LLOQs, LOQ standard solutions were prepared, used as the first level of each calibration curve, and validated by comparison of blank solutions. Validation parameters definition is given in the following paragraph.

To test the possible interferences at the analyte of interest's known retention time and *m/z*, a standard solution that had every analyte at a known concentration except for the one of interest, was analyzed. The selectivity % (Sel% = (Area at analyte retention time/Average area analyte in LLOQ) × 100) had to be ≤30%. The statistical parameter related to the linearity of calibration curve is the percentage difference (Diff% = (slope − average slope)/ average slope) × 100). Diff% had to be ≤25%.

Inaccuracy of lower limit of quantitation (LLOQ) had to be ≤20% and relative standard deviation % of accuracy of LLOQ ≤15%. Finally, to test recovery, matrices were spiked with a combined standard solution of amines at known concentration and processed as described in the Sample preparation section. The recovery (Rec = (Area at analyte retention time in spiked matrix/Average analyte area in spiked water solution) × 100) had to be between 85% and 120%. The validation parameters were respected in all cases.

Table 2. Validation parameters for calibration curves in (I) pH 2.5, 0.1 M phosphate buffer, (II) Atlantic bonito extraction solution, (III) European squid extraction solution. Diff% slope: Difference % of the slope of the calibration curve; RSD% LLOQ%: Relative standard deviation % of accuracy of lower limit of quantitation; BIAS% of LLOQ: Inaccuracy of lower limit of quantitation; LLOQ: Lower limit of quantitation.

Compound	Parameter	Matrices		
		(I)	(II)	(III)
DMA	Selectivity %	1.80	2.50	3.42
	Diff% slope	4.74	6.53	7.84
	RSD% LLOQ	14.7	15.0	14.9
	BIAS% LLOQ	19.0	20.3	19.5
	LLOQ (ng/mL)	25.0	45.0	45.0
	Recovery %	95.2	87.6	85.1
TMA	Selectivity %	0.50	1.32	1.97
	Diff% slope	3.60	5.19	6.41
	RSD% LLOQ	2.16	7.86	4.36
	BIAS% LLOQ	20.0	19.8	19.0
	LLOQ (ng/mL)	30.0	70.0	70.0
	Recovery %	93.6	89.3	86.0
TMAO	Selectivity %	0.05	0.90	0.85
	Diff% slope	4.07	14.6	8.2
	RSD% LLOQ	8.60	10.6	9.63
	BIAS% LLOQ	9.50	12.7	18.6
	LLOQ (ng/mL)	30.0	50.0	50.0
	Recovery %	102.5	91.4	90.7
CAD	Selectivity %	0.03	0.40	0.60
	Diff% slope	6.85	8.97	12.3
	RSD% LLOQ	9.80	15.3	14.3
	BIAS% LLOQ	16.0	17.6	19.0
	LLOQ (ng/mL)	20.0	40.0	40.0
	Recovery %	99.8	94.1	85.8

The UHPLC-tandem mass method was then applied to real fishery product samples of Atlantic bonito (*Sarda sarda)* and European squid (*Loligo vulgaris*): A) Freshly caught; B) freshly purchased in a local market; C) aged (four days at room temperature); and D) H_2O_2 treated (as described before).

The obtained results are summarized in Table 3. In freshly caught and freshly purchased samples the measured amount of amine was similar; in Table 3 the quantity of amines in freshly caught samples only was reported. High levels of TMAO were found in freshly caught Atlantic bonito (1700 ± 238 mg/kg) and European squid (1200 ± 336 mg/kg) samples. Conversely, in these samples the TMA amount was low, 170 ± 24 and 210 ± 29 mg/kg, respectively.

When fishery products initiated to degrade due to temperature (4 h at room temperature) or bacterial activities, the TMAO and TMA amounts reversed. Finally, when hydrogen peroxide was used as an illicit treatment as whitening and refreshing agents, exploiting its oxidant properties, the balance was shifted again towards a higher level of TMAO. In *Sarda sarda* the values after H_2O_2 treatment were TMAO 410 ± 57, TMA 720 ± 180 mg/kg; and in *Loligo vulgaris* were TMAO 850 ± 212, TMA 200 ± 28 mg/kg.

Table 3. Concentration values of trimethylamine-N-oxide (TMAO) and trimethylamine (TMA) in real fish samples of Atlantic bonito (*Sarda sarda*) and European squid (*Loligo vulgaris*). The amount is expressed in mg/kg. Fresh referred to freshly caught samples; aged to samples left at room temperature for 4 h; H_2O_2 treatment to samples treated with hydrogen peroxide (see Material and Methods).

Sample	Compound	Fresh (mg/kg)	Aged (mg/kg)	H_2O_2 Treatment (mg/kg)
Sarda sarda	TMAO	1700 ± 238	170 ± 24	410 ± 57
	TMA	170 ± 24	1250 ± 350	720 ± 180
Loligo vulgaris	TMAO	1200 ± 336	140 ± 20	850 ± 212
	TMA	210 ± 29	970 ± 242	200 ± 28

3.2. *Results of GC-MS Analysis of H_2O_2 Residues*

To confirm the illicit management of fishery products with hydrogen peroxide, we implemented a previously published GC-ECD assay based on peroxide detection by indirect oxidation of anisole to guaiacol (2-hydroxyanisole) (Scheme 1) [31].

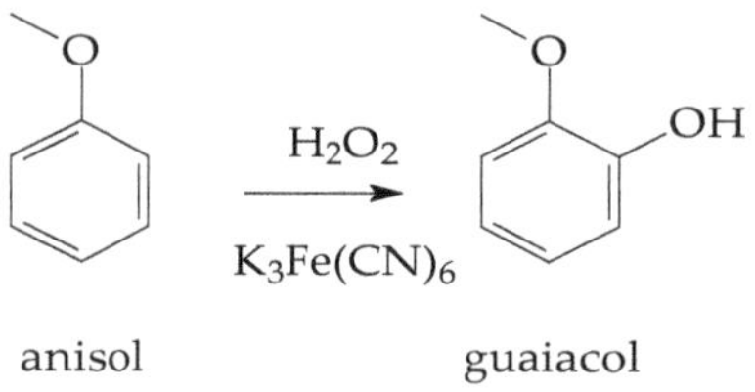

Scheme 1. Oxidation reaction with hydroxyl peroxide of anisole to guaiacol (2-hydroxyanisole) catalyzed by potassium ferricyanide.

We developed a headspace solid-phase microextraction (HS SPME) GC-MS methodology to improve sensitivity, easiness of operation, and reliability.

To quantify the residues of H_2O_2 we prepared a calibration curve with the standard addition method using fresh fish product samples of European squid and Atlantic bonito as matrices. The extracted solution of fresh fishery products was added with aliquots of hydrogen peroxide to give amounts of 0.0, 0.05, 0.1, 0.25, 0.5, and 1.0 μg/mL of H_2O_2. Then, 100 μL of potassium ferricyanide ($K_8Fe(CN)_6$) and 2 μL of anisole were added to the obtained solutions before fiber exposing for head space analysis. We measured the 2-hydroxyanisole peak area which was related to the H_2O_2 addition. Figure 3 shows a standard addition calibration curve of H_2O_2 in European squid samples. The obtained LLOQ was 50 ng/mL. When the concentration of hydrogen peroxide was zero, no 2-hydroxyanisole was detected.

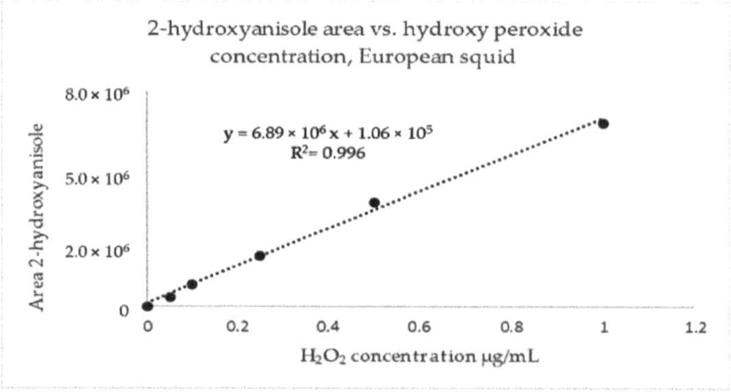

Figure 3. A standard addition curve of H_2O_2 in a fresh sample of European squid.

We then applied the developed HS SPME-GCMS method to five real fishery products samples, in particular squid which were the subject of illicit treatment with hydrogen peroxide due to its properties of whitening agent.

QC and real samples were prepared as described in the Sample Preparation section: The amount of H_2O_2 added for the redox reaction of anisole to guaiacol was set at 0.5 μg/mL. In these conditions we tested a blank solution without fishery products (no *Sarda sarda* or *Loligo vulgaris*), fresh (blank matrix), fresh and treated in controlled conditions (QC) and illicit treated fishery products samples of Atlantic bonito and European squid. All of the five samples coming from legal controls showed H_2O_2 values higher than 100 ppm. Table 4 shows the results. The peak area of 2-hydroxyanisole in the buffer without matrices confirmed the added amount of H_2O_2 of 0.5 ± 0.07 ppb. In the presence of matrices, the peak area decreased due to the matrix effect. In the case of fresh fish product samples, this effect was quantifiable in a loss of 12% of H_2O_2 amount (0.44 ± 0.11 μg/mL for Atlantic bonito and 0.43 ± 0.10 μg/mL for European squid). In fresh samples treated with hydrogen peroxide in controlled conditions (complete immersion of samples in a 0.8% H_2O_2 solution followed by rinsing with fresh water in laboratory) the amount of H_2O_2 was much higher than 1 ppm, the highest calibration curve point. However, extrapolating the results, it seemed larger than 100 ppm. The same was for the samples, especially squid, illegally treated with hydrogen peroxide.

Table 4. Quantitation of H_2O_2 in a 0.1 M phosphate buffer pH 2.5 solution and in fishery products samples. QC were quality control samples treated with hydrogen peroxide in controlled conditions (0.5 μg/mL of H_2O_2); five real samples were illegally treated with peroxide.

Matrix	Samples	H_2O_2 (μg/mL)
Buffer		0.5 ± 0.07
Sarda sarda *Loligo vulgaris*	fresh	0.44 ± 0.11 0.43 ± 0.10
Sarda sarda *Loligo vulgaris*	QC	>100
Sarda sarda *Loligo vulgaris*	real	>100

4. Discussion

The developed mass spectrometry based analytical methods were adequately selective to quantify the amount of peculiar molecules related to the illicit treatment of fishery products with hydrogen peroxide. In particular, to quantify trimethylamine, trimethylamine-N-oxide, dimethylamine, and 2-hydroxyanisole.

Many other analytical methods were proposed to measure the analytes [33–41]. The following paragraph gives a summary about was presented.

Bilgin's and Chung's research groups performed a quantitation of some amines in a large number of fish samples using HPLC coupled to a photodiode array detector [33] or other detectors such as chemiluminescent nitrogen, SPME-GC-MS, and spectrophotometric ones [34]. The first study regarded the determination of histamine, cadaverine, and tyramine after derivatization with dansyl chloride in 63 fish samples. The declared limit of quantitation was in the range between 0.010 and 0.100 μg/mL. Chung et al. determined TMAO, TMA, DMA, and FA in 266 fish samples. They were used respectively for TMAO, HPLC coupled to a chemiluminescent detector; for the TMA and DMA SPME-GC-MS method with a carboxen/divinylbenzen/polydimethylsiloxane fiber; and LC-visible analyzer for FA after derivatization with 2,4-dinitrophenylhydrazine. The stated LOQ were 25, 10, 10, 5 mg/kg for TMAO-DMA-TMA-FA, respectively [34]. Moreover, Soncin [40] and Chan et al. [35] in their studies used as analytical methodologies the headspace solid-phase extraction coupled with GC-MS. The first performed an identification and quantitation of markers of spoilage in fish and found as indicators four molecules. They defined the technique suitable for analyzing the volatile compounds [40]. Chan's study monitored the concentration of DMA and TMA in fish and the achieved limit of detection was 100 ppb [35].

Heude in his paper published on Food Analytical Methods in 2015 used a ^{1}H high resolution magic angle spinning NMR spectroscopy to determine the *K*-value [36] and trimethylamine nitrogen content as parameters of freshness and quality of fish products. They studied four species of fish and highlighted as great advantages the possibility to not process fish samples, no extraction was required for NMR analysis [36]. Feng Li et al. in their research used ion chromatography coupled to an unsuppressed conductivity detector to measure the amount of DMA, TMA, and TMAO in aquatic products. They obtained a good lower limit of detection (60, 80, 100 ng/mL for DMA-TMA-TMAO) and quantified the analytes in three real samples [38].

Finally, the remaining methods developed [37,39,41] were based on mass spectrometry techniques in order to determine the amines of interest. Le's group using a C18-PFP column and triple quadrupole quantify TMAO in human plasma samples as a potential indicator of cardiovascular health. The determined LOD and LOQ were 1 and 6 ng/mL [37]. Romero-Gonzáles et al. measured cadaverine and TMA with UHPLC-MS with a LLOQ of 25 and 60 ppb, respectively [39]. Finally, using a hydrophilic interaction liquid chromatography coupled to mass spectrometry Wu et al. quantified in fish meals five amines with satisfactory LOD and LOQ [41].

The present method is based on ultra-high liquid chromatography coupled to mass spectrometry and demonstrates to be sufficiently sensitive to quantify the four studied amines in fishery products in 5 min only. As previously indicated, the lower limits of quantification in matrices for DMA-TMA-TMAO and CAD were 45, 70, 50, and 40 ng/mL, respectively. The selectivity of the method was remarkable in spite of the low mass to the charge value of molecular protonated ions of the analytes. This region of low *m/z* ratio ($[M+H]^+$ 46, 60, 76, and 103 for DMA-TMA-TMAO, and CAD) is normally affected by a high background noise, but using tandem mass spectrometry the signal to noise ratio was reasonably good.

To highlight the illicit treatment of fishery products with hydrogen peroxide, we focused our attention on the ratio between TMAO and TMA. As shown in Table 3, subsequently to the prohibited washing with H_2O_2, the ratio between the measured compounds was shifted again towards a higher concentration of TMAO, one of the endogenous molecular marker of freshness. This ratio was not completely reversed because hydrogen peroxide acted only on the skin of the fishery products, besides the alkylamine is present in higher concentration in muscles. However, the treatment could make the product look younger and fresher.

As explained in the results section, the GC-MS method was an upgrade of the Tanaka et al. GC-ECD methodology for the quantitation of hydrogen peroxide in Chinese foods. The quantitation is possible thanks to the indirect measurement of 2-hydroxyanisole generated by a potassium ferricyanide catalyzed redox reaction of anisole with H_2O_2 [31]. In that paper, Tanaka et al. performed a deep investigation about reaction environment (pH range), catalyzer, and hydrogen peroxide concentrations in order to obtain the highest possible yield of redox reaction. For ECD quantitation, pentafluoro-benzoyl chloride was used as a derivatizing agent prior to analysis of the product of the reaction, 2-hydroxyanisole. The LLOQ by Tanaka et al. was 0.10 μg/mL.

With the present method based on mass spectrometry we aimed to improve the literature method both for sample preparation and sensitivity. It is well known that SPME followed by gas chromatography coupled to mass spectrometry is a highly sensitive method and the electron ionization source (EI) owns a high fragmentation repeatability.

The sample preparation herein described is free from the necessity of derivatization to detect the redox reaction product 2-hydroxyanisole. The HS SPME with carboxen/polydimethylsiloxane fiber was improved by testing various pH buffers, catalyzer concentrations, reaction times and fiber exposure times, stirring of solution.

We found that the best conditions to obtain the highest amount of 2-hydroxyanisole were: Solution buffering at pH 2.5, 0.1 M catalyzer concentration, 1 h at 60 °C time reaction, time of 30 min of exposure of the fiber without stirring. In these conditions a LLOQ value of 0.05 μg/mL was obtained.

With this HS SPME-GC-MS method we measured the amount of H_2O_2 in fishery products, comparing fresh and illicit treated samples. In the latter case, the concentration should be hypothesized by extrapolating over the calibration curve range because the signal was too abundant, and it was calculated larger than 100 ppm, as listed in Table 4.

5. Conclusions

In conclusion, the developed mass spectrometry based analytical methods show to be suitable to notice the illicit treatment of fishery products with hydrogen peroxide. Both the UHPLC-MS method and the HS SPME-GC-MS method are applicable to detect molecules related to the use of hydrogen peroxide solution to whiten and refresh aged fish food.

By LC-MS, low molecular mass amines were detected with high selectivity and good sensitivity. The TMAO and TMA ratio was shown to be reversed by the illicit treatment, simulating an unreal apparent freshness of fishery products foods.

The use of hydrogen peroxide on fish products was confirmed by the measurement of 2-hydroxyanisole with HS SPME-GC-MS after a redox reaction between anisole and residual H_2O_2 in the extracted solution, by exploiting potassium ferricyanide catalysis. After the optimization of sample preparation for headspace solid-phase microextraction and redox reaction parameters, the method was suitable to quantify the H_2O_2 residues in fish food matrices. When fishery products, especially squid, were whitened with hydrogen peroxide, its amount was shown to be easily detectable.

Author Contributions: Conceptualization, F.D.B., V.G., and C.M.; Methodology, F.D.B., M.Z., and R.A.; Validation, M.Z. and R.A.; Formal analysis, F.D.B.; Data curation, F.D.B.; Writing—original draft preparation, F.D.B.; Writing—review and editing, C.M. and R.A.; Supervision, C.M.; Project administration, C.M. All authors have read and agreed to the published version of the manuscript.

References

1. Regulation (EC) No 853/2004 of the European Parliament and of the Council of 29 April 2004 Laying Down Specific Hygiene Rules for Food of Animal Origin. Available online: http://data.europa.eu/eli/reg/2004/853/oj (accessed on 7 January 2020).
2. Council Regulation (EC) No 2406/96 of 26 November 1996 Laying Down Common Marketing Standards for Certain Fishery Products. Available online: http://data.europa.eu/eli/reg/1996/2406/oj (accessed on 7 January 2020).
3. Jacobsen, C.; Nielsen, H.H.; Jørgensen, B.; Nielsen, J. Chapter 15—Chemical processes responsible for quality deterioration in fish. In *Chemical Deterioration and Physical Instability of Food and Beverage*, 1st ed.; Skibsted, L.H., Andersen, M.L., Eds.; Elsevier: Amsterdam, the Netherlands, 2010; pp. 483–518. ISBN 978-1-84569-495-1.
4. Jack, L. Raw material selection: Fish. In *Chilled Fish: A Comprehensive Guide*, 3rd ed.; Brown, M., Ed.; Woodhead Publishing Limited: Cambridge, UK, 2008; Chapter 5; pp. 83–106.
5. Howgate, P. Tainting of farmed fish by geosmin and 2-methyl-iso-borneol: A review of sensory aspect and of uptake/depuration. *Aquaculture* **2004**, *234*, 155–181. [CrossRef]
6. Lin, J.K.; Hurng, D.C. Thermal conversion of trimethylamine-N-oxide to trimethylamine and dimethylamine in squids. *Food Chem. Toxicol.* **1985**, *23*, 579–583. [CrossRef]
7. Fu, X.; Xue, C.; Miao, B.; Liang, J.; Li, Z.; Cui, F. Purification and characterization of trimethylamine-N-oxide demethylase from jumbo squid (*Dosidicus gigas*). *J. Agric. Food Chem.* **2006**, *54*, 968–972. [CrossRef]
8. Fu, X.Y.; Xue, C.H.; Miao, B.C.; Li, Z.J.; Zhang, Y.Q.; Wang, Q. Effect of processing steps on the physico-chemical properties of dried-seasoned squid. *Food Chem.* **2007**, *103*, 287–294. [CrossRef]
9. Boziaris, I.S.; Stamatiou, A.P.; Nychas, G.J. Microbiological aspects and shelf life of processed seafood products. *J. Sci. Food Agric.* **2013**, *93*, 1184–1190. [CrossRef]
10. Ludewig, M.; Hohne, J.; Braun, P.G. Microbiological and sensory quality of salmon products from retail in Leipzig. *J. Food Saf. Food Qual.* **2019**, *70*, 10–16.

11. Junli, Z.; Jianrong, L.; Jia, J. Effects of thermal processing and various chemical substances on formaldehyde and dimethylamine formation in squid *Dosidicus gigas*. *J. Sci. Food Agric.* **2012**, *92*, 2436–2442.
12. Hansen, L.T.; Huss, H.H. Comparison of the microflora isolated from spoiled cold-smoked salmon from three smokehouses. *Food Res. Int.* **1998**, *31*, 703–711. [CrossRef]
13. Løvdal, T. The microbiology of cold smoked salmon. *Food Control* **2015**, *54*, 360–373. [CrossRef]
14. Chiou, T.; Chang, H.; Lo, L.; Lan, H.; Shiau, C. Changes in chemical constituents and physical indices during processing of dried-seasoned squid. *Fish. Sci.* **2000**, *4*, 708–715. [CrossRef]
15. Omura, Y.; Yamazawa, M.; Yamashita, Y.; Okazaki, E.; Watabe, S. Relationship between postmortem changes and browning of boiled, dried, and seasoned product made from Japanese Common Squid (*Tedarodes pacificus*) mantle muscle. *J. Food Sci.* **2007**, *72*, C044–C049. [CrossRef] [PubMed]
16. Peng, J.; Zheng, F.X.; Wei, L.; Lin, H.; Jiang, J.H.; Hui, G.H. Jumbo squid (*Dosidicus gigas*) quality enhancement using complex bio-preservative during cold storage. *J. Food Meas. Charact.* **2018**, *12*, 78–86. [CrossRef]
17. Zhu, J.; Wu, S.; Wang, Y.; Li, J. Quality changes and browning developments during storage of dried-seasoned squid (*Dosidicus gigas* and *Ommastrephes bartrami*). *J. Aquat. Food Prod. Technol.* **2016**, *25*, 1107–1119. [CrossRef]
18. Durmus, M.; Polat, A.; Oz, M.; Ozogul, Y.; Ucak, I. The effects of seasonal dynamics on sensory, chemical and microbiological quality parameters of vacuum-packed sardine (*Sardinella aurita*). *J. Food Nutr. Res.* **2014**, *53*, 344–352.
19. Manimaran, U.; Shakila, R.J.; Shalini, R.; Sivaraman, B.; Sumathi, G.; Selvaganapathi, R.; Jeyasekaran, G. Effect of additives in the shelflife extension of chilled and frozen stored Indian octopus (*Cistopus indicus*). *J. Food Sci. Technol.* **2016**, *53*, 1348–1354. [CrossRef]
20. Agustini, W.T.; Suzuki, M.; Suzuki, T.; Hagiwara, T.; Okouchi, S.; Takai, R. The possibility of using oxidation-reduction potential to evaluate fish freshness. *Fish. Sci.* **2001**, *67*, 547–549. [CrossRef]
21. Susanto, E.; Agustini, T.W.; Ritanto, E.; Dewi, E.N.; Swastawati, F. Changes in oxidation and reduction potential (Eh) and pH of tropical fish during storage. *J. Coast. Dev.* **2009**, *14*, 223–234.
22. Cheng, J.-H.; Sun, D.-W.; Zeng, X.-A.; Liu, D. Recent advances in methods and techniques for freshness quality determination and evaluation of fish and fish fillets: A review. *Crit. Rev. Food Sci. Nutr.* **2015**, *55*, 1012–1225. [CrossRef]
23. Zou, Q.; Bennion, B.J.; Daggett, V.; Murphy, K.P. The molecular mechanism of stabilization of proteins by TMAO and its ability to counteract the effects of urea. *J. Am. Chem. Soc.* **2002**, *124*, 1192–1202. [CrossRef]
24. Pillai, K.; Akhter, J.; Chua, T.C.; Morris, D.L. Mucolysis by ascorbic acid and hydrogen peroxide on compact mucin secreted in pseudomyxoma peritonei. *J. Surg. Res.* **2012**, *174*, 69–73. [CrossRef]
25. Jafarpour, A.; Sherkat, F.; Leonard, B.; Gorczyca, E.M. Colour improvement of common carp (Cyprinus carpio) fillets by hydrogen peroxide for surimi production. *Int. J. Food Sci. Technol.* **2008**, *43*, 1602–1609. [CrossRef]
26. Himonides, A.T.; Taylor, A.; Knowles, M.J. The improved whitening of cod and haddock flaps using hydrogen peroxide. *J. Sci. Food Agric.* **1999**, *79*, 845–850. [CrossRef]
27. Gale, J.P.; Yergey, A.L.; Duncan, M.W.; Yu, K. Quantifying small molecules by mass spectrometry. *LC GC N. Am.* **2015**, *33*, 34–41.
28. Kind, T.; Tsugawa, H.; Cajka, T.; Ma, Y.; Lai, Z.; Mehta, S.S.; Wohlgemuth, G.; Barupal, D.K.; Showalter, M.R.; Arita, M.; et al. Identification of small molecules using accurate mass MS/MS search. *Mass Spectrom. Rev.* **2018**, *37*, 513–532. [CrossRef]
29. De Vijlder, T.; Valkenborg, D.; Lemière, F.; Romijn, E.P.; Laukens, K.; Cuyckens, F. A tutorial in small molecule identification via electrospray ionization-mass spectrometry: The practical art of structural elucidation. *Mass Spectrom. Rev.* **2018**, *37*, 607–629. [CrossRef]
30. U.S. Food and Drug Administration. *Guidance for Industry: Bioanalytical Method Validation*; U.S. Department of Health and Human Services, FDA: Rockville, MD, USA, 2001.
31. Tanaka, A.; Ijima, M.; Kikuchi, Y. Determination of hydrogen peroxide in fish products and noodles (japanese) by gas-liquid chromatography with electron-capture detection. *J. Agric. Food Chem.* **1990**, *38*, 2154–2159. [CrossRef]
32. Dal Bello, F.; Santoro, V.; Scarpino, V.; Martano, C.; Aigotti, R.; Chiappa, A.; Davoli, E.; Medana, C. Antineoplastic drugs determination by HPLC-HRMS^n to monitor occupational exposure. *Drug Test. Anal* **2016**, *8*, 730–737.32. [CrossRef]
33. Bilgin, B.; Gençcelep, H. Determination of biogenic amines in fish products. *Food Sci. Biotechnol.* **2015**, *24*, 1907–1913. [CrossRef]

34. Chung, S.W.C.; Chan, B.T.P. Trimethylamine oxide, dimethylamine, trimethylamine and formaldehyde levels in main traded fish species in Hong Kong. *Food Addit. Contam. Part B Surveill.* **2009**, *2*, 44–51. [CrossRef]
35. Chan, S.T.; Yao, M.W.Y.; Wong, Y.C.; Wong, T.; Mok, C.S.; Sin, D.W.M. Evaluation of chemical indicators for monitoring freshness of food and determination of volatile amines in fish by headspace solid-phase microextraction and gas chromatography-mass spectrometry. *Eur. Food Res. Technol.* **2006**, *224*, 67–74. [CrossRef]
36. Heude, C.; Lemasson, E.; Elbayed, K.; Piotto, M. Rapid assessment of fish freshness and quality by 1H HR-MAS NMR spectroscopy. *Food Anal. Methods* **2015**, *8*, 907–915. [CrossRef]
37. Le, T.T.; Shafaei, A.; Genoni, A.; Christophersen, C.; Devine, A.; Lo, J.; Wall, P.L.; Boyce, M.C. Development and validation of a simple LC-MS/MS method for the simultaneous quantitative determination of trimethylamine-N-oxide and branched chain amino acids in human serum. *Anal. Bioanal. Chem.* **2019**, *411*, 1019–1028. [CrossRef] [PubMed]
38. Li, F.; Liu, H.; Xue, C.; Xin, X.; Xu, J.; Chang, Y.; Xue, Y.; Yin, L. Simultaneous determination of dimethylamine, trimethylamine and trimethylamine-n-oxide in aquatic products extracts by ion chromatography with non-suppressed conductivity detection. *J. Chromatogr. A* **2009**, *1216*, 5924–5926. [CrossRef] [PubMed]
39. Romero-González, R.; Alarcón-Flores, M.I.; Vidal, J.L.M.; Frenich, A.G. Simultaneous determination of four biogenic and three volatile amines in anchovy by ultra-high-performance liquid chromatography coupled to tandem mass spectrometry. *J. Agric. Food Chem.* **2012**, *60*, 5324–5329.
40. Soncin, S.; Chiesa, L.M.; Panseri, S.; Biondi, P.; Cantoni, C. Determination of volatile compounds of precooked prawn (*Penaeus vannamei*) and cultured gilthead sea bream (*Sparus aurata*) stored in ice as possible spoilage markers using solid phase microextraction and gas chromatography/mass spectrometry. *J. Sci. Food Agric.* **2009**, *89*, 436–442. [CrossRef]
41. Wu, T.H.; Bechtel, P.J. Screening for low molecular weight compounds in fish meal solubles by hydrophilic interaction liquid chromatography coupled to mass spectrometry. *Food Chem.* **2012**, *130*, 739–745. [CrossRef]

Hydrocolloid-Based Coatings with Nanoparticles and Transglutaminase Crosslinker as Innovative Strategy to Produce Healthier Fried Kobbah

Asmaa Al-Asmar [1,2], Concetta Valeria L. Giosafatto [1], Mohammed Sabbah [3] and Loredana Mariniello [1,*]

[1] Department of Chemical Sciences, University of Naples "Federico II", 80126 Naples, Italy; a.alasmar@najah.edu (A.A.-A.); giosafat@unina.it (C.V.L.G.)
[2] Analysis, Poison Control and Calibration Center (APCC), An-Najah National University, P.O. Box 7 Nablus, Palestine
[3] Department of Nutrition and Food Technology, An-Najah National University, P.O. Box 7 Nablus, Palestine; m.sabbah@najah.edu
* Correspondence: loredana.mariniello@unina.it

Abstract: This study addresses the effect of coating solutions on fried kobbah. Coating solutions were made of pectin (PEC) and grass pea flour (GPF), treated or not with transglutaminase (TGase) and nanoparticles (NPs)—namely mesoporous silica NPs (MSN) or chitosan NPs (CH–NPs). Acrylamide content (ACR), water, oil content and color of uncoated (control) and coated kobbah were investigated. Zeta potential, Z-average and in vitro digestion experiments were carried out. Zeta potential of CH–NPs was stable from pH 2.0 to pH 6.0 around + 35 mV but decreasing at pH > 6.0. However, the Z-average of CH–NPs increased by increasing the pH. All coating solutions were prepared at pH 6.0. ACR of the coated kobbah with TGase-treated GPF in the presence nanoparticles (MSN or CH–NPs) was reduced by 41.0% and 47.5%, respectively. However, the PEC containing CH–NPs showed the higher reduction of the ACR by 78.0%. Water content was higher in kobbah coated by PEC + CH–NPs solutions, while the oil content was lower. The color analysis indicated that kobbah with lower browning index containing lower ACR. Finally, in vitro digestion studies of both coating solutions and coated kobbah, demonstrated that the coating solutions and kobbah made by means of TGase or nanoparticles were efficiently digested.

Keywords: acrylamide; kobbah; transglutaminase; pectin; chitosan-nanoparticles; coatings; mesoporous silica nanoparticles; grass pea; HPLC-RP

---

1. Introduction

Kibbeh, kibbe, kobbah (also kubbeh, kubbah, kubbi) (pronunciation varies with region) is Eastern dish made of a ground bulgur (wheat-based food) mixed with minced beef meat formed as balls stuffed with cooked ground meat, onions, nuts and spices. They are usually cooked by deep frying for 8–10 min at high temperatures (160–180 °C), thus they have rough crust and thoroughly browned. They are home-made and consumed fresh or they are sold frozen in the super-markets and consumers can fry them at home [1].

Hydrocolloid materials are used for food protection, as well as for separating the different part of a food [2,3]. Coatings represent a thin layer of edible molecules that are laid on the surface of a food product and can be used to protect high perishable aliments. Pectin (PEC) is a polysaccharide present in the plant cell wall containing mainly galacturonic acid, but highly variable in composition, structure and molecular weight [4]. PEC is known as food additive (E440), useful for thickening mainly jam and marmalades and other products, since is provided with gelling properties [5]. Yossef [6], found that strawberry fruits dipped in PEC-based solutions retained physico-chemical properties as the same fruit protected by other hydrocolloid molecules, such as soy proteins, gluten or starch. Moreover, protein-based such as grass pea flour (GPF) was used for its high content in proteins. Grass pea (*Lathyrus sativus* L.) belongs to the leguminous family and is quite important in many Asian and African countries where it is cultivated either for animal feeding or human use. It is characterized by resistance to both abiotic and biotic stresses [7]. GPF containing proteins were able to act as transglutaminase substrates, giving arise to novel bioplastics endowed with improved technological properties than the ones cast without the enzyme [8–10].

Microbial transglutaminase (TGase) belongs to a family of enzymes (E.C. 2.3.2.13) (widely distributed in nature from microbes to animals and plants) capable of catalyzing iso–peptides bonds between endo-glutamine and endo-lysine residues belonging to proteins of different nature, giving arises to intra- and inter-molecular crosslinks [2,11]. TGase is widely used in the food industry as technological aid. Recently, Sabbah et al. [12], demonstrated that the proteins of *Nigella sativa* defatted cakes are act as TGase substrates and are responsible to enhance physico-chemical properties of the obtained films.

Using nanoparticles for developing of nanocomposite coatings is a way to improve their features and is of interest also for producing active packaging [13]. Mesoporous silica nanoparticles MSNs (Type MCM-41) are a kind of SiO_2–based nanoparticles that are promising materials for application in numerous aspects of biomedical and food purposes [14–16]. Recently, Fernandez-Bats et al. [17]; Giosafatto et al. [18]; Al-Asmar et al. [3], prepared and characterized the active protein, pectin and chitosan edible films grafted with MSNs, and they concluded that MSNs significantly influence the mechanical and permeability properties of the obtained materials. SiO_2 nanoparticles of different composition are labeled as E551, E554, E556 or E559 and used for instance as an anti-caking agent. The amount ingested daily is estimated to be 1.8 mg/kg (around 126 mg/day for a 70 kg person) [19]. Moreover, McCarthy [20] observed that SiO_2- based NPs with the size of 150 nm and 500 nm do not perform toxic effects on Calu-3 cells.

Chitosan nanoparticles (CH–NPs) are natural materials obtained from the marine byproducts, endowed with not able physico-chemical and antimicrobial characteristics, besides being sustainable and harmless for human health [21,22]. These properties suggest that CH–NPs can be used also as carrier for drug delivery. Lorevice et al. [23], obtained higher mechanical properties by adding CH–NPs to PEC films compared with control, allowing these novel materials to be an alternative to traditional food packaging production. Moreover, addition of small fractions of CH–NPs enhance mechanical and thermal stability of banana puree-based films [24]. Application of CH in foods is gaining interest, specifically after that shrimp chitin-derived CH has been recognized as Generally Recognized as Safe (GRAS) for common use in foods by the US Food and Drug Administration in 2011 [25,26].

Acrylamide (ACR) is a chemical that was discovered in foods in 2002 and it is present in a range of popular foods [27]. ACR is not present in raw foods, but it is formed from natural precursors during food treatment at high-temperature (>120 °C) following Maillard reaction [28]. ACR is formed during frying, baking roasting and toasting of the carbohydrate rich food and cereal products, as well as coffee. In particular, ACR occurs because of the reaction between the free amino acid asparagine and a carbonyl-containing compound. [29]. European Food Safety Authority (EFSA) scientists classified ACR is a 'probably carcinogenic to humans [30]. Hydrocolloid-based coatings recently become one of several strategies used to mitigate ACR formation and to improve the eating quality of different fried foods such as potato chips [31], bread [32] and fried banana [33]. Very recently, our research group demonstrated the effectiveness of hydrocolloid-based coatings prepared in the presence of TGase in reducing ACR content of French fries [34] and fried falafel, a typical Easter food [35]. In 2004, Al-Dmoor et al. [1] determined the ACR content in fried kobbah (food mostly eaten in Jordan, but also very popular in Palestine) finding values ranging from 2900 to 5300 $\mu g\ kg^{-1}$. Thus, kobbah contain ACR in values that impose attention to protect the health of consumers.

The aim of this study was to investigate the influence of different hydrocolloid-based coatings (containing PEC and GPF prepared in the presence or absence of MNS and/or TGase) on ACR, water content, oil content, digestibility and color of fried kobbah. The physicochemical properties of different coating solutions were also evaluated.

2. Materials and Methods

2.1. Materials

Methanol and ACR standard ≥99.8%, were purchased from Sigma–Aldrich Chemical Company (St. Louis, MO, USA). Acetonitrile HPLC analytical grade, n-hexane and formic acid were obtained from Carlo Erba reagents S.r.l. (Cornaredo, Milan, Italy). Oasis HLB 200 mg, 6 mL solid phase extraction (SPE) cartridges were from Waters (Milford, MA, USA). Syringe filters (0.45- and 0.22-μm PVDF) were from Alltech Associates (Deerfield, Italy). PEC of a low-methylated citrus peel (7%) (Aglupectin USP) was purchased from Silva Extracts s.r.l. (Gorle, Bergamo, Italy) and Activa®WM *Streptoverticillium* TGase was supplied by Ajinomoto Co (Tokyo, Japan). Sodium tripolyphosphate (TPP) and glycerol (GLY) was from Merck Chemical Company (Darmstadt, Germany). Chitosan (CH, mean molar mass of 3.7–104 g/mol) was procured from Professor R. Muzzarelli (University of Ancona, Ancona, Italy), with a degree of 9.0% N-acetylation. Grass pea seeds, corn oil, ground bulgur (wheat-based food), minced beef meat, onion, salt and spices were obtained from a local market in Naples (Italy).

2.2. Nanoparticles Preparation

MSN (MCM-41) were synthesized and characterized, as described in Fernandez-Bats et al. [17]. However, CH–NPs was prepared by using the ionic gelation method according to Chang et al. [36]. Briefly, adding the TPP 0.5% (*w/v*) dropwise to the CH solution (0.8%) by 40 min stirring the obtained suspension was centrifuged at 22,098× *g* for 10 min at 4 °C, then rinsed three times by Milli-Q water and dry at room temperature.

2.3. Kobbah Formulation and Manufacturing

Kobbah was made as described by Brazil et al. [37], soaking ground bulgur flour into hot water (80 °C) for 1 h, then the wheat flour, oil, salt and spices were added and mixed with the soaked ground bulgur. The dough was kept at the refrigerator for 1 h. Stuffing: onions, salt and spices were mixed with minced beef meat then cooked with olive oil. The dough was shaped into balls stuffed with cooked ground meat.

2.4. Preparation of the Coating Solutions

GPF was obtained according to Giosafatto et al. [8] and Al-Asmar et al. [34], in particular, the seeds were milled by a laboratory blender LB 20ES (Waring Commercial, Torrington, CT, USA) and the obtained flour was treated with a 425-μm stainless steel sieve (Octagon Digital Endecotts Limited, London, UK). A total of 83 g of GPF (containing 24% *w/w* proteins) were dissolved in 1 L Milli-Q water and the solutions were shacked for 1 h. the pH was brought to 9.0 with 1-M NaOH followed by centrifugation at 12,096× *g* for 10 min. After centrifugation, the supernatant was collected and used to prepare the GPF dipping solution. Nanoparticles, either MSN or CH–NP (1% *w/w* GPF proteins) were added to GPF at pH 6.0 than the solutions were mixed for 30 min at room temperature (GPF; GPF + MSN; GP + CH–NP). TGase (33 U/g of GPF proteins) was used to prepare the GPF + TGase; GPF + MSN + TGase and GPF + CH–NP + TGase, GLY was used as plasticizer (8% *w/w* GPF proteins) in all the GPF coating solutions, then incubated for 2 h at 37 °C. PEC-based solutions (1% *w/v*) prepared as described in Esposito et al. [38], were made from a PEC stock solution (2% *w/v*), then diluted with Milli-Q water. MSN or CH–NP (1% *w/w* PEC) were added to PEC and mixed for 30 min at room temperature (PEC; PEC + MSN; PEC + CH–NP). Each dipping solutions were adjusted to pH 6.0, then was used to coat kobbah before trying.

2.5. *Dipping and Frying Method*

Two hundred grams of kobbah (divided in 5 pieces) were immersed for 30 s into either in H_2O_d ("control" sample) or in one of these coating solutions: (1) GPF; (2) GPF reinforced with MSN (GPF + MSN); (3) GPF reinforced with CH–NP (GPF + CH–NP); (4) TGase-treated (GPF + TGase); (5) TGase-treated reinforced with MSN (GPF + MSN + TGase); (6) TGase-treated reinforced with CH–NP (GPF + CH–NP + TGase); (7) PEC; (8) PEC reinforced with MSN (PEC + MSN); and (9) PEC reinforced with CH–NP (PEC + CH–NP). Moreover, each sample was dripped for 2 min before frying to get rid of the excess of solutions. The frying conditions consisted in 2 L corn oil preheated (using a controlled temperature deep-fryer apparatus (GIRMI, Viterbo, Italy)) to the processing temperature (190 ± 5 °C), then the kobbah were deep fried for 4.5 min. The oil was replaced by new one for each different coating solutions. After frying, kobbah were drained for 2 min to remove oil excess [34,35,39].

2.6. *Zeta Potential and Z-Average of Coating Solutions*

The Zeta potential and particle size (Z-average) of the CH–NP solution (1 mg/mL) prepared at pH 2.0 were obtained by titration from pH 2.0 to pH 7.0, by means of Zetasizer Nano-ZSP (Malvern®, Worcestershire, UK) equipped with a He–Ne laser. All coating solutions used in this experiment were also tested for their Zeta potential and Z-average.

2.7. *Oil and Water Content*

The oil content was performed following frying and cooling of each processed samples around (3–5 g) in triplicate. The result was reported as a percentage on dry matter weight by n-hexane solvent extraction using the Soxhlet method [40].

Fried kobbah water content was obtained following the gravimetric method [41] in triplicate.

2.8. *Acrylamide Detection*

2.8.1. Preparation of Acrylamide Standard

ACR standard stock solution (1.0 mg/mL) was obtained as described in Al-Asmar et al. [34,35]. In particular 10 mg of the ACR standard were dissolved in 10 mL of Milli-Q water. From the stock solution, different concentrations of calibration standards (100, 250, 500, 1000, 2000, 3000, 4000 and 5000 μg/L), were prepared, respectively. All series of standard solutions were kept in glass dark bottles at 4 °C until used.

2.8.2. Acrylamide Extraction

ACR extraction was performed as described in Al-Asmar et al. [34,35]. Briefly, about 200 g of fried kobbah, were put in n-hexane for 30 min to get rid of the oil [42]. After that, n-hexane was let to evaporate under fume hood at room temperature, then samples were subjected for ACR extraction. The treated fried kobbah samples were milled at 1300 rpm for 1 min by means of a rotary mill Grindomix GM200, (Retsch GmbH, Haan, Germany). Freeze drying was used to dry the samples before ACR extraction that was carried out by following the procedure of Wang et al. [43]. Briefly, two different tubes were set up for each sample, one for detecting ACR formed in kobbah samples, and the second one to carry out the "Recovery test for ACR in all kobbah types (in each sample 150 μg/L of ACR standard were added". In both tubes, 1.0 g (dry weight) of sample, was placed in both tubes and only in the second one there was the ACR standard added. Carrez reagent potassium salt and Carrez reagent zinc salts (50 μL) were included in each sample. In each tube, 10 mL of HPLC water were finally added. The samples were put in an incubated shaker for 30 min at 25 °C and 170 rpm, then centrifuged at 7741× *g* for 10 min at 4 °C. The supernatant was filtered with 0.45-μm syringe filter for the clean-up of the Oasis HLB SPE cartridges. The SPE cartridge was preventively conditioned with 2.0 mL of methanol followed by washing with 2 mL of water before loading 2.0 mL of the filtered supernatant, the first 0.5 mL was discarded and the remaining elute collected (~1.5 mL; exact volume was measured by weight and converted by means of density). All extracts were kept in dark glass vials at 4 °C before analysis. The clean sample extracts were further filtered through 0.2-µm nylon syringe filters before HPLC-UV (ultra violet) analysis of fried kobbah [34]. Each determination was performed in triplicate.

2.8.3. HPLC-UV Analysis

HPLC-UV analysis was used to determine the ACR, by using the RP-HPLC method on Beckman Gold HPLC instrument equipped with a dual pump and a diode array detector [34]. The column Synergi™ 4-µm Hydro-RP 80 Å HPLC Column 250 × 3 mm (Phenomenex, Torrance, CA, USA) was used [44].

The operating conditions described also in Al-Asmar et al. [34] were the following: the wavelength detection was 210 nm, a gradient elution of 0.1% formic acid (*v/v*) in water: acetonitrile (97:3, *v/v*) was used. Solvent A was water and Solvent B was acetonitrile, both solvents containing 0.10% (*v/v*) formic acid; flow rate, 1 mL/min. The gradient elution program was applied as follows: 97% A (3% B) for 10 min, increased to 20% A (80% B) from 10 to 12 min and kept at 20% A (80% B) for 5.0 min, increased to 95% B (5% A) from 17 to 19 min and kept at 95% B for 5 min, increased to 97% A (3% B) from 24 to 26 min and kept for 4 min. The injection volume was equal to 20 µL. The total chromatographic runtime was 30 min for each sample and the temperature was kept at 30 °C (GECKO 2000 "HPLC column heater", Spectra Lab Scientific, Inc., Markham, ON, Canada) to ensure optimal separation. In all samples (ACR standard and fried kobbah-derived), the ACR retention time was 4.9 min.

2.9. Color Analysis

Color measurement of food products was considered as an indirect measure of other quality features such as flavor and contents of pigments [45]. Chroma Meter Konica Minolta CR-400 (Japan) was utilized to determine L*, a*, b* values of fried kobbah samples. L* a* b* is an international standard for color measurement adopted by the Commission Internationale d'Eclairage (CIE) in 1976. L* is the lightness component, which ranges from 0 to 100 and parameters a* (from green to red) and b* (from blue to yellow) are the two chromatic components, which range from −120 to 120 [46]. Total color difference to the control sample (ΔE) indicates the magnitude of color difference between coated kobbah and uncoated control kobbah and it was obtained by the following equation [33,45]:

$$\Delta E = \sqrt{(L* - L'*)^2 + (a* - a'*)^2 + (b* - b'*)} \quad (1)$$

where L'*, a'* and b'* are the parameters of treated kobbah and L*, a* and b* the ones of the control (uncoated fried kobbah).

The browning index (BI) allowed to define the overall changes in browning color [33,47]. BI of the fried kobbah was calculated by the following equation:

$$BI = \frac{100\,(x - 0.31)}{0.17} \quad (2)$$

where:

$$X = \frac{(a* + 1.75\,L*)}{(5.645\,L* + a* - 3.012b*)} \quad (3)$$

2.10. Sodium Dodecyl Sulfate Polyacrylamide Gel Electrophoresis (SDS-PAGE) and In Vitro Digestion

SDS-PAGE, performed as described in Lemmli [48], was carried out at a constant voltage (80 V for 2–3 h). The protein bands were stained with Coomassie Brilliant Blue R250 (Bio-Rad, Segrate, Milan, Italy). Bio-Rad Precision Protein Standards were used as molecular weight markers.

GPF-based FFSs and fried kobbah either treated or not by TGase (33 U/g protein) reinforced or not by MSN, were subjected to in vitro gastric digestion (IVD) by using an adult model [49–51]. Then,

100 mg of each sample was incubated with 4 mL of simulated salivary fluid (SSF, 150 mM of NaCl, 3 mM of urea, pH 6.9) containing 75 U of amylase enzyme/g protein for 5 min at 37 °C at 170 rpm [35]. The amylase activity was stopped by adjusting the pH at 2.5. Afterwards, the samples were subjected to IVD as described by Giosafatto et al. [49] and Al-Asmar et al. [35], with some modifications. Briefly, 100 μL of simulated gastric fluid (SGF, 0.15 M of NaCl, pH 2.5) were placed in 1.5 mL microcentrifuge tubes and added to 100 μL of oral phase and then incubated at 37 °C. Thereafter, 50 μL of pepsin (0.1 mg/mL dissolved in SGF) were added to initiate the digestion. At intervals of 1, 2, 5, 10, 20, 40 and 60 min, 40 μL of the 0.5 M of ammonium bicarbonate (NH_4HCO_3) were added to each vial to stop the pepsin reaction. The control was set up by incubating the sample for 60 min without the protease. The samples were then analyzed by SDS-PAGE (12%) procedure described above.

2.11. Statistical Analysis

The statistical analysis was performed by means of JMP software 10.0 (SAS Institute, Cary, NC, USA), Two-way ANOVA and the *t*-student test for mean comparisons were used. Differences were considered significant at $p < 0.05$

3. Results and Discussion

3.1. Chitosan Nanoparticles, Mesoporous Silica Nanoparticles and Film Forming Solutions Characterization

Zeta potential is an important value to indicate the stability of solutions. Moreover, Z-average shows the size of the particles. Figure 1 shows Zeta potential (panel A) and Z-average (panel B) of the CH–NP in the function of pH. The results indicate that Zeta potential of CH–NP was stable at +35 mV started from pH 2.0 to pH 6.0 and decreases to +20 mV when the pH is equal to 7.0. This finding is in accordance with other authors' results [23,52,53]. The Z-average of CH–NPs at pH 2.0 was around 99.5 d.nm and increased at higher pH to reach 800 d.nm at pH 7.0. The obtained results were in agreement with those from Ali et al. [52], who explained that at pH higher than 6.0 the protonated amino groups start to lose protons and the ionic bonds start decreasing. Thus, the rises of Z-average together with the reduction in Zeta potential at pH 6.0 is because of the particle aggregation at this pH, rather than the additional increase of individual particle size [52,54]. In addition, we synthesized the MSN according to Fernandez-Bats et al. [17], with the very similar Z-average. These authors have analyzed MSN also by TEM and evidenced an average size of 143 ± 26 nm. MSN were used to improve the physio-chemical of PEC and CH films the results reported in Giosafatto et al. [18].

The coating solutions used during this study were also characterized for their stability. Zeta potential and particle size results are reported in Table 1. The results showed that stability was significantly increased after treatment of GPF-based solutions (−13.7 mV) with MSN (−16.8 mV) or TGase (−19.8 mV), also when the enzymatic crosslinking was carried out in of GPF-solutions nanoreinforced either with MSN (−18.4 mV) or CH–NPs (−18.2 mV). However, no significant effect on Zeta potential were found by adding MSN or CH–NPs on PEC FFSs stability. On the contrary, the particle size of GPF solutions was 201.3 d.nm, but this value increased significantly when CH–NP were incorporated with or without TGase. No significant change on the Z-average of FFS after adding MSN on the GPF was observed and these results are similar to those published previously by Fernandez-Bats et al. [17]. Adding TGase as crosslinker to the GPF together with MSN or CH–NP showed a significant increasing on the Z-average of FFSs. In addition, PEC FFS Z-average was (3198 d.nm) and it rises significantly to (3421 d.nm) after the addition of CH–NPs.

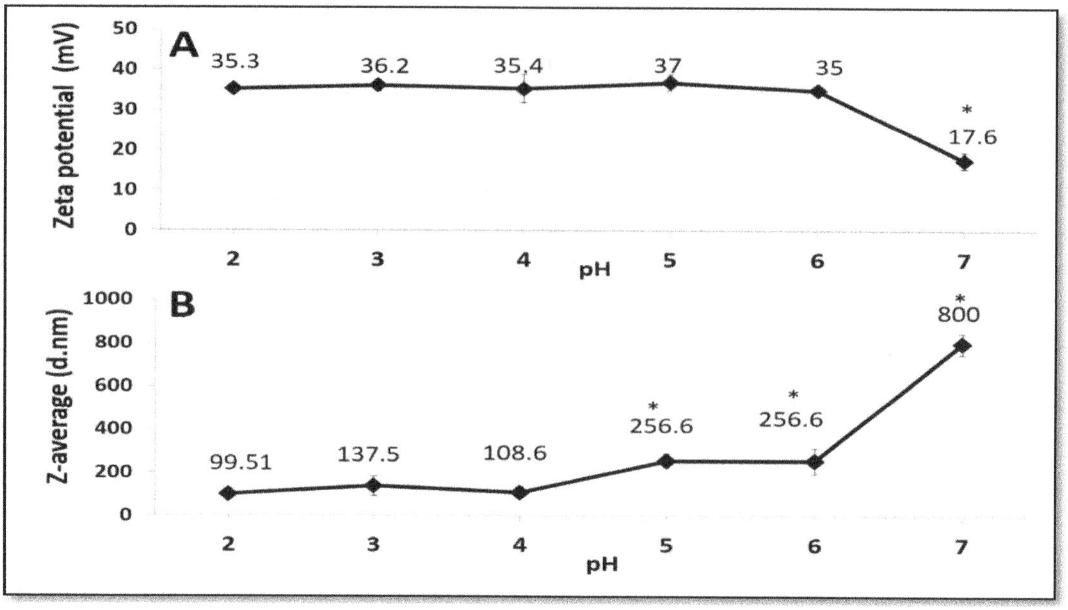

Figure 1. pH effect on Zeta potential (Panel **A**) and Z-average (Panel **B**) of 1 mg/mL chitosan nanoparticles (CH–NPs). Values marked with * were significantly different respect to the value at pH 2.0.

Table 1. Effect of 1% mesoporous silica nanoparticles (MSN) or 1% chitosan nanoparticles (CH–NPs) on Zeta potential and Z-average on either grass pea flour (GPF)-based (without or with transglutaminase (TGase) (33 U/g protein)) or pectin (PEC)-based film forming solutions at pH 6.

FFSs	Zeta Potential (mV)	Z-Average (d.nm)
GPF	−13.7 ± 0.6	201.3 ± 11.1
GPF + MSN	−16.8 ± 0.9 ^{a}	191.4 ± 14.2
GPF + CH–NP	−14.1 ± 0.8 ^{b}	385.6 ± 28.2 ^{a,b}
GPF + TGase	−19.8 ± 1.2 ^{a,b}	240.9 ± 14.4 ^{a,b}
GPF + MSN + TGase	−18.4 ± 0.5 ^{a}	333.0 ± 22.3 ^{a,b,c}
GPF + CH–NP + TGase	−18.2 ± 0.9 ^{a}	507.7 ± 18.9 ^{a,b,c,d}
PEC	−33.7 ± 2.1	3198 ± 79
PEC + MSN	−31.8 ± 2.9	3110 ± 77
PEC + CH–NP	−32.4 ± 3.2	3421 ± 63 *

The value significantly different from GPF FFSs are indicated by "a", the value indicated by "b" were significantly different from GPF + MSN film forming solution (FFS), whereas the value indicated by "c" were significantly different from GPF + TGase FFS, the value indicated by "d" was significantly different from GPF + MSN + TGase FFSs, the value indicated by "*" was significantly different respect to the PEC and PEC + MSN FFSs. Data represent the average values of three repetitions using (2-way ANOVA, $p < 0.05$ for mean comparison). Additional details are reported in the main text.

3.2. Influence on Nanoreinforced and TGase-Crosslinked Hydrocolloid Coating Solutions on Acrylamide Content

Kobbah is an ethnic food consumed dispersed among all the world not only in the Arab region. The main aim of this study consisted in studying the effect of the different coating solutions to decrease the ACR content that is formed during frying. The ACR content was performed by RP-HPLC and the results reported in Figure 2. Two main different dipping solutions (GPF and PEC), reinforced by means of two different nanoparticles (1% MSN and 1% CH–NP (*w/w*)) were used to coat the kobbah prior to frying. The GPF was enzymatically crosslinked by means of TGase in the presence or absence of NPs. The control sample was the kobbah dipped into distilled water. Figure 2 shows that control exhibited the highest ACR content reaching a value of 3039.7 µg/kg. On the contrary, all used coating materials were able to significantly reduce ACR content. Kobbah dipped into GPF solution showed about 22.5% reduction in ACR content, while PEC-based coating solution reduced the ACR to 55.5%. The previous work about potato French Fries showed that PEC alone reduced ACR formation about 48% [34]. Coating solutions containing NPs (either MSN or CH–NP) in addition to GPF provoked slightly significant reduction of ACR comparing to the GPF-based coating sample not containing NPs. Higher significant reduction was observed when even MSN or CH–NP were mixed with PEC. The lowest ACR content was detected in the kobbah coated by PEC solutions containing CH–NP. In fact, in these samples the ACR content was 678 μg/kg with the 78.0% ACR reduction in comparison to the control. Recently, Mekawi et al. [31] discovered that the addition of pomegranate peel NP extracts, to the sunflower oil during deep fat frying is responsible for ACR reduction to about (54%) in potato chips. The addition of the enzyme (33U TGase/g GPF protein) into nanoreinforced GPF (even with MSN or CH–NP) reduced the ACR formation significantly (about 41.0% and 47.5%, respectively) in respect to the nanoreinforced GPF prepared without TGase (Figure 2). The obtained data may indicate a potential synergistic effect between NPs and TGase which reduces the Maillard reaction. The ACR recovery was between 93% and 108% (Table 2).

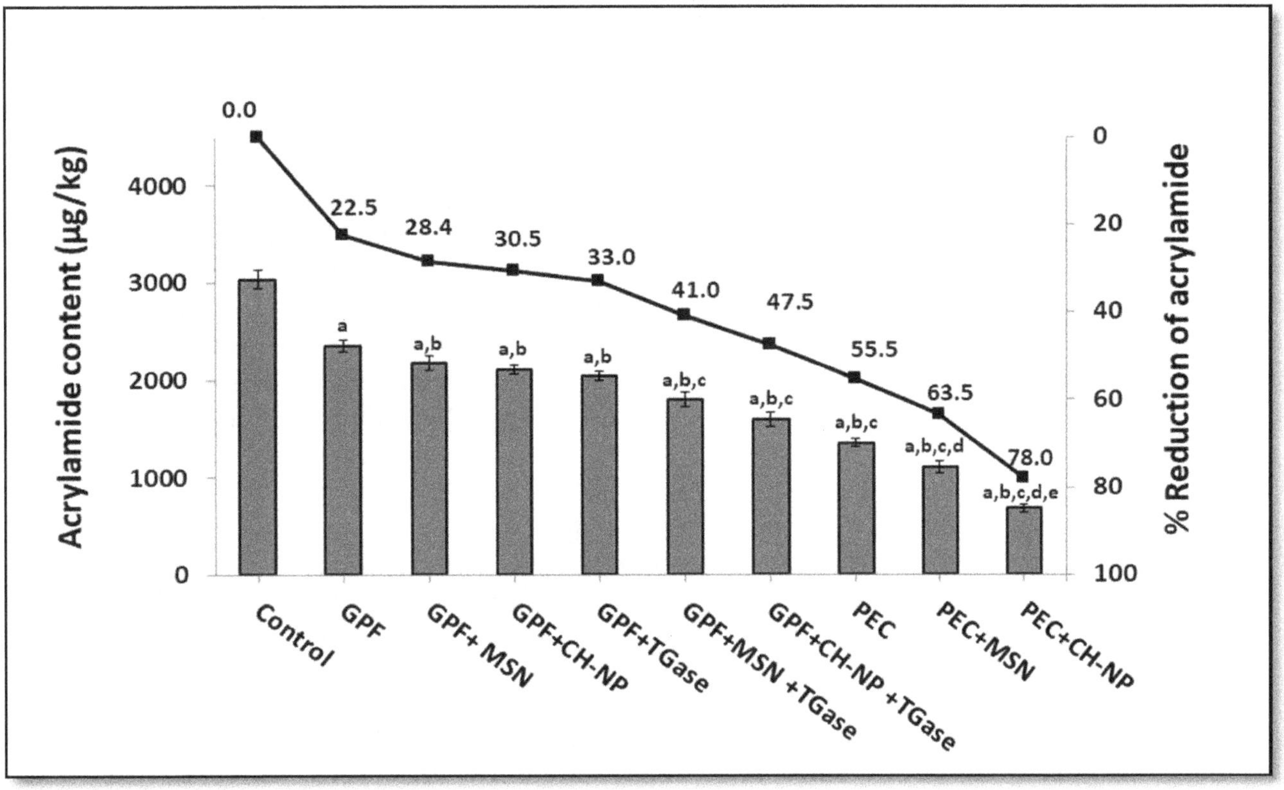

Figure 2. Influence of different hydrocolloid coatings on acrylamide content of fried kobbah (y-axis on the left based on fat-free dry matters (FFDM)) and% acrylamide reduction (y-axis on the right). "Control" represents kobbah sample dipped in distilled water. Columns "a" indicate values significantly different from the control sample; columns "b" indicate values significantly different from grass pea flour (GPF)-coated kobbah; columns "c" indicate values significantly different from GPF + mesoporous silica nanoparticles (MSN)-coated kobbah or GPF + transglutaminase (TGase)-coated kobbah; columns "d" indicate values significantly different from PEC-coated kobbah; columns "e" indicate values significantly different from pectin (PEC) + MSN-coated kobbah. Additional details are reported in the main text.

Table 2. Acrylamide content (ACR) recovery in all kobbah samples (in each sample 150 µg/L of ACR standard were used).

Kobbah Types	ACR Content in Spiked Sample (µg/Kg)	Recovery (%)
Control	3186 ± 61	98
Dipped in GPF	2511 ± 135 [a]	103
Dipped in GPF + MSN	2329 ± 103 [a,b]	101
Dipped in GPF + CH–NP	2255 ± 51 [a,b]	96
Dipped in GPF + TGase	2186 ± 48 [a,b]	98
Dipped in GPF + MSN + TGase	1934 ± 70 [a,b,c]	93
Dipped in GPF + CH–NP + TGase	1744 ± 49 [a,b,c]	99
Dipped in PEC	1495 ± 39 [a,b,c]	95
Dipped in PEC + MSN	1250 ± 50 [a,b,c,d]	95
Dipped in PEC + CH–NP	841 ± 37 [a,b,c,d,e]	108

Values significantly different from those obtained for the controls are indicated by "a", the value signed with "b" were significantly different from kobbah coated only by grass pea flour (GPF), whereas the value indicated by "c" were significantly different from kobbah coated with GPF in the presence of nanoparticles or TGase alone, the value indicated by "d" were significantly different from kobbah coated only by pectin (PEC) and the value indicated by "e" was significantly different respect to the kobbah coated by PEC + mesoporous silica nanoparticles (MSN). Data reported are the average values of three repetitions using (2-way ANOVA, $p < 0.05$ for mean comparison). Additional details are reported in the main text.

3.3. Influence of Nanoreinforced and Crosslinked Hydrocolloid Coating Solutions on Water and Oil Content

Water content of the kobbah (coated or not) was evaluated and the results reported in Figure 3. The obtained data have shown that the water content significantly increases in kobbah coated with any of the different hydrocolloid solutions used in this research. In fact, the lowest water content was found in the control sample (equal to 18%), while water content in coated kobbah by PEC-based solutions was (32.0%), significantly higher compared to the kobbah coated by GPF-based solutions (21.0%). Nanoreinforcement by using either MSN or CH–NP in both GPF-based or PEC-based solutions, provokes the increasing in water content of the kobbah significantly higher in comparison to samples coated by solutions made of only GPF or PEC. Our findings are supported by Osheba et al. [55], that concluded that CH–NP rise the moisture content of fish fingers up to 52.7%, while the uncoated samples exhibits 34.6% moisture. Regarding the use of TGase, our results prove that the enzyme action in both GPF-based and GPF + NP-based solutions show a significant higher water content respect to the kobbah coated without TGase. Comparable effects were observed by Rossi Marquez et al. [39], where TGase-mediated cross-links are responsible of the reduction of the water evaporation during frying. Moreover, the results demonstrate that the water content of kobbah coated by GPF + CH–NP + TGase is significantly higher compared to the kobbah coated by only GPF + TGase and GPF + MSN + TGase (Figure 3). Recently, Castelo Branco Melo et al. [56] found out that CH–NP led the delaying of the ripening process of the grapes as evident from the decreased weight loss, soluble solids and increased moisture retention.

One of the main health problems is the highest oil content of fried foods. Several studies concluded that coating the fried foods before frying by hydrocolloids materials reduced the oil uptake during frying [39,57]. Figure 4 shows the oil content of kobbah just dipped into water (and used as control) or the ones coated with different solutions. Coating significantly reduces the oil content in comparison to the control, which shows the highest oil content (36.9%), whereas the lowest value was obtained in the fried kobbah coated by PEC + CH–NP (15.2%). There was not any significant difference between the GPF coated kobbah and the kobbah protected by GPF nanoreinforced with MSN or CH–NP. On the other hand, significantly difference in oil content of the fried kobbah were observed between PEC-coated samples and PEC + NP-coated samples. Enzymatically cross-linking of GPF, without and with NPs, demonstrated a significant oil uptake reduction in the coated fried kobbah compared to kobbah coated by GPF or in the presence of NPs (Figure 4). PEC-based coating materials containing NPs (either MSN or CH–NP) induced a significant reduction in oil content of the coated kobbah (18.1%

and 15.2%, respectively) compared to kobbah coated with PEC (20.8%). Moreover, using CH–NPs for coating the fish fingers, Osheba et al. [55] have demonstrated a significant reduction of oil uptake which changed from 16.4% in uncoated fish fingers to 4.5% in coated ones.

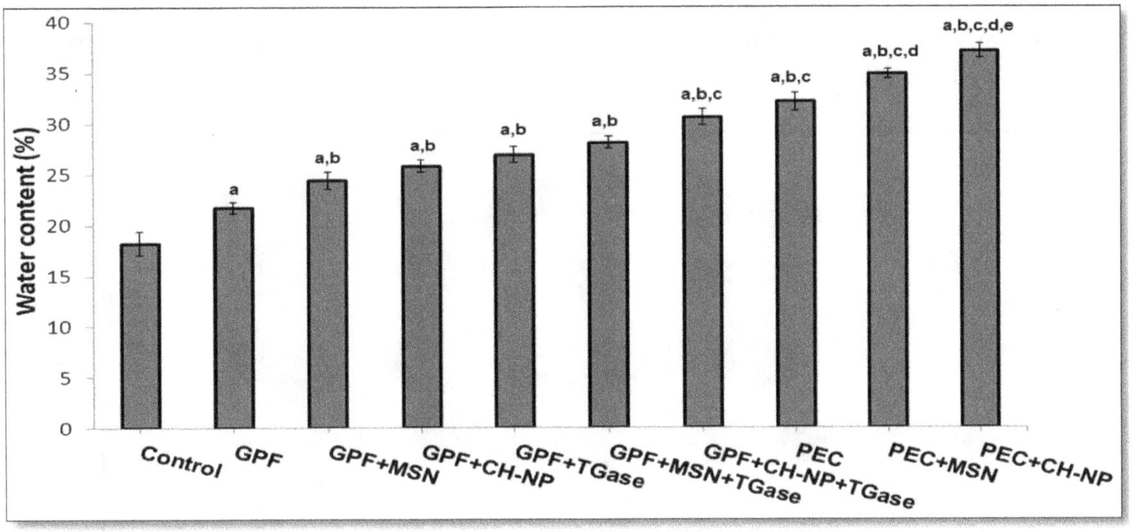

Figure 3. Effect of different hydrocolloid coatings on fried kobbah water content. "Control" represents the kobbah sample dipped in distilled water. Columns "a" indicate values significantly different from the control sample; columns "b" report values significantly different from grass pea flour (GPF)-coated kobbah; columns "c" indicate values significantly different from GPF + mesoporous silica nanoparticles (MSN)-coated kobbah or GPF + transglutaminase (TGase)-coated kobbah; columns "d" indicate values significantly different from pectin (PEC)-coated kobbah; columns "e" indicate values significantly different from PEC + MSN-coated kobbah. Additional details are reported in the main text.

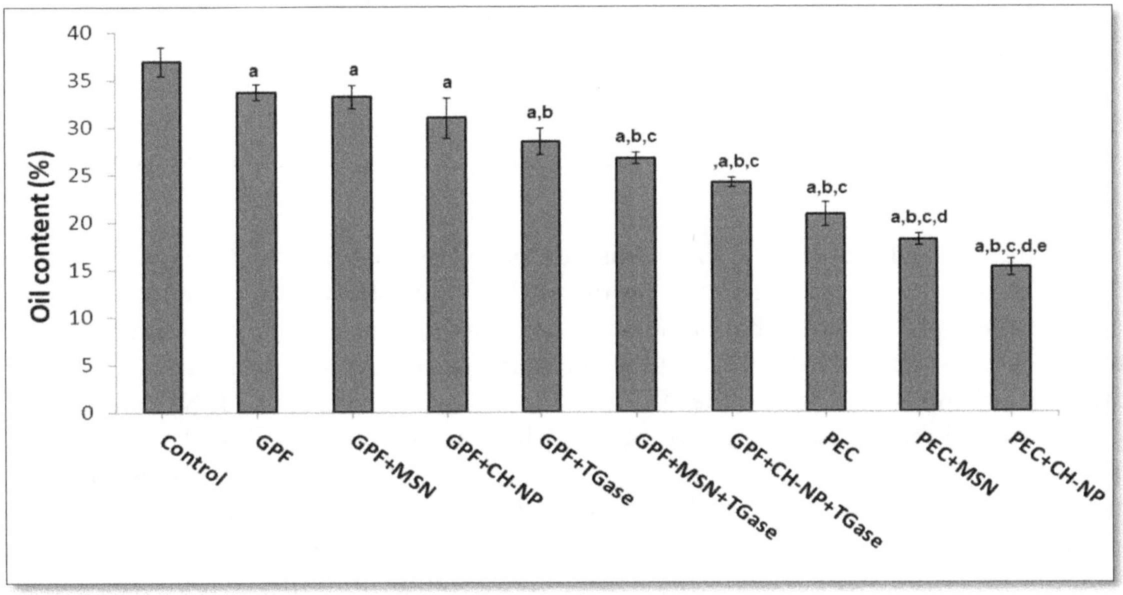

Figure 4. Influence of different hydrocolloid coatings on oil content of fried kobbah. Columns "a" indicate values significantly different from the control sample; columns "b" report values significantly different from grass pea flour (GPF)-coated kobbah; columns "c" indicate values significantly different from GPF + mesoporous silica nanoparticles (MSN)-coated kobbah or GPF + transglutaminase (TGase)-coated kobbah; columns "d" indicate values significantly different from pectin (PEC)-coated kobbah; columns "e" indicate values significantly different from PEC + MSN-coated kobbah. Additional details are reported in the main text.

3.4. Influence of Nanoreinforced and Crosslinked Hydrocolloid Coating Solutions on the Kobbah Color

Food color is important for the industries, as consumers are highly influenced by this feature. The color is dependent by several processes occurring during food processing [45]. Figure 5 shows the aspect of all the kobbah samples obtained in this study, while the results of color analysis are reported in Table 3, together with L*, a*, b* values and their derivatives, such as total color difference to control (ΔE) and Browning Index (BI).

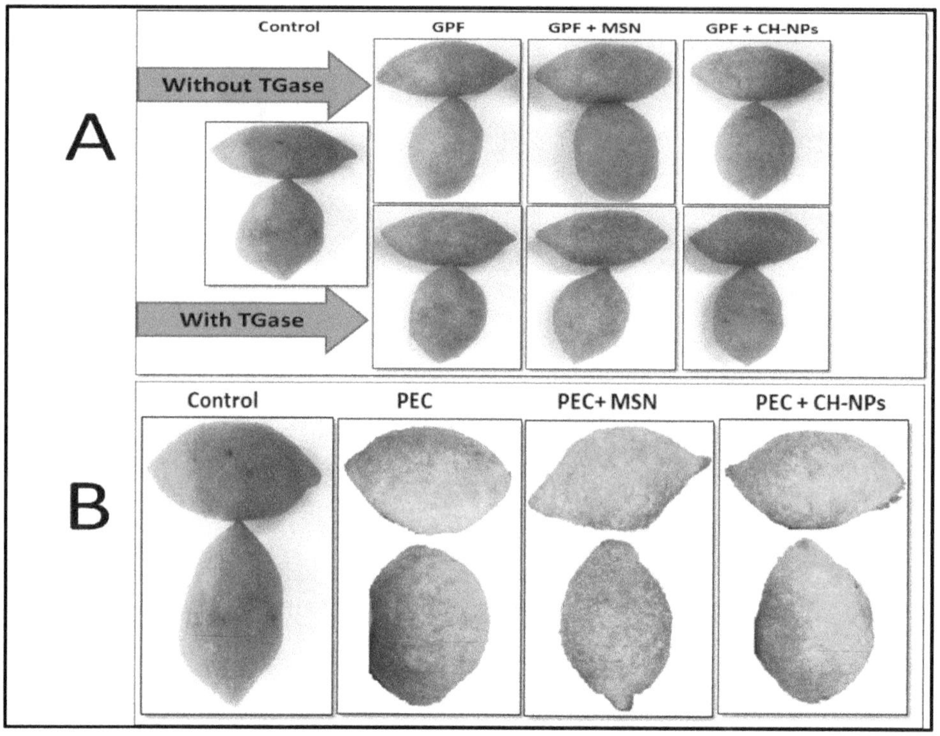

Figure 5. Images of kobbah samples coated by hydrocolloid coatings made of grass pea flour (GPF), GPF + mesoporous silica nanoparticles (MSN), GPF + chitosan nanoparticles (CH–NP), GPF + transglutaminase (TGase); GPF + MSN + TGase and GPF + CH–NP + TGase (Panel **A**); pectin (PEC), PEC + MSN and PEC + CH–NP (Panel **B**). "Control" represents the kobbah sample dipped in distilled water.

It was found that color of fried kobbah was influenced by coating which as a consequence could change the color of the final products. The lightness (L*) value showed that the lowest value was in the control samples (49.25 ± 0.68), that is uncoated and the highest value was founded in the kobbah coated by PEC + CH–NP (60.78 ± 1.02) and these results was conformed to Figure 5.

Moreover, kobbah coated with either PEC containing or not MSN or CH–NPs dipping solutions showed significant higher L* value comparing to the kobbah coated with GPF alone or with TGase or nanoparticles. The a* values showed a significant reduction after treated kobbah by different coating solutions the lowest value was in the kobbah coated with PEC + CH–NP (3.32 ± 1.35). GPF containing nanoparticles either with or without TGase showed significant reduction in the a* value comparing to the kobbah coated by only GPF. The b* value showed no significant different between the untreated and the treated kobbah except the coated kobbah by PEC + CH–NP was significant higher comparing to the kobbah coated by GPE.

Total color difference to control (ΔE) showed the highest value was in the kobbah coated by PEC–NP about (12.7 ± 0.98). However, the ΔE of the PEC coating solutions coating nanoparticles was significantly higher comparing to the kobbah coated with GPF solutions. The results indicated that the kobbah coated by the GPF containing CH–NP alone or with TGase was significantly higher than kobbah coated by only GPF (Table 3). In contrast, the highest BI was in the control kobbah (110.09 ± 3.54) and it decreased significantly after coating the kobbah by different coating solutions and the lowest value was detected into the kobbah coated by PEC + CH–NP equal to 79.91 ± 1.72. This result is in correlation with the acrylamide results that indicated that the lowest ACR was in the kobbah coated by PEC–NPs. Jackson and Al-Taher. [58] and EFSA report [30], concluded that the surface color is highly correlated to acrylamide levels, where higher BI means higher ACR content. This demonstrates that the surface browning degree could be an indicator of ACR formation during cooking of kobbah product.

Table 3. Color properties of fried kobbah coated with different hydrocolloid-based solutions.

Kobbah Types	L*	a*	b*	ΔE	BI
Control	49.25 ± 0.68	8.28 ± 0.19	31.96 ± 0.76	0.0 ± 0.00	110.09 ± 3.54
Dipped in GPF	51.04 ± 0.34 *	7.77 ± 0.12	31.26 ± 0.57	2.02 ± 0.46 *	100.80 ± 2.99 *
Dipped in GPF + MSN	52.08 ± 1.01 *	6.69 ± 0.33 *,a	33.14 ± 0.18	3.48 ± 0.93 *	104.20 ± 3.36
Dipped in GPF CH–NP	53.23 ± 0.97 *,a	5.96 ± 0.16 *,a	31.88 ± 1.62	4.82 ± 0.76 *,a	95.00 ± 8.61 *
Dipped in GPF + TGase	54.93 ± 1.11 *,a	6.10 ± 0.82 *,a	33.01 ± 0.54	6.20 ± 1.30 *,a	95.16 ± 4.36 *,a
Dipped in GPF + MSN + TGase	55.35 ± 1.08 *,a	5.97 ± 0.31 *,a	31.55 ± 1.80	6.70 ± 1.10 *,a	88.68 ± 8.02 *,a
Dipped in GPF + CH–NP + TGase	55.70 ± 1.00 *,a	5.81 ± 0.42 *,a	32.06 ± 1.29	7.01 ± 0.84 *,a	89.44 ± 6.35 *,a
Dipped in PEC	56.56 ± 0.69 *,a,b	5.44 ± 0.31 *,a	32.58 ± 0.36	7.87 ± 0.75 *,a,b	88.75 ± 2.02 *,a
Dipped in PEC + MSN	58.03 ± 0.77 *,a,b	5.01 ± 0.25 *,a	32.99 ± 0.38	9.36 ± 0.77 *,a,b,c	86.78 ± 3.18 *,a
Dipped in PEC + CH–NP	60.78 ± 1.02 *,a,b,c	3.32 ± 1.35 *,a,c	33.34 ± 1.00 a	12.70 ± 0.98 *,a,b,c	79.91 ± 1.72 *,a,c

Columns significantly different from those obtained by analyzing the control are indicated by "*", the columns indicated by "a" were significantly different from kobbah coated only by grass pea flour (GPF), whereas the columns with "b" were significantly different from kobbah coated by GPF with transglutaminase (TGase) alone, the columns indicated by "c" were significantly different from kobbah coated only by pectin (PEC) or PEC in presence of mesoporous silica nanoparticles (MSN). The results represent the average values of three repetitions using (2-way ANOVA, $p < 0.05$ for mean comparison). Additional details are reported in the main text.

3.5. Effect of Nanoreinforced and Crosslinked Hydrocolloid Coating Solutions on the Digestibility of Fried Kobbah

In order to verify whether the coating composition could affect digestibility of the fried food, IVD experiments were performed by a protocol set up within the INFOGEST Cost Action [59]. According to INFOGEST protocol, IVD experiments were set up under physiological conditions, followed by SDS-PAGE (12%) analysis as shown in Figure 6. Samples "C" represents the control since such samples were treated with SGF prepared without pepsin. To study the digestibility rate two different kinds of bands were observed: 25 kDa band for samples containing GPF and GPF + MSN and 250 kDa band for samples set up in the presence of the enzyme (GPF + TGase and GPF + MSN + TGase). By visual inspection of the SDS-PAGE patterns of Figure 6, it is possible to assess that MSN do not affect digestibility (comparing Panel B to Panel A and Panel D to Panel C). However, looking at 250 kDa band present in TGase-treated samples (Figure 6, Panels C and D) it is not possible to note significative differences among different samples following pepsin treatment. Thus, densitometry analysis was performed, and the results reported in Figure 7, Panel A. It is possible to note that a significant rate of digestibility of 250 kDa band present in TGase-treated FFS samples, was observed after 10 min pepsin incubation. Similar results were obtained studying digestibility of GPF-based bioplastics crosslinked by means of TGase [8]. Densitometry analysis results of 25 kDa (present in FFS samples not treated with the enzyme) confirmed what was observed by visual inspection, namely an higher digestibility rate already after 1 min pepsin incubation (Figure 7, Panel A). Comparable data were reported by Romano et al. [9].

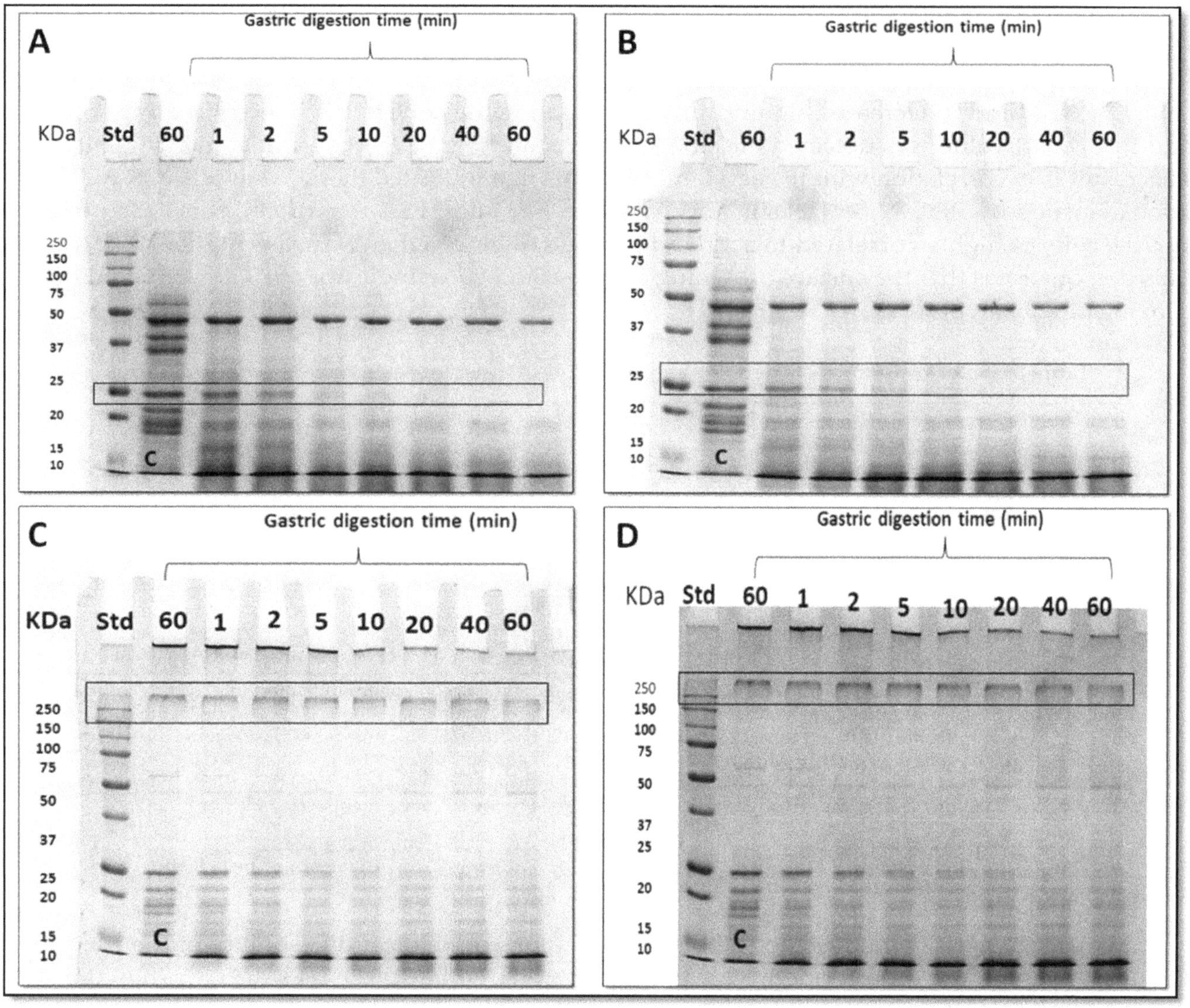

Figure 6. Sodium dodecyl sulfate polyacrylamide gel electrophoresis (SDS-PAGE) of grass pea flour (GPF) film forming solutions (FFSs) after in vitro digestion (IVD) experiments. (Panel **A**): GPF; (Panel **B**): GPF + mesoporous silica nanoparticles (MSN); (Panel **C**): GPF + transglutaminase (TGase 33 *U*/g); (Panel **D**): GPF + MSN + TGase (33 *U*/g). The bands in the rectangle are those chosen for densitometry analysis. C is control sample incubated with simulated gastric fluid (SGF) prepared without pepsin. Std, Molecular mass standards, Bio-Rad.

IVD experiments were performed also using kobbah dipped in GPF or GPF containing MSN FFSs-treated (Figure 8). IVD treatment was effective on protein component of kobbah, mainly proteins present in kobbah ingredients (i.e., mostly bulgur flour, beef meat). The ~45 kDa band of samples not treated with TGase was subjected to densitometry analysis, while in enzyme–treated samples the 250 kDa was analyzed.

Densitometry analysis of those bands are observed in Figure 7, Panel B. The digestion seems to be slower in the food coated by protein crosslinked by means of TGase enzyme. However, all the proteins were completely digested by pepsin at the longest incubation time in all different coated kobbah (Figure 7, Panel B).

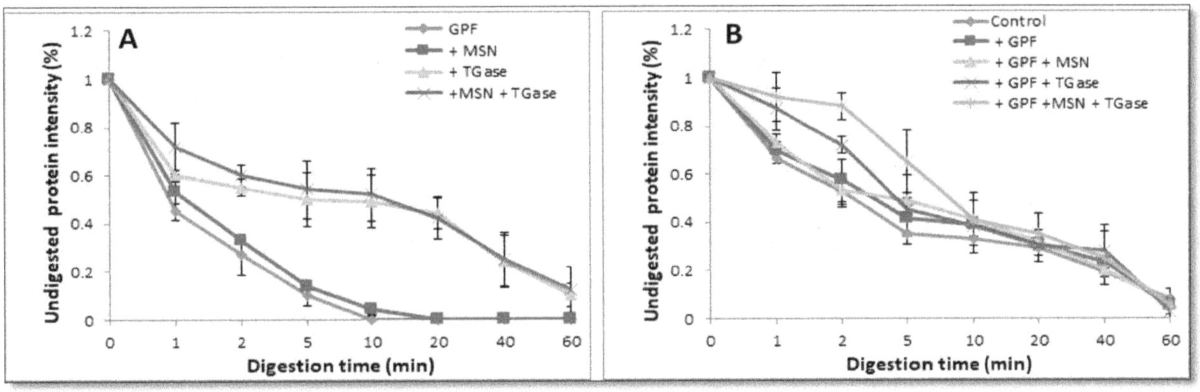

Figure 7. Intensity of the protein framed bands in gels of Figures 6 and 8, obtained after in vitro gastric digestion (IVD). Both grass pea flour (GPF)-based film forming solutions (FFSs) (Panel **A**) and fried kobbah coated with all GPF-based FFSs (Panel **B**), were subjected to densitometry analysis.

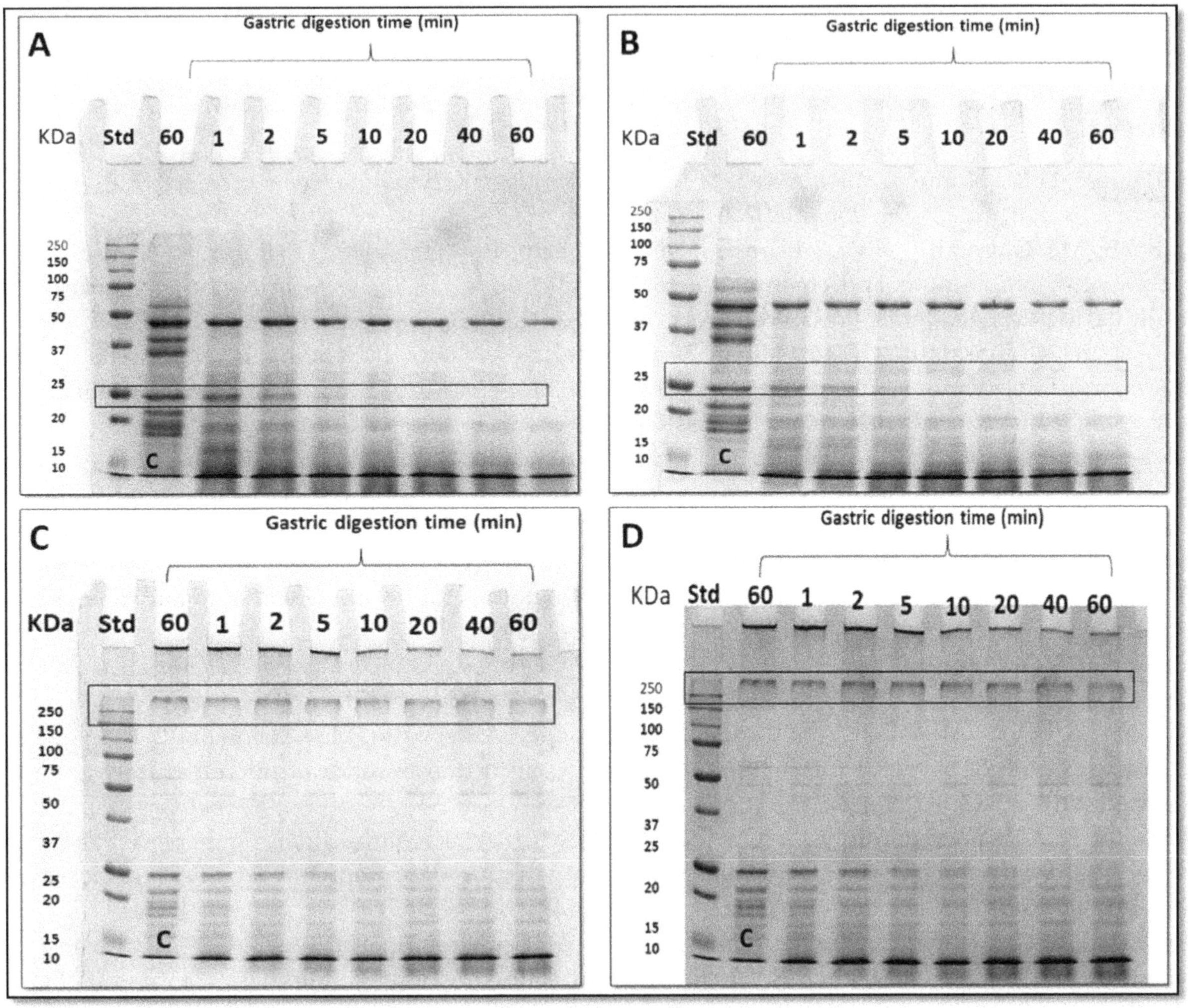

Figure 8. Sodium dodecyl sulfate polyacrylamide gel electrophoresis (SDS-PAGE) of fried kobbah digested by in vitro gastric digestion (IVD) experiments. (Panel **A**): kobbah dipped in water (control); (Panel **B**): kobbah dipped in grass pea flour (GPF); (Panel **C**): kobbah dipped in GPF + mesoporous silica nanoparticles (MSN); (Panel **D**): kobbah dipped in GPF + transglutaminase (TGase 33 *U*/g); (Panel **E**): kobbah dipped in GPF + MSN + TGase (33 *U*/g). The bands in the rectangle are those chosen for densitometry analysis. C is control sample incubated with simulated gastric fluid (SGF) prepared without pepsin. Std, Molecular mass standards, Bio-Rad.

4. Conclusions

Healthier fried kobbah was successfully obtained by dipping method using either GPF or PEC-based solutions. TGase-treated GPF in the presence of nanoparticles was demonstrated to have also important function to reduce ACR formation. The best coating solution that significantly reduced ACR was the one made of PEC nanoreinforced with CH–NP. From the obtained results we conclude that increasing water content inside the fried food by coating is an effective way to mitigate ACR formation and oil content. Reducing the browning index of the fried kobbah is a key indicator to the healthier kobbah. Moreover, the gastric digestion results showed that TGase-mediated modification fairly decreased the rate of digestion in both coating solutions and fried kobbah, even though protein component was completely digested at the end of the longest incubation time.

Author Contributions: Conceptualization, L.M. and A.A.-A.; methodology, A.A.-A., C.V.L.G. and M.S.; software, A.A.-A., C.V.L.G., M.S.; validation, A.A.-A., C.V.L.G. and L.M.; formal analysis, A.A.-A., C.V.L.G. and L.M.; investigation, A.A.-A. and L.M.; resources, L.M.; data curation, L.M.; writing—original draft preparation, A.A.-A.; writing—review and editing, L.M.; visualization, A.A.-A. and L.M.; supervision, L.M.; project administration, L.M.; funding acquisition, L.M. All authors have read and agreed to the published version of the manuscript.

Acknowledgments: Authors would like to thank Mrs. Maria Fenderico for SDS-PAGE support.

References

1. Al-Dmoor, H.M.; Humeid, M.A.; Alawi, M.A. Investigation of acrylamide levels in selected fried and baked foods in Jordan. *Food Agric. Environ.* **2004**, *2*, 157–165.
2. Sabbah, M.; Giosafatto, C.V.L.; Esposito, M.; Di Pierro, P.; Mariniello, L.; Porta, R. Transglutaminase cross-linked edible films and coatings for food applications. In *Enzymes in Food Biotechnology*, 1st ed.; Kuddus, M., Ed.; Academic Press: New York, NY, USA, 2019; pp. 369–388.
3. Al-Asmar, A.; Giosafatto, C.V.L.; Sabbah, M.; Sanchez, A.; Villalonga Santana, R.; Mariniello, L. Effect of mesoporous silica nanoparticles on the physicochemical properties of pectin packaging material for strawberry wrapping. *Nanomaterials* **2020**, *10*, 52. [CrossRef] [PubMed]
4. Lara-Espinoza, C.; Carvajal-Millán, E.; Balandrán-Quintana, R.; López-Franco, Y.; Rascón-Chu, A. Pectin and pectin-based composite materials: Beyond food texture. *Molecules* **2018**, *23*, 942. [CrossRef] [PubMed]
5. Padmaja, N.; Bosco, S.J.D. Preservation of jujube fruits by edible Aloe *vera* gel coating to maintain quality and safety. *Indian J. Sci. Res. Technol.* **2014**, *3*, 79–88.
6. Yossef, M.A. Comparison of different edible coatings materials for improvement of quality and shelf life of perishable fruits. *Middle East J. Applied Sci.* **2014**, *4*, 416–424.
7. Campbell, C.G. *Grass Pea, Lathyrus sativus L.; Promoting the Conservation and Use of Underutilized and Neglected Crops*; IPGRI: Rome, Italy, 1997; Volume 18, pp. 1–91.
8. Giosafatto, C.V.L.; Al-Asmar, A.; D'Angelo, A.; Roviello, V.; Esposito, M.; Mariniello, L. Preparation and characterization of bioplastics from grass pea flour cast in the presence of microbial transglutaminase. *Coatings* **2018**, *8*, 435. [CrossRef]
9. Romano, A.; Giosafatto, C.V.L.; Al-Asmar, A.; Aponte, M.; Masi, P.; Mariniello, L. Grass pea (*Lathyrus sativus*) flour: Microstructure, physico-chemical properties and in vitro digestion. *Eur. Food Res. Technol.* **2019**, *245*, 191–198. [CrossRef]
10. Romano, A.; Giosafatto, C.V.L.; Al-Asmar, A.; Masi, P.; Romano, R.; Mariniello, L. Structure and in vitro digestibility of grass pea (*Lathyrus sativus* L.) flour following transglutaminase treatment. *Eur. Food Res. Technol.* **2019**, *245*, 1899–1905. [CrossRef]
11. Giosafatto, C.V.L.; Al-Asmar, A.; Mariniello, L. Transglutaminase protein substrates of food interest. In *Enzymes in Food Technology: Improvement and Innovation*; Kuddus, M., Ed.; Springer Nature; Singapore Pte Ltd.: Singapore, 2018; pp. 293–317.
12. Sabbah, M.; Altamimi, M.; Di Pierro, P.; Schiraldi, C.; Cammarota, M.; Porta, R. Black edible films from protein-containing defatted cake of *Nigella sativa* seeds. *Int. J. Mol. Sci.* **2020**, *21*, 832. [CrossRef]
13. Zink, J.; Wyrobnik, T.; Prinz, T.; Schmid, M. Physical, chemical and biochemical modifications of protein-based films and coatings: An extensive review. *Int. J. Mol. Sci.* **2016**, *17*, 1376. [CrossRef]
14. Liu, J.; He, D.; Liu, Q.; He, X.; Wang, K.; Yang, X.; Shangguan, J.; Tang, J.; Mao, Y. Vertically ordered

mesoporous silica film-assisted label-free and universal electrochemiluminescence aptasensor platform. *Anal. Chem.* **2016**, *88*, 11707–11713. [CrossRef] [PubMed]

15. Slowing, I.I.; Vivero-Escoto, J.L.; Trewyn, B.G.; Lin, V.S.-Y. Mesoporous silica nanoparticles: Structural design and applications. *J. Mater. Chem.* **2010**, *20*, 7924–7937. [CrossRef]
16. Tyagi, V.; Sharma, A.; Gupta, R.K. Design and properties of polysaccharide based: Silica hybrid packaging material. *Int. J. Food Sci. Nutr.* **2017**, *2*, 140–150.
17. Fernandez-Bats, I.; Di Pierro, P.; Villalonga-Santana, R.; Garcia-Almendarez, B.; Porta, R. Bioactive mesoporous silica nanocomposite films obtained from native and transglutaminase-crosslinked bitter vetch proteins. *Food Hydrocolloid.* **2018**, *82*, 106–115. [CrossRef]
18. Giosafatto, C.V.L.; Sabbah, M.; Al-Asmar, A.; Esposito, M.; Sanchez, A.; Villalonga-Santana, R.; Cammarota, M.; Mariniello, L.; Di Pierro, P.; Porta, R. Effect of mesoporous silica nanoparticles on glycerol plasticized anionic and cationic polysaccharide edible films. *Coatings* **2019**, *9*, 172. [CrossRef]
19. Dekkers, S.; Krystek, P.; Peters, R.J.B.; Lankveld, D.P.K.; Bokkers, B.G.H.; van Hoeven-Arentzen, P.H.; Bouwmeester, H.; Oomen, A.G. Presence and risks of nanosilica in food products. *Nanotoxicology* **2011**, *5*, 393–405. [CrossRef]
20. McCarthy, J.; Inkielewicz-Stępniak, I.; Corbalan, J.J.; Radomski, M.W. Mechanisms of toxicity of amorphous silica nanoparticles on human lung submucosal cells in vitro: Protective Effects of Fisetin. *Chem. Res. Toxicol.* **2012**, *25*, 2227–2235. [CrossRef]
21. Malmiri, H.J.; Jahanian, M.A.G.; Berenjian, A. Potential applications of chitosan nanoparticles as novel support in enzyme immobilization. *Am. J. Biochem. Biotechnol.* **2012**, *8*, 203–219.
22. Divya, K.; Jisha, M.S. Chitosan nanoparticles preparation and applications. *Environ. Chem. Lett.* **2018**, *16*, 101–112. [CrossRef]
23. Lorevice, M.V.; Otoni, C.G.; de Moura, M.R.; Mattoso, L.H.C. Chitosan nanoparticles on the improvement of thermal, barrier, and mechanical properties of high-and low-methyl pectin films. *Food Hydrocoll.* **2016**, *52*, 732–740. [CrossRef]
24. Martelli, M.R.; Barros, T.T.; de Moura, M.R.; Mattoso, L.H.C.; Assis, O.B.G. Effect of chitosan nanoparticles and pectin content on mechanical properties and water vapor permeability of banana puree films. *J. Food Sci.* **2013**, *78*, N98–N104. [CrossRef] [PubMed]
25. Radhakrishnan, Y.; Gopal, G.; Lakshmanan, C.C.; Nandakumar, K.S. Chitosan nanoparticles for generating novel systems for better applications: A Review. *J. Mol. Genet. Med.* **2015**, S4-005. [CrossRef]
26. Hu, Z.; Gänzle, M.G. Challenges and opportunities related to the use of chitosan as a food preservative. *J. Appl. Microbiol.* **2018**, *126*, 1318–1331. [CrossRef] [PubMed]
27. Tareke, E.; Rydberg, P.; Karlsson, P.; Eriksson, S.; Törnqvist, M. Analysis of acrylamide, a carcinogen formed in heated foodstuffs. *J. Agric. Food Chem.* **2002**, *50*, 4998–5006. [CrossRef]
28. Mottram, D.S.; Wedzicha, B.L.; Dodson, A.T. Food chemistry: Acrylamide is formed in the Maillard reaction. *Nature* **2002**, *419*, 448–449. [CrossRef] [PubMed]
29. Zyzak, D.V.; Sanders, R.A.; Stojanovic, M.; Tallmadge, D.H.; Eberhart, B.L.; Ewald, D.K.; Gruber, D.C.; Morsch, T.R.; Strothers, M.A.; Rizzi, G.P.; et al. Acrylamide formation mechanism in heated foods. *J. Agric. Food Chem.* **2003**, *51*, 4782–4787. [CrossRef]
30. EFSA CONTAM Panel (EFSA Panel on Contaminants in the Food Chain). Scientific opinion on acrylamide in food. *EFSA J.* **2015**, *13*, 4104. [CrossRef]
31. Mekawi, E.M.; Sharoba, A.M.; Ramadan, M.F. Reduction of acrylamide formation in potato chips during deep-frying in sunflower oil using pomegranate peel nanoparticles extract. *J. Food Meas. Characteriz.* **2019**. [CrossRef]
32. Liu, J.; Liu, X.; Man, Y.; Liu, Y. Reduction of acrylamide content in bread crust by starch coating. *J. Sci. Food Agric.* **2017**, *98*, 336–345. [CrossRef]
33. Suyatma, N.E.; Ulfah, K.; Prangdimurti, E.; Ishikawa, Y. Effect of blanching and pectin coating as pre-frying treatments to reduce acrylamide formation in banana chips. *Int. Food. Res. J.* **2015**, *22*, 936–942.
34. Al-Asmar, A.; Naviglio, D.; Giosafatto, C.V.L.; Mariniello, L. Hydrocolloid-based coatings are effective at reducing acrylamide and oil content of French fries. *Coatings* **2018**, *8*, 147. [CrossRef]
35. Al-Asmar, A.; Giosafatto, C.V.L.; Panzella, L.; Mariniello, L. The effect of transglutaminase to improve the quality of either traditional or pectin-coated falafel (Fried Middle Eastern Food). *Coatings* **2019**, *9*, 331. [CrossRef]
36. Chang, P.R.; Jian, R.; Yu, J.; Ma, X. Fabrication and characterisation of chitosan nanoparticles/plasticised-starch

composites. *Food Chem.* **2010**, *120*, 736–740. [CrossRef]
37. Brasil, T.A.; Capitani, C.D.; Takeuchi, K.P.; de Castro Ferreira, T.A.P. Physical, chemical and sensory properties of gluten-free kibbeh formulated with millet flour (*Pennisetum glaucum* (L.) R. Br.). *Food Sci. Technol.* **2015**, *35*, 361–367. [CrossRef]
38. Esposito, M.; Di Pierro, P.; Regalado-Gonzales, C.; Mariniello, L.; Giosafatto, C.V.L.; Porta, R. Polyamines as new cationic plasticizers for pectin-based edible films. *Carbohydr. Polym.* **2016**, *153*, 222–228. [CrossRef]
39. Rossi Marquez, G.; Di Pierro, P.; Esposito, M.; Mariniello, L.; Porta, R. Application of transglutaminase-crosslinked whey protein/pectin films as water barrier coatings in fried and baked foods. *Food Bioprocess. Technol.* **2013**, *7*, 447–455. [CrossRef]
40. *Association of Official Analytical Chemists (AOAC) Official Method 960.39 Fat (Crude) or Ether Extract in Meat First Action 1960 Final Action*; AOAC International: Arlington, MA, USA, 2006.
41. *Association of Official Analytical Chemists (AOAC) Official Method 950.46 (39.1.02) Moisture (M)*; AOAC International: Arlington, MA, USA, 2006.
42. Zeng, X.; Cheng, K.-W.; Du, Y.; Kong, R.; Lo, C.; Chu, I.K.; Chen, F.; Wang, M. Activities of hydrocolloids as inhibitors of acrylamide formation in model systems and fried potato strips. *Food Chem.* **2010**, *121*, 424–428. [CrossRef]
43. Wang, H.; Feng, F.; Guo, Y.; Shuang, S.; Choi, M.M.F. HPLC-UV quantitative analysis of acrylamide in baked and deep-fried Chinese foods. *J. Food Compos. Anal.* **2013**, *31*, 7–11. [CrossRef]
44. Michalak, J.; Gujska, E.; Kuncewicz, A. RP-HPLC-DAD studies on acrylamide in cereal-based baby foods. *J. Food Compos. Anal.* **2013**, *32*, 68–73. [CrossRef]
45. Pathare, P.B.; Opara, U.L.; Al-Said, F.A.-J. Colour measurement and analysis in fresh and processed foods: A Review. *Food Bioprocess Technol.* **2012**, *6*, 36–60. [CrossRef]
46. Papadakis, S.E.; Abdul-Malek, S.; Kamdem, R.E.; Yam, K.L. A versatile and inexpensive technique for measuring colour of foods. *Food Technol.* **2000**, *54*, 48–51.
47. Palou, E.; Lopez-Malo, A.; Barbosa-Canovas, G.V.; Welti-Chanes, J.; Swanson, B.G. Polyphenoloxidase activity and colour of blanched and high hydrostatic pressure treated banana puree. *J. Food Sci.* **1999**, *64*, 42–45. [CrossRef]
48. Laemmli, U.K. Cleavage of structural proteins during the assembly of the head of Bacteriophage T4. *Nature* **1970**, *227*, 680–985. [CrossRef] [PubMed]
49. Giosafatto, C.V.L.; Rigby, N.M.; Wellner, N.; Ridout, M.; Husband, F.; Mackie, A.R. Microbial transglutaminase-mediated modification of ovalbumin. *Food Hydrocoll.* **2012**, *26*, 261–267. [CrossRef]
50. Bourlieu, C.; Ménard, O.; Bouzerzour, K.; Mandalari, G.; Macierzanka, A.; Mackie, A.R.; Dupont, D. Specificity of Infant Digestive Conditions: Some Clues for Developing Relevant In Vitro Models. *Crit. Rev. Food Sci. Nutr.* **2014**, *54*, 1427–1457. [CrossRef]
51. Minekus, M.; Alminger, M.; Alvito, P.; Balance, S.; Bohn, T.; Bourlieu, C.; Carrière, F.; Boutrou, R.; Corredig, M.; Dupont, D.; et al. A standardised static in vitro digestion method suitable for food—An international consensus. *Food Funct.* **2014**, *5*, 1113–1124. [CrossRef]
52. Ali, S.W.; Joshi, M.; Rajendran, S. Synthesis and characterization of chitosan nanoparticles with enhanced antimicrobial activity. *Int. J. Nanosci.* **2011**, *10*, 979–984. [CrossRef]
53. Antoniou, J.; Liu, F.; Majeed, H.; Zhong, F. Characterization of tara gum edible films incorporated with bulk chitosan and chitosan nanoparticles: A comparative study. *Food Hydrocoll.* **2015**, *44*, 309–319. [CrossRef]
54. Ali, S.W.; Joshi, M.; Rajendran, S. Modulation of size, shape and surface charge of chitosan nanoparticles with reference to antimicrobial activity. American Scientific Publishers. *Adv. Sci. Lett.* **2010**, *3*, 452–460. [CrossRef]
55. Osheba, S.A.; Sorour, A.M.; Abdou, E.S. Effect of chitosan nanoparticles as active coating on chemical quality and oil uptake of fish fingers. *J. Agri. Environ. Sci.* **2013**, *2*, 1–14.
56. Melo, N.F.C.B.; de MendonçaSoares, B.L.; Diniz, K.M.; Leal, C.F.; Canto, D.; Flores, M.A.; da Costa Tavares-Filho, J.H.; Galembeck, A.; Stamford, T.L.M.; Stamford-Arnaud, T.M.; et al. Effects of fungal chitosan nanoparticles as eco-friendly edible coatings on the quality of postharvest table grapes. *Postharvest Biol. Technol.* **2018**, *139*, 56–66. [CrossRef]
57. Angor, M.K.M.; Ajo, R.; Al-Rousan, W.; Al-Abdullah, B. Effect of starchy coating films on the reduction of fat uptake in deep-fat fried potato pellet chips. *Ital. J. Food Sci.* **2013**, *25*, 45–50.
58. Jackson, L.S.; Al-Taher, F. Effects of Consumer Food Preparation on Acrylamide Formation. In *Chemistry and Safety of Acrylamide in Food. Advances in Experimental Medicine and Biology*; Friedman, M., Mottram, D., Eds.;

Springer: Boston, MA, USA, 2005; Volume 561, pp. 447–465.
59. FA-1005—Improving Health Properties of Food by Sharing Our Knowledge on the Digestive Process (INFOGEST). Available online: https://www.cost.eu/actions/FA1005/#tabs\T1\textbar{}Name:overview (accessed on 17 February 2020).

Permissions

All chapters in this book were first published by MDPI; hereby published with permission under the Creative Commons Attribution License or equivalent. Every chapter published in this book has been scrutinized by our experts. Their significance has been extensively debated. The topics covered herein carry significant findings which will fuel the growth of the discipline. They may even be implemented as practical applications or may be referred to as a beginning point for another development.

The contributors of this book come from diverse backgrounds, making this book a truly international effort. This book will bring forth new frontiers with its revolutionizing research information and detailed analysis of the nascent developments around the world.

We would like to thank all the contributing authors for lending their expertise to make the book truly unique. They have played a crucial role in the development of this book. Without their invaluable contributions this book wouldn't have been possible. They have made vital efforts to compile up to date information on the varied aspects of this subject to make this book a valuable addition to the collection of many professionals and students.

This book was conceptualized with the vision of imparting up-to-date information and advanced data in this field. To ensure the same, a matchless editorial board was set up. Every individual on the board went through rigorous rounds of assessment to prove their worth. After which they invested a large part of their time researching and compiling the most relevant data for our readers.

The editorial board has been involved in producing this book since its inception. They have spent rigorous hours researching and exploring the diverse topics which have resulted in the successful publishing of this book. They have passed on their knowledge of decades through this book. To expedite this challenging task, the publisher supported the team at every step. A small team of assistant editors was also appointed to further simplify the editing procedure and attain best results for the readers.

Apart from the editorial board, the designing team has also invested a significant amount of their time in understanding the subject and creating the most relevant covers. They scrutinized every image to scout for the most suitable representation of the subject and create an appropriate cover for the book.

The publishing team has been an ardent support to the editorial, designing and production team. Their endless efforts to recruit the best for this project, has resulted in the accomplishment of this book. They are a veteran in the field of academics and their pool of knowledge is as vast as their experience in printing. Their expertise and guidance has proved useful at every step. Their uncompromising quality standards have made this book an exceptional effort. Their encouragement from time to time has been an inspiration for everyone.

The publisher and the editorial board hope that this book will prove to be a valuable piece of knowledge for researchers, students, practitioners and scholars across the globe.

List of Contributors

Worraprat Chaisuwan and Apisit Manassa
Interdisciplinary Program in Biotechnology, Graduate School, Chiang Mai University, Chiang Mai 50200, Thailand
Faculty of Agro-Industry, Chiang Mai University, 155 Moo 2, Mae Hia, Mueang, Chiang Mai 50100, Thailand

Yuthana Phimolsiripol, Kittisak Jantanasakulwong, Thanongsak Chaiyaso and Phisit Seesuriyachan
Faculty of Agro-Industry, Chiang Mai University, 155 Moo 2, Mae Hia, Mueang, Chiang Mai 50100, Thailand
Cluster of Agro Bio-Circular-Green Industry (Agro BCG), Chiang Mai University, Chiang Mai 50200, Thailand

Wasu Pathom-aree
Department of Biology, Faculty of Science, Chiang Mai University, Chiang Mai 50200, Thailand

SangGuan You
Department of Marine Food Science and Technology, Gangneung-Wonju National University, Gangneung, Gangwon 210-702, Korea

Bernadette-Emőke Teleky
Institute of Life Sciences, University of Agricultural Sciences and Veterinary Medicine, Calea Manas, tur 3-5, 400372 Cluj-Napoca, Romania

Gheorghe Adrian Martau and Dan Cristian Vodnar
Institute of Life Sciences, University of Agricultural Sciences and Veterinary Medicine, Calea Manas, tur 3-5, 400372 Cluj-Napoca, Romania
Faculty of Food Science and Technology, University of Agricultural Sciences and Veterinary Medicine, Calea Manas, tur 3-5, 400372 Cluj-Napoca, Romania

Mathilde Hirondart, Natacha Rombaut, Anne Sylvie Fabiano-Tixier and Farid Chemat
Avignon University, INRAE, UMR408, GREEN Team Extraction, F-84000 Avignon, France
ORTESA, LabCom Naturex-Avignon University, F-84000 Avignon, France

Antoine Bily
ORTESA, LabCom Naturex-Avignon University, F-84000 Avignon, France
Naturex-Givaudan, 250 rue Pierre Bayle, BP 81218, CEDEX 9, F-84911 Avignon, France

Renata Raina-Fulton and Aisha A. Mohamad
Department of Chemistry & Biochemistry, Trace Analysis Facility, University of Regina; 3737 Wascana Parkway, Regina, SK S4S 0A2, Canada

Aleksandar Ž. Kostić, Danijel D. Milinčić, Sladjana P. Stanojević, Miroljub B. Barać and Mirjana B. Pešić
Chemistry and Biochemistry, Faculty of Agriculture, University of Belgrade, Nemanjina 6, 11080 Belgrade, Serbia

Tanja S. Petrović
Preservation and Fermentation, Faculty of Agriculture, University of Belgrade, Nemanjina 6, 11080 Belgrade, Serbia

Vesna S. Krnjaja
Institute for Animal Husbandry, Autoput 16, 11080 Belgrade, Serbia

Živoslav Lj. Tešić
Analytical Chemistry, Faculty of Chemistry, University of Belgrade, Studentski Trg 12-16, 11158 Belgrade, Serbia

Jan Teipel, Maren Hegmanns, Thomas Kuballa, Stephan G. Walch and Dirk W. Lachenmeier
Chemisches und Veterinäruntersuchungsamt (CVUA) Karlsruhe, Weissenburger Strasse 3, 76187 Karlsruhe, Germany

Steffen Schwarz
Coffee Consulate, Hans-Thoma-Strasse 20, 68163 Mannheim, Germany

Carmen M. Breitling-Utzmann
Chemisches und Veterinäruntersuchungsamt Stuttgart, Schaflandstr. 3/2, 70736 Fellbach, Germany

Tsai-Li Kung
Taoyuan District Agricultural Research and Extension Station, Council of Agriculture, Executive Yuan, Taoyuan 327, Taiwan

Yi-Ju Chen, Louis Kuoping Chao, Chin-Sheng Wu and Hsin-Chun Chen
Department of Cosmeceutics, China Medical University, Taichung 404, Taiwan

Li-Yun Lin
Department of Food Science and Technology, Hungkuang University, Taichung 433, Taiwan

Saira Sultan, Gabriele Netzel, Michael E. Netzel, Mukan Yin, Rafat Al Jassim and Mary T. Fletcher
Queensland Alliance for Agriculture and Food Innovation (QAAFI), The University of Queensland, Health and Food Sciences Precinct, Coopers Plains, QLD 4108, Australia

Cindy Giles
Department of Agriculture and Fisheries, Queensland Government, Health and Food Sciences Precinct, Coopers Plains, QLD 4108, Australia

Everaldo Attard
Division of Rural Sciences and Food Systems, Institute of Earth Systems, University of Malta, Msida MSD 2080, Malta

Daniele Naviglio and Martina Ciaravolo
Department of Chemical Sciences, University of Naples Federico II, via Cintia; Monte S. Angelo Complex, 80126 Naples, Italy

Monica Gallo
Department of Molecular Medicine and Medical Biotechnology, University of Naples Federico II, via Pansini 5, 80131 Naples, Italy

Alex O. Okaru
Department of Pharmaceutical Chemistry, University of Nairobi, Nairobi, Kenya
Chemisches und Veterinäruntersuchungsamt (CVUA) Karlsruhe, Weissenburger Strasse 3, 76187 Karlsruhe, Germany

Chen Chen
College of Biological Science and Engineering, Shaanxi University of Technology, Hanzhong 723000, China

Long Xu
College of Biological Science and Engineering, Shaanxi University of Technology, Hanzhong 723000, China
Centre of Molecular and Environmental Biology, University of Minho, Department of Biology, Campus de Gualtar, 4710-057 Braga, Portugal

Qi Zhang
College of Biological Science and Engineering, Shaanxi University of Technology, Hanzhong 723000, China
College of Veterinary Medicine, Northwest A&F University, Yangling 712100, China

Xiao-ying Zhang
College of Biological Science and Engineering, Shaanxi University of Technology, Hanzhong 723000, China
Centre of Molecular and Environmental Biology, University of Minho, Department of Biology, Campus de Gualtar, 4710-057 Braga, Portugal
College of Veterinary Medicine, Northwest A&F University, Yangling 712100, China

Simone A. Osborne
Commonwealth Scientific and Industrial Research Organisation, Agriculture and Food, St. Lucia, QLD 4067, Australia

Christian Vella
Department of Pharmacy, Faculty of Medicine and Surgery, University of Malta, Msida MSD 2080, Malta

Pierpaolo Scarano
Department of Science and Technology, University of Sannio, Via Port'Arsa 11, 82100 Benevento, Italy

Xiao-yi Suo and Xin-ping Li
College of Veterinary Medicine, Northwest A&F University, Yangling 712100, China

Eddie T. T. Tan
Queensland Alliance for Agriculture and Food Innovation (QAAFI), The University of Queensland, Health and Food Sciences Precinct, Coopers Plains, QLD 4108, Australia
Alliance of Research and Innovation for Food (ARIF), Faculty of Applied Sciences, Universiti Teknologi MARA, Cawangan Negeri Sembilan, Kuala Pilah Campus, Negeri Sembilan 72000, Malaysia

Ken W. L. Yong
Department of Agriculture and Fisheries, Health and Food Sciences Precinct, Coopers Plains, QLD 4108, Australia

Elena Fattore, Renzo Bagnati, Andrea Colombo, Roberto Fanelli and Enrico Davoli
Department of Environmental Health Sciences, Istituto di Ricerche Farmacologiche Mario Negri IRCCS, 20156 Milano, Italy

Roberto Miniero, Gianfranco Brambilla and Alessandro Di Domenico
Toxicological Chemistry Unit, Environment Department, Italian National Institute of Health, 00161 Rome, Italy

Alessandra Roncarati
School of Biosciences and Veterinary Medicine, University of Camerino, I-62024 Matelica, Italy

Pan Gao and Tianying Yan
College of Information Science and Technology, Shihezi University, Shihezi 832000, China
Key Laboratory of Oasis Ecology Agriculture, Shihezi University, Shihezi 832003, China

Wei Xu
College of Agriculture, Shihezi University, Shihezi 832003, China
Xinjiang Production and Construction Corps Key Laboratory of Special Fruits and Vegetables Cultivation Physiology and Germplasm Resources Utilization, Shihezi 832003, China

Xin Lv
Key Laboratory of Oasis Ecology Agriculture, Shihezi University, Shihezi 832003, China
College of Agriculture, Shihezi University, Shihezi 832003, China

Chu Zhang and Yong He
College of Biosystems Engineering and Food Science, Zhejiang University, Hangzhou 310058, China
Key Laboratory of Spectroscopy Sensing, Ministry of Agriculture and Rural Affairs, Hangzhou 310058, China

Cheryl M. Armstrong, Joseph A. Capobianco, Terence P. Strobaugh Jr. and Andrew G. Gehring
Molecular Characterization of Foodborne Pathogens Research Unit, United States Department of Agriculture, Eastern Regional Research Center, Wyndmoor, PA 19038, USA

Leah E. Ruth and Fernando M. Rubio
Abraxis, Inc., Warminster, PA 18974, USA

Federica Dal Bello, Riccardo Aigotti, Michael Zorzi and Claudio Medana
Molecular Biotechnology and Health Sciences Dept. Università degli Studi di Torino, Via Pietro Giuria 5, 10125 Torino, Italy

Valerio Giaccone
Department of Animal Medicine, Productions and Health, Università degli Studi di Padova, Viale dell'Università 16, 35020 Legnaro, PD, Italy

Asmaa Al-Asmar
Department of Chemical Sciences, University of Naples "Federico II", 80126 Naples, Italy
Analysis, Poison Control and Calibration Center (APCC), An-Najah National University, Palestine

Concetta Valeria L. Giosafatto and Loredana Mariniello
Department of Chemical Sciences, University of Naples "Federico II", 80126 Naples, Italy

Mohammed Sabbah
Department of Nutrition and Food Technology, An-Najah National University, Palestine

Index

Printed in the USA
CPSIA information can be obtained
at www.ICGtesting.com
JSHW062138251023
50683JS00026B/84